Integrated Circuits and Systems

Series Editor
Anantha P. Chandrakasan
Massachusetts Institute of Technology
Cambridge, Massachusetts

More information about this series at http://www.springer.com/series/7236

Patrick P. Mercier • Anantha P. Chandrakasan
Editors

Ultra-Low-Power Short-Range Radios

Springer

Editors
Patrick P. Mercier
University of California
San Diego
La Jolla, CA, USA

Anantha P. Chandrakasan
Massachusetts Institute
of Technology
Cambridge, MA, USA

ISSN 1558-9412
Integrated Circuits and Systems
ISBN 978-3-319-14713-0 ISBN 978-3-319-14714-7 (eBook)
DOI 10.1007/978-3-319-14714-7

Library of Congress Control Number: 2015942638

Springer Cham Heidelberg New York Dordrecht London

Printed on acid-free paper

Springer International Publishing AG Switzerland is part of Springer Science+Business Media (www.springer.com)

Preface

Wireless communication is becoming increasingly ubiquitous today, with many consumer, industrial, medical, and military devices leveraging newly found untethered freedoms to enable exciting new applications and opportunities. To date, the cellular and computer industries have been the primary driver of advances in wireless functionality, exemplified by the popularity and continuing advances in 4G carrier aggregation and next-generation WiFi products. While these (and other) standards support the high throughput and range necessary for cellular and computing applications, there are emerging classes of applications including sensor networks, Internet of Things (IoT), and body-area networks that have very different requirements. Specifically, devices used in these applications are not necessarily looking for the fastest throughput, but instead focus on achieving *sufficient* throughput under ultra-low-power budgets. To reduce power, these devices typically operate over shorter ranges, pushing the power requirements of long-haul communications to more energy-rich gateways. It is expected that these ultra-low-power, short-range radios will begin to comprise a large fraction of total volume of radios as these emerging application spaces mature.

To support these application spaces, there are currently a number of standards specifically responsive to the needs of low-power short-range communications. Examples include Bluetooth, Bluetooth Low Energy, and IEEE 802.15.6 (Wireless Body Area Networks), with a few other standards that support low-power though not necessarily short-range (e.g., Zigbee and 802.11ah). However, in many cases strictly adhering to standards can limit design creativity, potentially resulting in solutions with higher-than-desired power. Thus, the majority of this book does not focus on developing radios for specific standards, but instead focuses on circuit and system techniques that achieve ultra-low-power while not sacrificing too much on other important metrics. Many of the discussed techniques can then be applied to standards-based radios, though some techniques are naturally better suited for custom, proprietary solutions.

In either case, the purpose of this book is not to act as a textbook or a design manual for specific radios, but should instead be used by engineers who have a background in RF to better understand the challenges, requirements, circuits, and system-level techniques that can be used to design ultra-low-power short-range radios.

Organization of the Book

This book is organized into 12 chapters. To set the tone, the first chapter begins with an overview of general trends in low-power radio design, including a benchmarking section that covers state-of-the-art performance in this space. Then, chapter "Channel Modeling for Wireless Body Area Networks" discusses channel modeling, with a specific emphasis on the body channel, in order to help the reader better understand the requirements placed on ultra-low-power short range radios. Following these discussions, the book dives into the details of specific use-cases and implementations of ultra-low-power short-range radios, starting with narrowband radios, then moving to alternative forms of wireless communication, and concluding with very short-range communications and energy management.

On the topic of narrowband radios, chapter "Circuit Techniques for Ultra-Low Power Radios" discusses design techniques that, through crystal-based injection-locking approaches, enable the achievement of ultra-low-power operation without the use of phase-locked loops (PLLs). As an alternative approach, chapter "Architectures for Ultra-Low-Power Multi-Channel Resonator-Based Wireless Transceivers" introduces a similar concept, but in this case using higher frequency high-Q resonators as PLL and front-end filter replacements. Chapter "Ultra-Low Power Wake-Up Radios" then considers low-power wake-up radios for applications that have asynchronous, event-driven communication needs. Chapter "Commercially Viable Ultra-Low Power Wireless" reviews trends and techniques that are appropriate for commercial, standards-driven radios, and chapter "Synchronization Clocks for Ultra-Low Power Wireless Networks" discusses ultra-low-power timing circuits necessary for synchronization amongst radios in both standards-based and custom radio networks.

In applications that require ultra-high throughput or secure, non-radiating communication, techniques other than narrowband communication may offer superior performance. To this end, chapter "Pulsed Ultra-Wideband Transceivers" discussed ultra-wideband (UWB) circuits and systems for use in ultra-low-power short-range applications. For on-body communication applications, chapter "Human Body Communication Transceiver for Energy Efficient BAN" reviews the history and recent trends regarding human-body communications (HBC) and its application to wearable, biomedical monitoring devices.

Chapters "Centimeter-Range Inductive Radios" and "Near-Field Wireless Power Transfer" then introduce near-field communication (NFC) and power delivery concepts for applications where communication over a few centimeters is desired.

Specifically, chapter "Centimeter-Range Inductive Radios" focuses on increasing the throughput of NFC links, while chapter "Near-Field Wireless Power Transfer" reviews resonant coupling theory and derives formulae describing optimal conditions for efficient or maximal delivery of wireless power. Finally, chapter "Energy Harvesting Opportunities for Low-Power Radios" concludes the book with a discussion on energy harvesting and energy management circuits and devices that are appropriate to extend operational lifetime of many ultra-low-power short-range wireless devices.

Acknowledgements

Putting together a book of this size and scope would not have been possible without the help of many people. We would first and foremost like to thank all of the contributing authors for their insightful content that comprises the majority of the book—your efforts are greatly appreciated. We would also like to thank the editing and support staff at Springer, especially Charles Glaser and Jessica Lauffer.

San Diego, CA, USA Patrick P. Mercier
Cambridge, MA, USA Anantha P. Chandrakasan

About the Authors

Patrick P. Mercier is an Assistant Professor of Electrical and Computer Engineering, and the Associate Director of the Center for Wearable Sensors, both at the University of California, San Diego (UCSD). His research interests include the design of energy-efficient microsystems, focusing on the design of RF circuits, power converters, and sensor interfaces for mobile electronics and biomedical applications. Prior to joining UCSD, he completed his Ph.D. degree in Electrical Engineering and Computer Science at the Massachusetts Institute of Technology (MIT). Prof. Mercier has received numerous awards include the *IEEE International Solid-State Circuits Conference (ISSCC)* Jack Kilby Award for Outstanding Student Paper, a Beckman Young Investigator Award, the Hellman Foundation Award, a UCSD ECE Graduate Teaching Award, an Intel Ph.D. Fellowship, a Natural Sciences and Engineering Council of Canada (NSERC) Julie Payette and Post Graduate fellowships, amongst others. He has over 70 publications and invited presentations at venues such as *ISSCC, IEEE Journal of Solid-State Circuits, and Nature Biotechnology.* Prof. Mercier currently serves as an Associated Editor for the *IEEE Transactions on Biomedical Circuits and Systems and the IEEE Transactions on VLSI.*

Anantha P. Chandrakasan is the Joseph F. and Nancy P. Keithley Professor of Electrical Engineering at the Massachusetts Institute of Technology, Cambridge. He is the Head of the MIT EECS Department. He has received several awards including the 2009 Semiconductor Industry Association (SIA) University Researcher Award and the 2013 IEEE Donald O. Pederson Award in Solid-State Circuits. His research interests include micro-power digital and mixed-signal integrated circuit design, wireless microsensor system design, portable multimedia devices, energy efficient radios, and emerging technologies. He has served as the Conference Chair for the *IEEE International Solid-State Circuits Conference* (*ISSCC*) since 2010.

Contents

Introduction to Ultra Low Power Transceiver Design

Dhongue Lee and Patrick P. Mercier

Abstract Design of radios with ultra-low-power consumption can enable many new and exciting applications ranging from wearable healthcare to Internet of Things devices and beyond. Achieving low power operation is usually an exercise in trading-off important performance metrics with power. This chapter presents an overview of state-of-the-art narrowband architectures and techniques that achieve ultra-low-power operation, and concludes with a section that benchmarks recent state-of-the-art designs in order to illustrate power-performance trade-offs.

Keywords Low power transceiver • Ultra low power radio design

1 Introduction

Recent advancements in integrated radio design have enabled many new applications ranging from wearable healthcare or fitness monitors to Internet of Things (IoT) devices, structural integrity monitors, and beyond. In many of these applications, device size and battery life are of critical importance. Since radios often consume a significant portion of the power budget in small sensing nodes [1], reducing radio power consumption can be an impactful way to effectively decrease device size or increase operational lifetime. Reducing radio power can be challenging, however, as there are important tradeoffs between power consumption and performance metrics such as radiated output power, linearity, sensitivity, channelization capabilities, and interference sensitivity. Low-power radio designs often sacrifice one or more of these metrics in the pursuit of low overall power consumption.

The purpose of this chapter is to briefly introduce the main challenges facing narrowband ultra-low-power (ULP) design. This chapter will begin by first defining what is meant by ULP, will introduce common architectures that achieve low-power operation, and will conclude with some benchmarking data across various

D. Lee • P.P. Mercier (✉)
Department of Electrical and Computer Engineering, University of California, San Diego (UCSD), La Jolla, CA, USA
e-mail: dhl034@ucsd.edu; pmercier@ucsd.edu

P.P. Mercier, A.P. Chandrakasan (eds.), *Ultra-Low-Power Short-Range Radios*, Integrated Circuits and Systems, DOI 10.1007/978-3-319-14714-7_1

architectures. More detailed descriptions of circuit techniques for narrowband ULP radios can be found in chapters "Circuit Techniques for Ultra-Low Power Radios", "Architectures for Ultra-Low-Power Multi-Channel Resonator-Based Wireless Transceivers", "Ultra-Low Power Wake-Up Radios", "Commercially Viable Ultra-Low Power Wireless", and "Synchronization Clocks for Ultra-Low Power Wireless Networks", while chapters "Pulsed Ultra-Wideband Transceivers", "Human Body Communication Transceiver for Energy Efficient BAN", and "Centimeter-Range Inductive Radios" focus on non-narrowband applications. A more detailed overview of ULP transmitters that includes PA design challenges can be found in [2], from which the first part of this chapter is based.

2 Link Budgeting

Wireless sensor networks (WSN) and *body area networks* (BAN) are two common areas that generally require short-range ULP transceivers. The purpose of this section is to briefly review path loss models and link budgets for these representative applications to derive minimum required output power. A more detailed discussion on path loss models is presented in chapter "Channel Modeling for Wireless Body Area Networks".

2.1 Wireless Sensor Networks

To come up with a general WSN link budget, consider a *representative* WSN system operating with a carrier frequency of 2.4 GHz at a communication distance of 10 m. Most WSN nodes operate in peer-to-peer ad-hoc networks, where each node can potentially act as a relay between other nodes. Consequently, WSN transceivers must balance power specifications evenly between transmit and receive modes in order to optimize system-level energy efficiency. As a result, receivers are typically designed to have an input sensitivity close to −90 dBm. The minimum transmitter output power can then be calculated by preparing a link budget using the Friis equation for free space as a baseline case:

$$P_{transmit} = P_{receiver} - G_{antenna} - 20\,\log_{10}\frac{\lambda}{4\pi D},$$

where λ is the carrier wavelength and D is the distance between nodes. This equation tells us that, in free space, a 10 m link suffers from 60 dB of path loss at 2,450 MHz. A typical surface-mount antenna at 2,450 MHz has a gain of 0 dBi, which leads to the minimum transmit power of −30 dBm under ideal condition. However, a WSN transceiver working in a hostile environment could experience as much as 30 dB of additional loss, for a total of 90 dB of path loss [3]. Therefore, a WSN transmitter should have a maximum output power of 0 dBm, our definition of a ULP transmitter.

2.2 *Body-Area Networks*

On the other hand, BAN applications have much lower transmission distances: 1–2 m is often sufficient to communicate information around the human body. This should theoretically result in a lower path loss than in WSNs: 40–46 dB in free space at 2.4 GHz. Unfortunately, the presence of the human body in BANs adds significant attenuation, resulting in a measured path loss that range from 40 to 80 dB [4]. As an added complication, this path loss is highly variable and depends not only on the carrier frequency and the distance between nodes, but also on the relative position of the body and its surrounding environment (discussed in [4] and chapter "Channel Modeling for Wireless Body Area Networks"). Fortunately, the frequency of this variation is limited by the response time of a human (hundreds of milliseconds), enabling relatively low-complexity automatic gain control loops to compensate for such variation. Additionally, while the channel itself has high losses, it can generally be modeled as a non-frequency selective channel with no resolvable multi-path, eliminating the requirement for complex multi-path cancellation schemes [5]. Other studies have shown slightly lower path loss results (by 10–15 dB) at 900 and 400 MHz [6] as a result of lower tissue conductivity and higher relative permeability. Operating at these frequencies, however, reduces the radiation efficiency of electrically small antennas, which may negate the path loss advantage when computing the system-level energy efficiency. As a result, there is no clear rule-of-thumb regarding carrier frequency selection in BANs, as the available size and location of the antenna affects this decision dramatically.

To calculate the generally required PA output power range in a BAN, we first exploit a natural property of the system: most BAN users will be wearing a smartphone or smartwatch platform that is energy-rich, at least in comparison to a wearable or implantable sensor node. Thus, we can utilize these smart devices in an energy-asymmetric star topology network, where the smart watch/phone platform acts as a highly-sensitive centralized base-station. Assuming a base-station receiver sensitivity of −100 dBm, as typically encountered in commercial Bluetooth receivers, along with path loss of 40–80 dB and 10 dB link margin, the most efficient PA implementation would dynamically alter its output power between −10 and −50 dBm depending on instantaneous channel conditions; −10 dBm is also the recommended transmit power according to IEEE802.15.6 BAN standard [7].

To put these number in perspective, recall that −10 dBm corresponds to 100 μW output power. It is very challenging to design all downstream blocks to consume well under 100 μW in order to limit the overall system power consumption. The rest of this chapter will thus review architectures and circuits that help address this problem.

3 Modulation Schemes

Achieving ULP operation requires optimization across all design layers, ranging from modulation schemes and architectures, to circuits and devices. One of the most direct ways to reduce power is to reduce the complexity of the modulation schemes, which usually also reduces the overall architectural complexity.

Figure 1 shows an example of a conventional superheterodyne transceiver architecture that adopts quadrature amplitude modulation (QAM) or phase-shift keying (PSK) modulation. These modulations schemes are spectrally efficient, and are therefore appealing in high throughput applications and/or in crowded frequency bands. They do, however, bring additional complexities into the architecture such as the requirement of a phase-locked loop (PLL) and a linear PA, and therefore generally have a higher power consumption than if simpler modulation schemes were employed. For example, low-power PLLs in the 900-MHz range typically consume 800 μW [8, 9], which approaches the total power budget of typical ULP transceivers. Furthermore, the long start-up time of most PLLs discourages their use in aggressively duty-cycled systems. Additionally, linearization techniques associated with the PA add additional power consumption, making QAM modulation difficult to incorporate without significant power overhead.

Adopting a non-coherent modulation scheme such as on-off keying (OOK) or frequency-shift keying (FSK) can greatly simplify the architecture, potentially eliminating the PLL and the requirement of a quadrature path, as illustrated in Fig. 2. These simple modulation schemes are also compatible with nonlinear but highly efficient PAs, enabling a further reduction in power consumption. Thus, designers of ULP transceivers generally favor direct-modulation transmitter architectures utilizing OOK or FSK modulation schemes (or in some cases, binary PSK (BPSK) if a high-stability local oscillator (LO) is used), instead of architectures that feature more spectrally efficient modulation schemes. For the same reason, receiver designers often favor envelope-detector and super-regenerative architectures for ULP

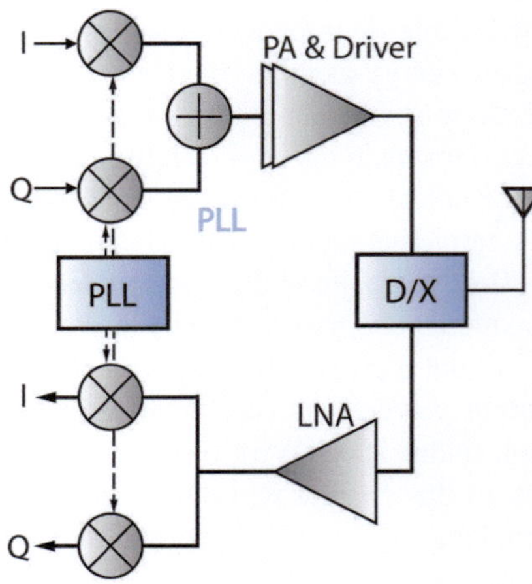

Fig. 1 A generic IQ transceiver architecture

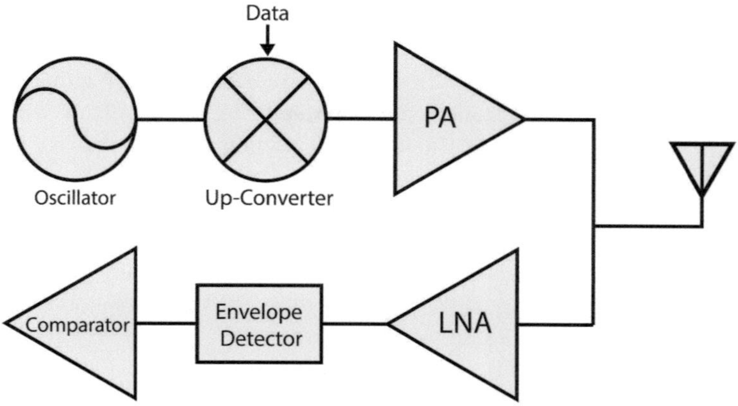

Fig. 2 A generic low-complexity transceiver architecture

applications. Unfortunately, adopting non-coherent modulation schemes and low-complexity architectures directly trade-off performance (in terms of throughput, spectral efficiency, etc.) for ULP operation. The next generation of ULP radios must find a way to achieve both high throughput/spectral efficiency at ultra-low powers in order to satisfy the throughput and efficiency demands of next-generation applications.

The method of generating an LO usually defines the architecture and ultimate power limits of a transmitter, and thus, various architectures that aim to minimize the power consumed in LO generation are discussed in Sect. 4. Section 5 will subsequently discuss ULP receiver architectures.

4 Low Power TX Architectures

There are four popular methods to synthesize the carrier frequency in ULP transmitters (TXs):

1. **Phase-locked loops**. The most popular LO generation method, at least at higher output powers, is to employ a PLL to synthesize a carrier frequency that is locked to an on-board reference circuit such as a crystal [10, 11]. Although this is the most robust carrier synthesis solution, continuously operating a PLL is too energy expensive for many ULP applications. Section 4.1 will discuss methods to duty-cycle the PLL for reduced power consumption.
2. **Free-running DCOs**. In applications employing non-coherent signaling with relatively wide bandwidths, it may not be necessary to achieve low phase noise or ultra-high LO precision. In such cases, employing a free-running Digitally Controlled Oscillator (DCO) may be an appropriate solution. Section 4.2 will present an example free-running DCO.

3. **Slow frequency correction loops**. In cases where good phase noise and/or BPSK modulation is required, it is possible to periodically calibrate what is an otherwise free-running DCO using a slow frequency-correction loop from either on-board reference or from a base-station, thereby stabilizing the oscillator without the power/complexity overhead of a full PLL. Section 4.3 will discuss two such possibilities.
4. **High-Q resonators**. Alternatively, a stable and low noise LO can be generated directly from a high-Q resonator without requiring a PLL, generally at the expense of frequency tunability. Section 4.4 will discuss methods to use low-frequency (e.g., crystal) or high-frequency (e.g., FBAR) resonators to generate low-power and stable RF LOs.

4.1 Duty-Cycled PLLs

Although they are power hungry, PLLs can robustly synthesize arbitrary RF carrier frequencies with low phase noise. One possible method to utilize a PLL in an energy-constrained system is to duty-cycle its operation as shown in [12, 13]. Figure 3 shows a representative binary FSK (BFSK) transmitter topology where a PLL is locked to an oscillator only prior to transmitting data [12].

With the loop open, the oscillation frequency of the voltage controlled oscillator (VCO—or the analog version of a DCO) can experience frequency-pulling by both noise and strong signals in adjacent channels. In the absence of the latter, the frequency drift can be minimized at 2.5 Hz/μs for a low-voltage multi-gigahertz VCO in a modern CMOS process [12, 14]. The system is designed with an assumption that only one TX is expected to operate in the area. However, in an area with a congested bandwidth, frequency pulling by a nearby strong interferer may become an issue and require multiple calibrations per data transmission.

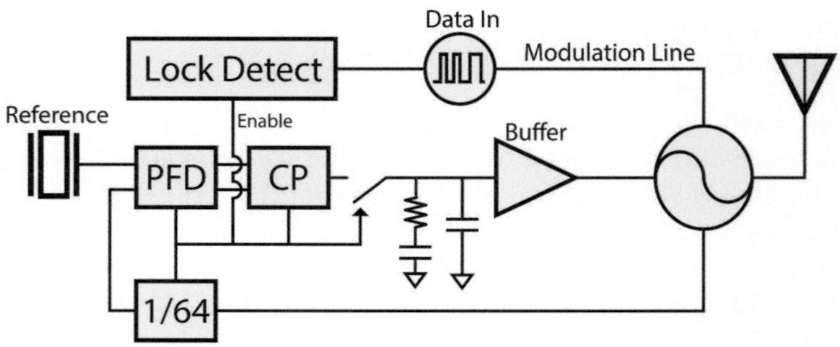

Fig. 3 A generic direct modulation transmitter architecture (from [12])

4.2 Free-Running DCOs

Free-running DCOs offer the ultimate in terms of a low-complexity design, and as a result can achieve extremely low active- and standby-mode power consumption [15–18]. Typically, using a free-running oscillator restricts modulation schemes to non-coherent FSK or OOK, and can only realistically be used when operating with a sufficiently wide bandwidth such that precise setting of phase and even frequency is not required. Since the human body is an excellent temperature regulator [19], implanted devices often exploit this architecture along with a well-regulated supply voltage to help stabilize the oscillator [15, 19, 20].

A representative free-running oscillator architecture is described in [15] and achieves an average power consumption of 78 pW through aggressive duty-cycling and ultra-low-leakage design. This particular transmitter was designed as part of a system that operated off of harvested energy from the endocochlear potential located within the inner-ear of mammals [21]. Figure 4 shows a simplified schematic of the single-stage, direct-RF transmitter consisting of a cross coupled NMOS pair loaded by an on-board inductive element that acts as both a resonant element and an antenna, further limiting the overall transmitter complexity and minimizing the number of stages that must operate at RF for minimal standby power consumption.

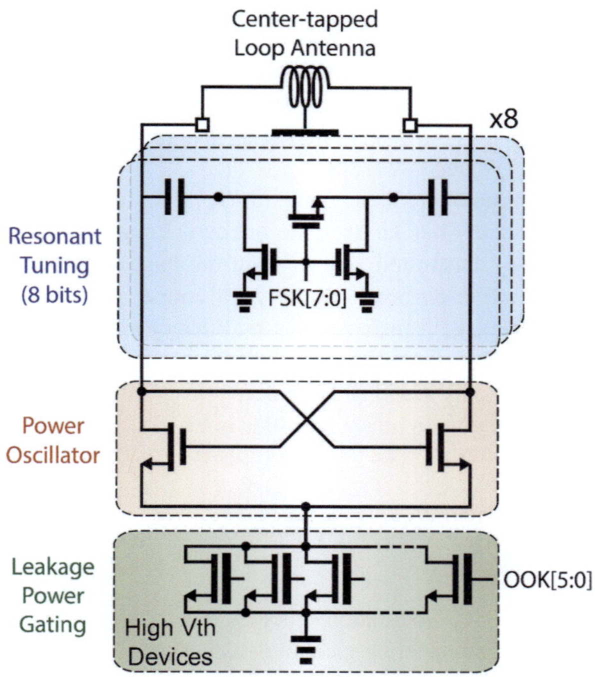

Fig. 4 Power oscillator architecture with a free running DCO (from [15])

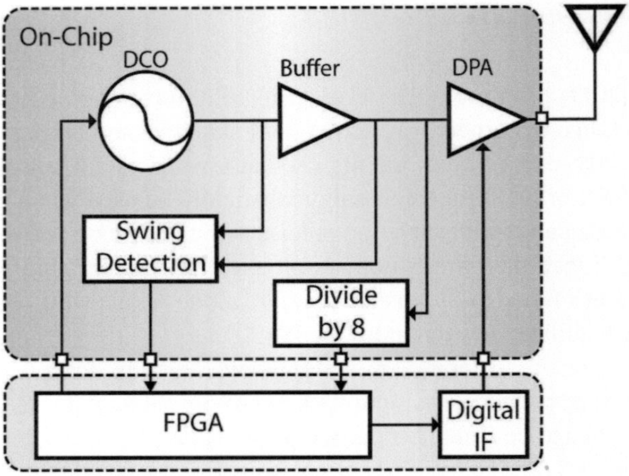

Fig. 5 Frequency calibration through FPGA (from [22])

4.3 Slow Frequency Correction Loops

In nonimplanted or temperature/voltage-regulated environments, the frequency of a free-running DCO is susceptible to frequency shifts, and may require periodic calibration. While such calibrations will not improve phase noise, they may be necessary to avoid drifting into an adjacent channel. An on-board microcontroller, if available, can be used for calibration as suggested in [22] and shown in Fig. 5. Here, a swing-detector and a divide-by-8 circuit are used to provide digital information to the FPGA for calibration purposes [22].

An alternative method of frequency calibration in star networks first recognizes that the base station likely has an accurate onboard timing reference. The sensor node can thus receive transmitted signals from the base station to extract timing information for calibration purposes [23, 24]. Of course, this requires an onboard receiver (which not all ULP sensing nodes may have) and time/power overhead. A representative FSK transmitter operating at 920 MHz is presented in [23]. This technique is nominally only reasonable in networks with a base station (e.g., star networks) and is not suitable in mesh or ad-hoc networks.

4.4 High-Q Resonator-Based TX

Utilizing higher-order PSK modulation schemes in ULP transmitter applications is often desirable due to their superior spectral density compared with OOK and FSK, while maintaining a constant envelop necessary to still incorporate efficient, non-linear PAs. However, the architectures presented in the preceding two subsections are not well suited to PSK modulation, in part due to the lack of a well-defined

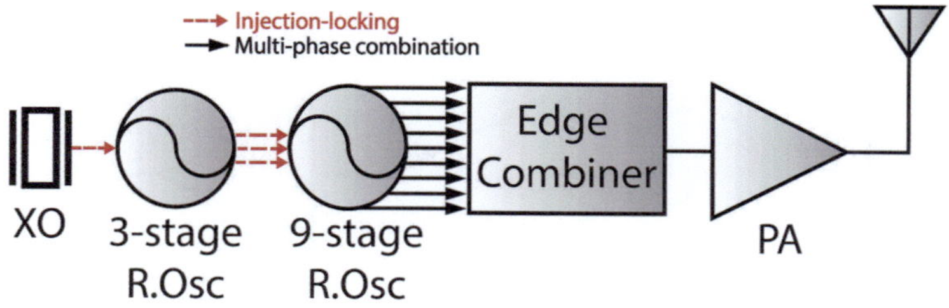

Fig. 6 Injection-locking-based transmitter architecture

LO phase throughout an entire packet. To combat this without requiring the use of a PLL, several architectures have proposed free-running LOs synthesized directly from high-Q resonators.

Figure 6 shows a high-level block diagram of an injection-locking based transmitter that was first introduced in [25] and also implemented with some modifications using sub-harmonic injection in [26, 27]. A radio-frequency identification (RFID) tag transmitter based on the same principle can be seen in [28].

In [25], a three-stage ring oscillator is injection-locked to an external crystal oscillator. Injection locking allows the ring oscillator, which can have very low power especially in advanced process nodes, to inherent the good phase noise performance of the crystal oscillator at low power. The three phases of the ring oscillator are used as an injection signal to lock a 9-stage ring oscillator, producing 9 phases that, when edge combined, multiply the frequency by 9×, thereby producing the desired RF carrier. It should be noted, however, that the limited tunability of crystal resonators prevents this architecture from being used in applications that demand multi-channel access, which is a very significant drawback in congested environments.

Another promising way to achieve good LO frequency stability without requiring a PLL is to use a high-Q resonator operating directly at RF. For example, Film Bulk Acoustic Resonator (FBAR) or Surface Acoustic Wave resonator (SAW) have high quality factors and operate directly at common RF frequencies. Thus, they can be used in simple direct-modulation architectures, as shown in Fig. 7 [29–32]. Due to their limited tunability, multiple resonators may be needed to achieve channel selection, as demonstrated in [31] and illustrated in Fig. 8. In this architecture, the MSK (Minimum Shift Keying: an FSK modulation with minimum spacing between frequencies) transmitter multiplexes the FBARs by using transmission gates and a buffer.

The MSK transmitter achieves data rate of 1 Mbps and output power up to −2.5 dBm while consuming 550 μW. The use of high-Q resonators enable a phase noise of −132 dBc/Hz at 1 MHz offset, but also comes with 4 μs start-up time and may not be compatible with deeply duty-cycled systems. More details about this architecture are described in chapter "Architectures for Ultra-Low-Power Multi-Channel Resonator-Based Wireless Transceivers".

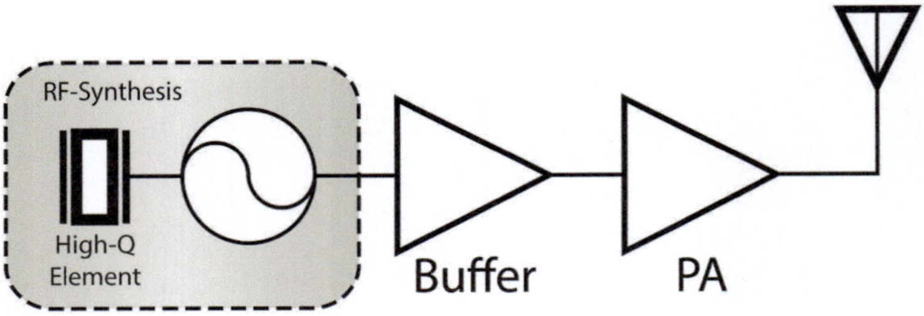

Fig. 7 FBAR-based transmitter architecture

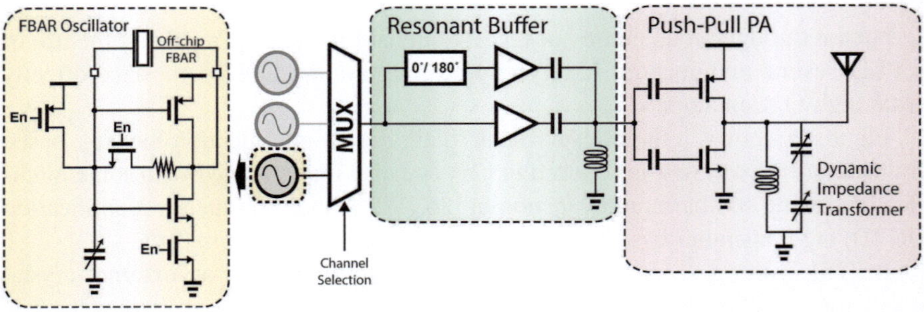

Fig. 8 Schematics of multichannel FBAR-based transmitter (from [31] and discussed in detail in chapter "Architectures for Ultra-Low-Power Multi-Channel Resonator-Based Wireless Transceivers")

5 Low Power RX Architectures

As was the case for transmitters, the power consumption of a receiver (RX) is highly influenced by the modulation scheme chosen, which then defines the overall receiver architecture. Receivers can be broadly categorized by the following demodulation schemes:

1. **Clocked demodulation**. Broadly, RF energy is first mixed down to a lower frequency (or two) before demodulation occurs. In general, a low phase noise LO is desired, though not strictly required. This category can be further sub-categorized into the following architectures:

   a. *Super-heterodyne/heterodyne RX*. Although the traditional dual down-conversion architecture is extremely robust, as seen in Fig. 1, it requires two LOs and an additional mixer and thus has difficulty achieving ULP operation. Section 5.1 discusses several techniques to minimize the power consumption in such architectures.
   b. *Homodyne (zero-IF) RX*. It is generally more energy-efficient to operate at baseband than at RF, and for this reason, homodyne architectures save power by directly down-converting the input signal to baseband. However, homodyne receivers are well-known to suffer from DC offsets, flicker noise,

LO leakage, and other issues. Thus, low-power methods to address these issues are required, several of which are also discussed in Sect. 5.1.

2. **Energy/Envelope detection**. A PLL can be amongst the most power hungry blocks in ULP receivers, and as a result, receiver architectures that eliminate the requirement of a PLL can more easily achieve ULP operation. Eliminating the PLL, however, generally either reduces LO precision, making coherent demodulation difficult, or through high-Q resonators precludes multi-channel operation. Instead, it is possible to perform non-coherent demodulation by observing the signal's energy level either directly at RF, or after down conversion to an imprecise intermediate frequency. Naturally, doing so relies on less spectrally efficiency modulation schemes (e.g., OOK), and has difficulty dealing with blockers. Section 5.2 describes methods to perform envelope/energy detection in more detail.
3. **Super-regenerative receiver**. A super-regenerative receiver (SRR) achieves ultra-high gain using a low-complexity unstable network in an efficient and controlled manner. While most SRRs indeed have envelope/energy detectors, SRRs have sufficiently different requirements to consider them separately. Section 5.3 describes the basic operation of a super-regenerative receiver and presents examples from the recent literature.

5.1 Clocked Demodulator

The power consumption of a receiver is normally dominated by frequency synthesis, RF amplification, and the LO buffer. Although the dual down-conversion architecture is very robust for demodulating data while rejecting unwanted signals, the requirement of multiple down-conversion mixers and two reference signals is often prohibitively expensive from a power perspective in ULP applications. Consequently, low-IF and zero-IF architectures have gained popularity in ULP radio design due to their low implementation complexity (i.e., the minimal number of blocks that consume power). Image rejection problems associated with low-IF receiver architectures can be solved by using high-Q resonators as image rejection filters. Furthermore, certain low power standards, such as ZigBee, require very loose specifications of image rejection and channel filtering [33], potentially saving implementation complexity, though at the expense of an increased chance of interference. Additional power can be saved by adopting simpler modulation schemes such as OOK, FSK, and low-index PSKs, though at the expense of reduce spectral efficiency.

While such system-level simplifications can decrease power consumption, they may not be sufficient to meet ULP power budget constraints. Thus, efforts have been made to further lower the power consumption in such architectures by replacing a PLL with a clever method of frequency synthesis [34–36], lowering the supply voltage [17, 37, 38], and replacing the LNA with a passive mixer front end [17].

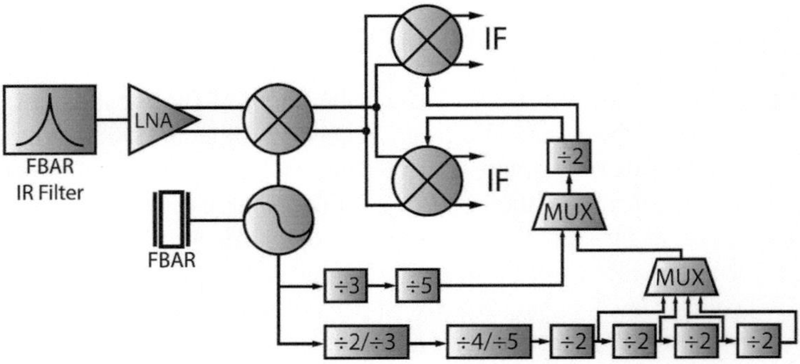

Fig. 9 PLL-less FBAR-based super-heterodyne RX schematics [36]

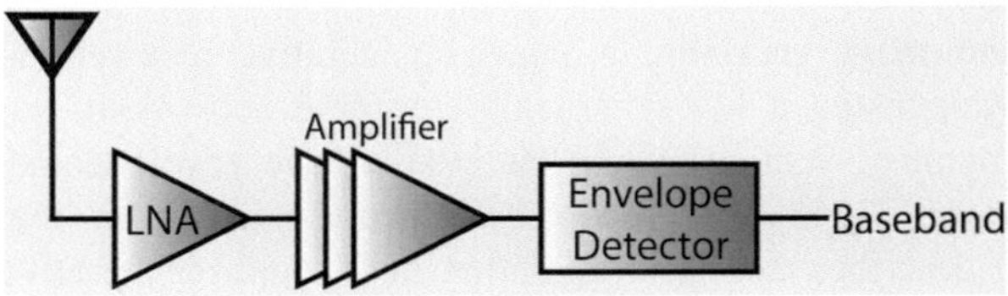

Fig. 10 Schematics of envelope-detector-based RX frontend

Figure 9 shows an example of a representative FBAR-based multi-channel super-heterodyne receiver architecture [36]. Here, multi-channel operation has been achieved by first down-converting the whole channel band to a wideband IF (5–80 MHz), while using multiple frequency dividers driven by the resonator in order to define all necessary channels in the 2.4 GHz ISM band, as illustrated in Fig. 9.

5.2 *Energy/Envelope Detector-Based RX*

The high power consumption associated with coherent demodulation often steers designers to choose simpler non-coherent modulation scheme such as OOK or FSK at the cost of reduced spectral efficiency. Energy or envelope detector-based receiver architectures further save power consumption by removing the need for an accurate high frequency clock. The key difference between energy and envelope detection is whether a self-mixer or an envelope detector is used, where the former implements an actual squaring operation.

Figure 10 shows a schematic of a generic envelope-detector-based receiver. Envelop and energy detectors generally offer extremely low-power operation compared to all other demodulators, though they suffer from poor blocker rejection and SNR due to translation of blockers to DC and minimum detectable signals set

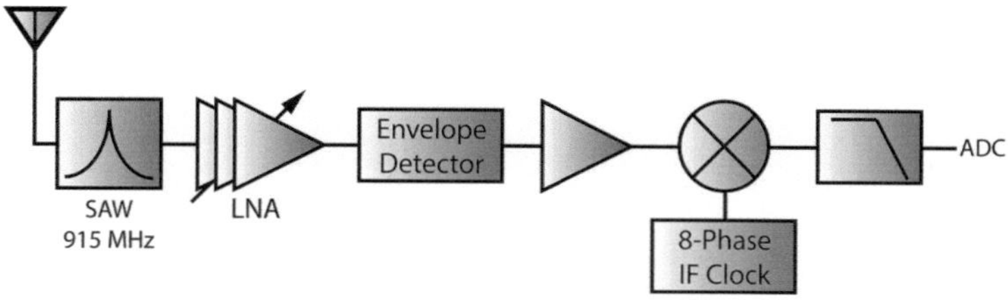

Fig. 11 Schematics of 2-tone FSK receiver [22]

by non-linear elements. These problems can potentially be mitigated by using a 2-tone modulation scheme with a high-gain LNA [22], or by using an uncertain-IF architecture [39].

An example 2-tone receiver architecture is shown in Fig. 11. Unlike a traditional envelope detector that down-converts any signal (potentially including interferers) to DC, in a 2-tone system the signal is transmitted at two separate frequencies with a known frequency offset, such that the intermodulation between these two signals lies at a known-IF, which can then be filtered and demodulated with substantial blocker rejection. The implementation in [22] uses the best phase-aligned LO signal amongst 8 phases in order to demodulate the signal without using a quadrature path.

An uncertain-IF architecture utilizes an imprecise and thus power-efficient LO (e.g., a ring oscillator) to down-convert the desired signal to a wide, uncertain IF to take advantage of the large gain attainable at such frequencies. However, down-conversion introduces image content, and a high-Q RF filter, such as bulk acoustic wave resonators (BAW), must be used as image rejection filters [39], which in some cases may be cost or area prohibitive.

5.3 Super-Regenerative Receiver

Super-regenerative receivers (SRR) use an oscillator with a variable bias current to steer the two complex poles of the oscillator from the left half *s*-plane to the right half plane, effectively "oscillating" and "quenching" the system in a non-linear fashion. During the start-up of the super-regenerative oscillator (SRO), any small signal and noise in the vicinity of the oscillator's natural frequency is exponentially amplified, thereby achieving enormous gain—much higher gain than an open loop amplifier. A representative SRR is shown in Fig. 12. Most SRRs use an envelope detector to demodulate OOK or other amplitude modulated signals. Unlike other envelope-detector-based receivers, an SRR can have an arbitrarily large gain (limited by the quench period and power supply rail) and does not nominally suffer from an envelope detector's low detection threshold.

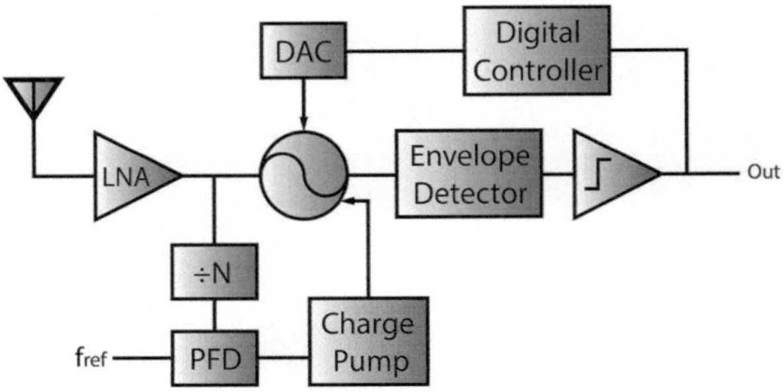

Fig. 12 Schematics of super-regenerative receiver with BW calibration [41]

While super-regenerative amplifiers achieve the largest gain-to-current ratio of any competing amplifier topology by exploiting the positive feedback growth characteristics of a building oscillation, any blocker in a nearby channel can force an oscillation independent of the presence of a signal at the desired band. Consequently, SRR architectures are very susceptible to blockers, and the susceptibility is inversely proportional to the rate at which the transconductance of the oscillator, or the quench signal, grows [40]. An effort has been made to mitigate this problem at the cost of additional complexity and data rate by calibrating the quench signal with a digital feedback loop as shown in Fig. 12 [41].

6 Benchmarking

In contrast to traditional high-output-power systems where PAs dominate system power budgets and maximizing their efficiency is imperative, the absolute efficiency of PAs in ULP transmitters is not as critically important. For example, if LO generation, mixing, and baseband circuits require 1 mW, the difference in overall system power consumption for a PA radiating −10 dBm (100 μW) at a PA efficiency of 40 versus 50 % is approximately 4 %. Thus, a more important figure of merit in ULP transmitter design is transmitter efficiency: $\eta_{TX} = P_{rad,out}/P_{total}$, where $P_{rad,out}$ is the radiated output power, and P_{total} is the total transmitter power consumption. Figure 13 plots this metric for recently published ULP transmitters and two representative commercial transmitters [42, 43], illustrating the difficulty of achieving high transmitter efficiency at low output powers. Next-generation designs should thus endeavor to further minimize the power consumption of the downstream blocks to further increase the overall transmitter efficiency.

Another metric of critical importance when designing energy-constrained systems is the energy required to transmit a bit of information: $E_{bit} = P_{total}/DR$, where DR is the data rate. While providing useful information to system designers, E_{bit} does not facilitate a fair or direct comparison between transmitter designs, as output power and bandwidth can both have a substantial effect on E_{bit}. For

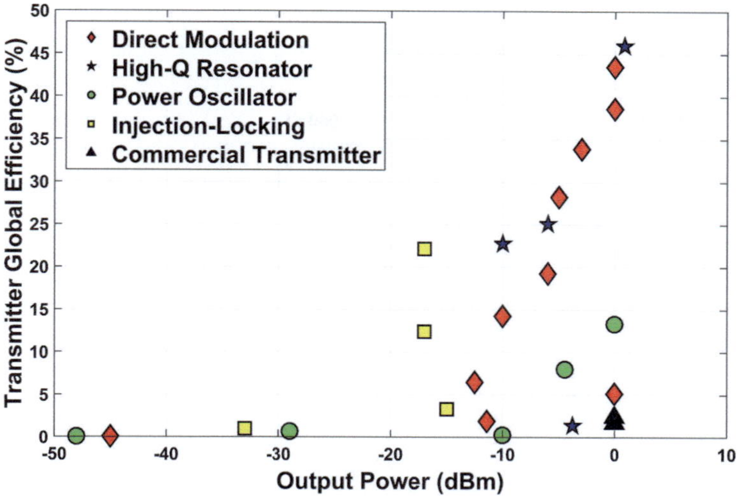

Fig. 13 A survey of the global transmitter efficiency of recently-published ULP transmitters

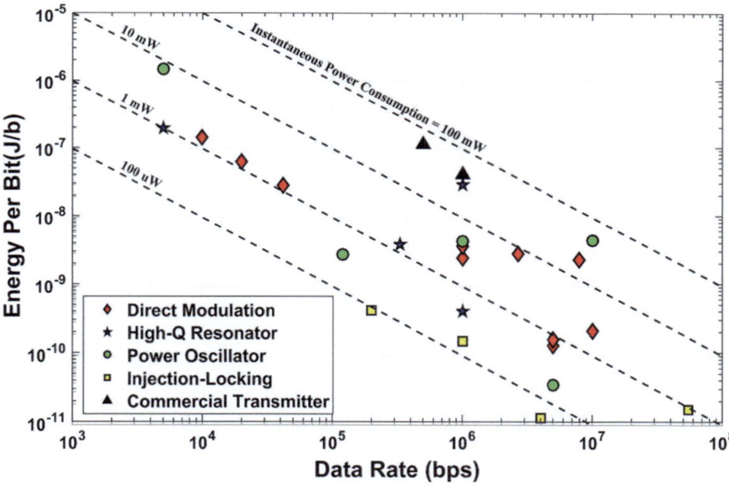

Fig. 14 A survey of the energy efficiency of recently-published ULP transmitters

example, a low output power and wideband TX will have a very low E_{bit}; however, communication distance will be short, and the radiated spectrum may violate standards, limiting FDMA opportunities, and increasing the possibility of interference.

To illustrate this, Fig. 14 shows a plot of E_{bit} versus data rate for recently published transmitters. Since E_{bit} is inversely proportional to data rate, it is generally possible to increase the data rate with little power overhead to achieve a lower E_{bit}. For example, it does not cost much in power to increase the modulation rate from

1 to 10 Mbps in an OOK transmitter. Doing so, however, significantly increases the occupied bandwidth, and may not be practical in multi-user scenarios. Thus, when interpreting an E_{bit} transmitter plot, one should draw a vertical line at the data rate compatible with the spectral resource available in the given application/standard, and only look at transmitter to the left of this line.

The required data rate in many WSN and BAN cases is often much lower than the maximum achievable radio data rates. Does this mean such applications fundamentally require higher energy-per-bit? The answer is decidedly no, since the above E_{bit} plot illustrates active-mode energy only, at the instantaneous data rate reported in the published work. Duty-cycling can be employed to effectively shift a data point to the left—that is, a radio that is efficient at 1 Mbps can likely also be efficient at 100 kbps, provided it is capable of rapid and/or efficient duty-cycling. Figure 15 illustrates this by plotting duty-cycled energy-per-bit, $E_{bit,DC}$, versus a very wide range of duty-cycled data rates for representative transmitters that publish standby power numbers [15, 16, 20, 42–44]. Specifically, the plot is constructed by computing:

$$E_{bit,DC} = E_{bit} + \frac{P_{standby}}{DR_{DC}}$$

where E_{bit} is the active-mode energy-per-bit ($= P_{active}/DR_{instantaneous}$), $P_{standby}$ is the transmitter's standby power, and DR_{DC} is the duty-cycled data rate. For simplicity, the overhead for duty-cycling is assumed to be negligible, though this may not be the case in practice, particularly in designs that have long start-up times (e.g., those with PLLs). Most of the published ULP transmitters that specify standby power

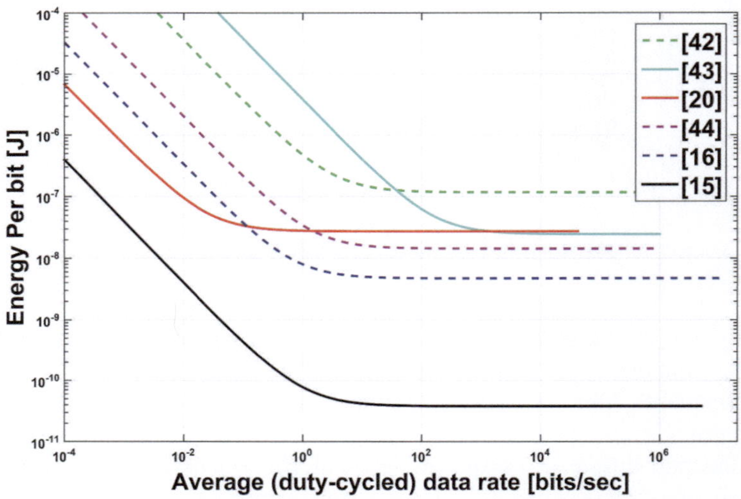

Fig. 15 Energy per bit vs. average (duty-cycled) data rate

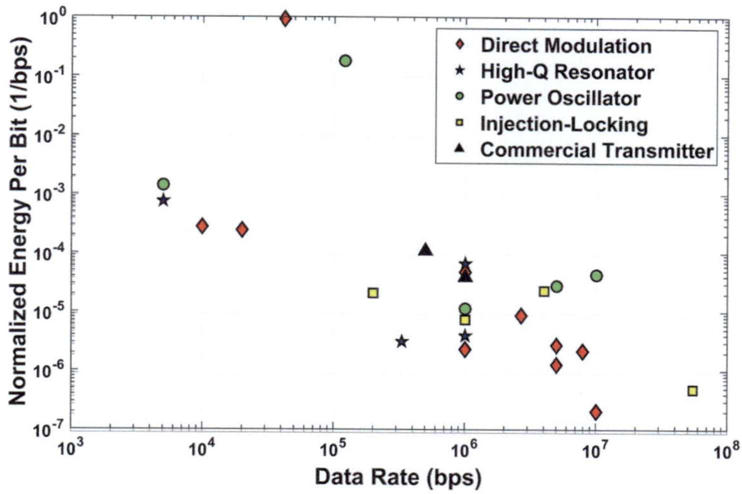

Fig. 16 Normalized energy per bit vs. data rate for representative transmitters

scale nicely down to low data rates, though eventually quiescent and leakage power, integrating over very long time intervals, result in increased energy-per-bit at very low average data rates.

Duty-cycling, however, still does not address the issue that low output power designs naturally achieve superior E_{bit}. To facilitate a more fair comparison between competing designs, E_{bit} can be normalized by the radiated output power to create the following normalized energy-per-bit figure of merit [13, 45]:

$$FOM_{E_{nom}} = \frac{E_{bit}}{P_{rad,out}}.$$

A lower $FOM_{E_{nom}}$ is preferable, as less energy is required to transmit a bit of information at a higher radiated output power. Figure 16 illustrates this FOM for representative transmitters. The same data rate caveat applies here as above.

When it comes to ULP receivers, the specifications of interests are power consumption, data rate, and sensitivity. For this reason, energy-per-bit and sensitivity are plotted against data rate in Figs. 16 and 17 respectively. Receivers are categorized into three architectures: Clocked-demodulators [13, 34–36, 46–54], envelope-detectors [3, 22, 23, 30, 39, 55], and super-regenerative [19, 26, 32, 41, 56, 57].

Figure 17 shows that some clocked-demodulator-based receivers consume just as low power as envelope-detector-based and super-regenerative receivers by replacing a PLL with a high-Q resonator [34], or turning off the PLL during the receiver operation [13]. Figure 18 shows that super-regenerative receiver and clocked demodulator architectures attain similar levels of performance given the trade-off

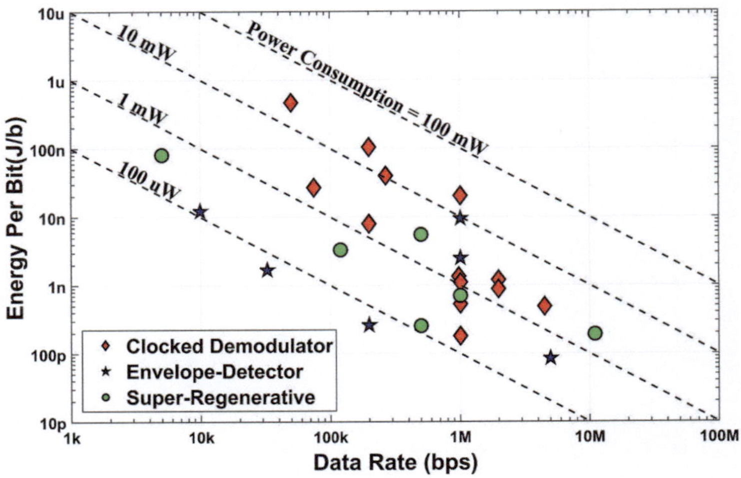

Fig. 17 A survey of the energy efficiency of recently-published ULP receivers

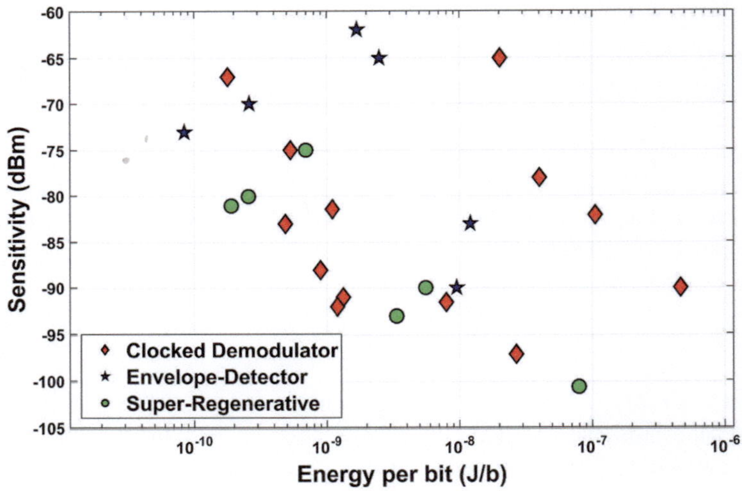

Fig. 18 A survey of the sensitivity of recently-published ULP receivers

between sensitivity and energy-per-bit. Envelope/energy detectors consume the least amount of power, but also generally exhibit inferior sensitivity than the other architectures. Consequently, an envelope detector can be used at the expense of non-linearity, blocker rejection, and sensitivity when the absolute power consumption is the most important constraint. However, a more favorable trade-off between power consumption and sensitivity can be achieved by using super-regenerative or high-Q resonator-based clocked demodulation architectures for a given energy-per-bit.

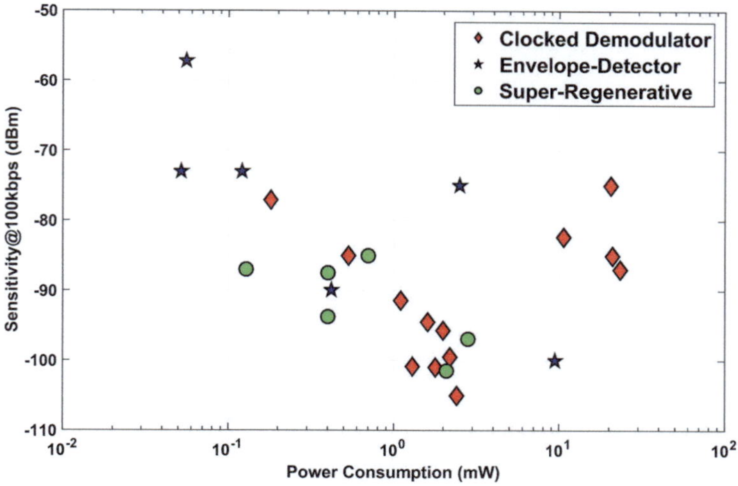

Fig. 19 A survey of the scaled sensitivity @ 100 kbps of recently-published ULP receivers

To foster a better comparison between the architectures, receiver sensitivity can be normalized to the same data rate, in this example arbitrarily chosen as 100 kbps. The normalized sensitivity can be derived for each receiver since data rate scales linearly with bandwidth and thus inversely with sensitivity [58]:

$$Sensitivity_{@100kbps} = Sensitivity - 10\log_{10}\frac{Data\ Rate}{100\ kbps}$$

For example, a 10 times increase in data rate would result in 10 dB higher sensitivity. Figure 19 shows a plot of normalized sensitivity at 100 kbps versus power consumption. Here it can be seen more clearly that envelope-detectors suffer a slight degradation in sensitivity compared to the best performance of other receiver architectures from the recent literature.

7 Conclusions

Ultra-low-power narrowband radios can open up many unique applications ranging from Internet-of-Things and industrial sensor networks, to wearable sensors, healthcare devices, and beyond. Achieving ultra-low-power consumption while maintaining robust operation involves difficult trade-offs between output power, data rate, bandwidth, channel selectivity, sensitivity, and energy efficiency that must be overcome through a combination of innovative circuit design, novel architectures, and system-level considerations. This chapter has introduced typical architectures used in ULP radios, discussed their relative merits, and provided some benchmarking data to help identify what architectures might make the most sense given system-level specifications. While optimal implementations depend strongly on the given application, in general the most efficient radios employ low-complexity modulation schemes (e.g., OOK, FSK, and possibly BPSK), and are run by an efficient LO stabilized without a PLL.

That being said, this chapter was just an introduction. Chapters "Circuit Techniques for Ultra-Low Power Radios", "Architectures for Ultra-Low-Power Multi-Channel Resonator-Based Wireless Transceivers", "Ultra-Low Power Wake-Up Radios", "Commercially Viable Ultra-Low Power Wireless", and "Synchronization Clocks for Ultra-Low Power Wireless Networks" describe trade-offs and techniques in narrowband radios in more detail. Before doing so, chapter "Channel Modeling for Wireless Body Area Networks" describes channels models typically found in BAN applications in order to better appreciate the communication requirements in representative ULP environments.

References

1. J.M. Rabaey, J. Ammer, T. Karalar, S. Li, B. Otis, M. Sheets, T. Tuan, PicoRadios for wireless sensor networks: the next challenge in ultra-low power design, in *IEEE International Digest of Technical Papers. Solid-State Circuits Conference*, 2002, pp. 2001–2002
2. D.-G. Lee, L.G. Salem, P.P. Mercier, Ultra-low-power transmitter design. *Microwave Magazine*, April 2015
3. S. Sayilir, W.-F. Loke, J. Lee, H. Diamond, B. Epstein, D.L. Rhodes, B. Jung, A −90 dBm sensitivity wireless transceiver using VCO-PA-LNA-switch-modulator co-design for low power insect-based wireless sensor networks. IEEE J. Solid-State Circuits **49**(4), 996–1006 (2014)
4. P.S. Hall, Y.I. Nechayev, A. Alomainy, C.C. Constantinou, C. Parini, M.R. Kamarudin, T.Z. Salim, D.T.M. Hee, R. Dubrovka, A.S. Owadally, A. Serra, P. Nepa, M. Gallo, M. Bozzetti, Antennas and propagation for on-body communication systems. IEEE Antenn. Propag. Mag. **49**(3), 41–58 (2007)
5. D. Smith, D. Miniutti, L. Hanlen, A. Zhang, D. Lewis, D. Rodda, B. Gilbert, Power delay profiles for dynamic narrowband body area network channels. *IEEE 802.15.6 standard*, 2009
6. A. Fort, F. Keshmiri, G.R. Crusats, C. Craeye, C. Oestges, A body area propagation model derived from fundamental principles: analytical analysis and comparison with measurements. IEEE Trans. Antenn. Propag. **58**(2), 503–514 (2010)
7. IEEE Standard for Local and metropolitan area networks – Part 15.6: Wireless Body Area Networks. *IEEE Std 802.15.6*, 2012
8. G. Devita, A.C.W. Wong, N. Kasparidis, P. Corbishley, A. Burdett, P. Paddan, A 0.9 mW PLL integrated in an ultra-low-power SoC for WPAN and WBAN applications, in *2010 Proceedings of ESSCIRC*, 2010, pp. 158–161
9. W. Deng, D. Yang, T. Ueno, T. Siriburanon, S. Kondo, K. Okada, A. Matsuzawa, 15.1 A 0.0066 mm^2 780 μW fully synthesizable PLL with a current-output DAC and an interpolative phase-coupled oscillator using edge-injection technique, in *IEEE International Solid-State Circuits Conference Digest of Technical Papers (ISSCC)*, 2014, pp. 266–267
10. J. Gil, J.-H. Kim, C.S. Kim, C. Park, J. Park, H. Park, H. Lee, S.-J. Lee, Y.-H. Jang, M. Koo, J.-M. Gil, K. Han, Y.W. Kwon, I. Song, A fully integrated low-power high-coexistence 2.4-GHz ZigBee transceiver for biomedical and healthcare applications. IEEE Trans. Microw. Theory Tech. **62**(9), 1879–1889 (2014)

11. A. Molnar, B. Lu, S. Lanzisera, B. W. Cook, K.S.J. Pister, An ultra-low power 900 MHz RF transceiver for wireless sensor networks, in *Proceedings of the IEEE 2004 Custom Integrated Circuits Conference (IEEE Cat. No.04CH37571)*, 2004, pp. 401–404
12. V. Karam, P.H.R. Popplewell, A. Shamim, J. Rogers, C. Plett, A 6.3 GHz BFSK transmitter with on-chip antenna for self-powered medical sensor applications, in *IEEE Radio Frequency Integrated Circuits (RFIC) Symposium*, 2007, pp. 101–104
13. M. Vidojkovic, X. Huang, P. Harpe, S. Rampu, C. Zhou, L. Huang, J. van de Molengraft, K. Imamura, B. Busze, F. Bouwens, M. Konijnenburg, J. Santana, A. Breeschoten, J. Huisken, K. Philips, G. Dolmans, H. de Groot, A 2.4 GHz ULP OOK single-chip transceiver for healthcare applications. IEEE Trans. Biomed. Circuits Syst. **5**(6), 523–534 (2011)
14. A. Yamagishi, M. Ugajin, T. Tsukahara, A 1-V 2.4-GHz PLL synthesizer with a fully differential prescaler and a low-off-leakage charge pump, in *IEEE MTT-S International Microwave Symposium Digest*, 2003, vol. 2, pp. 733–736
15. P.P. Mercier, S. Bandyopadhyay, A.C. Lysaght, K.M. Stankovic, A.P. Chandrakasan, A sub-nW 2.4 GHz transmitter for low data-rate sensing applications. IEEE J. Solid-State Circuits **49**(7), 1463–1474 (2014)
16. G. Chen, H. Ghaed, R. Haque, M. Wieckowski, Y. Kim, G. Kim, D. Fick, D. Kim, M. Seok, K. Wise, D. Blaauw, D. Sylvester, A cubic-millimeter energy-autonomous wireless intraocular pressure monitor, in *2011 IEEE International Solid-State Circuits Conference*, 2011, pp. 310–312
17. B.W. Cook, A. Berny, A. Molnar, S. Lanzisera, K.S.J. Pister, Low-power 2.4-GHz transceiver with passive RX front-end and 400-mV supply. IEEE J. Solid-State Circuits **41**(12), 2757–2766 (2006)
18. X. Huang, P. Harpe, X. Wang, G. Dolmans, H. de Groot, A 0 dBm 10 Mbps 2.4 GHz ultra-low power ASK/OOK transmitter with digital pulse-shaping, in *IEEE Radio Frequency Integrated Circuits Symposium*, 2010, pp. 263–266
19. J.L. Bohorquez, A.P. Chandrakasan, J.L. Dawson, A 350 μW CMOS MSK transmitter and 400 μW OOK super-regenerative receiver for medical implant communications. IEEE J. Solid-State Circuits **44**(4), 1248–1259 (2009)
20. E.Y. Chow, S. Chakraborty, W.J. Chappell, P.P. Irazoqui, Mixed-signal integrated circuits for self-contained sub-cubic millimeter biomedical implants, in *IEEE International Solid-State Circuits Conference – (ISSCC)*, 2010, pp. 236–237
21. P.P. Mercier, A.C. Lysaght, S. Bandyopadhyay, A.P. Chandrakasan, K.M. Stankovic, Energy extraction from the biologic battery in the inner ear. Nat. Biotechnol. **30**(12), 1240–1243 (2012)
22. X. Huang, A. Ba, P. Harpe, G. Dolmans, H. de Groot, J.R. Long, A 915 MHz, ultra-low power 2-tone transceiver with enhanced interference resilience. IEEE J. Solid-State Circuits **47**(12), 3197–3207 (2012)
23. J. Bae, L. Yan, H.-J. Yoo, A low energy injection-locked FSK transceiver with frequency-to-amplitude conversion for body sensor applications. IEEE J. Solid-State Circuits **46**(4), 928–937 (2011)
24. G. Papotto, F. Carrara, A. Finocchiaro, G. Palmisano, A 90-nm CMOS 5-Mbps crystal-less RF-powered transceiver for wireless sensor network nodes. IEEE J. Solid-State Circuits **49**(2), 335–346 (2014)
25. J. Pandey, B.P. Otis, A sub-100 μW MICS/ISM band transmitter based on injection-locking and frequency multiplication. IEEE J. Solid-State Circuits **46**(5), 1049–1058 (2011)
26. C. Ma, C. Hu, J. Cheng, L. Xia, P.Y. Chiang, A near-threshold, 0.16 nJ/b OOK-transmitter with 0.18 nJ/b noise-cancelling super-regenerative receiver for the medical implant communications service. IEEE Trans. Biomed. Circuits Syst. **7**(6), 841–850 (2013)
27. M.M. Izad, C.-H. Heng, A 17 pJ/bit 915 MHz 8PSK/O-QPSK transmitter for high data rate biomedical applications, in *Proceedings of the IEEE 2012 Custom Integrated Circuits Conference*, 2012, pp. 1–4
28. F. Zhang, M.A. Stoneback, B.P. Otis, A 23 μA RF-powered transmitter for biomedical applications, in *2011 IEEE Radio Frequency Integrated Circuits Symposium*, 2011, pp. 1–4

29. Y. Chee, A. Niknejad, J. Rabaey, A 46% efficient 0.8 dBm transmitter for wireless sensor networks, in *Symposium on VLSI Circuits, 2006. Digest of Technical Papers*, 2006, pp. 43–44
30. D.C. Daly, A.P. Chandrakasan, An energy-efficient OOK transceiver for wireless sensor networks. IEEE J. Solid-State Circuits **42**(5), 1003–1011 (2007)
31. A. Paidimarri, P.M. Nadeau, P.P. Mercier, A.P. Chandrakasan, A 2.4 GHz multi-channel FBAR-based transmitter with an integrated pulse-shaping power amplifier. IEEE J. Solid-State Circuits **48**(4), 1042–1054 (2013)
32. B. Otis, Y.H. Chee, J. Rabaey, A 400 μW-RX, 1.6 mW-TX superregenerative transceiver for wireless sensor networks, in *ISSCC. IEEE International Digest of Technical Papers. Solid-State Circuits Conference*, 2005, pp. 396–398
33. I. Nam, K. Choi, J. Lee, H.-K. Cha, B.-I. Seo, K. Kwon, K. Lee, A 2.4-GHz low-power low-IF receiver and direct-conversion transmitter in 0.18-μm CMOS for IEEE 802.15.4 WPAN applications. IEEE Trans. Microw. Theory Tech. **55**(4), 682–689 (2007)
34. P.M. Nadeau, A. Paidimarri, P.P. Mercier, A.P. Chandrakasan, Multi-channel 180 pJ/b 2.4 GHz FBAR-based receiver, in *IEEE Radio Frequency Integrated Circuits Symposium*, 2012, pp. 381–384
35. A. Heragu, D. Ruffieux, C. Enz, A 2.4-GHz MEMS-based PLL-free multi-channel receiver with channel filtering at RF, in *Proceedings of the ESSCIRC (ESSCIRC)*, 2012, pp. 137–140
36. K. Wang, J. Koo, R. Ruby, B. Otis, 21.7 A 1.8 mW PLL-free channelized 2.4 GHz ZigBee receiver utilizing fixed-LO temperature-compensated FBAR resonator, in *IEEE International Solid-State Circuits Conference Digest of Technical Papers (ISSCC)*, 2014, pp. 372–373
37. N. Stanic, A. Balankutty, P.R. Kinget, Y. Tsividis, A 2.4-GHz ISM-band sliding-IF receiver with a 0.5-V supply. IEEE J. Solid-State Circuits **43**(5), 1138–1145 (2008)
38. Z. Lin, P.-I. Mak, R. Martins, 9.4 A 0.5 V 1.15 mW 0.2 mm^2 sub-GHz ZigBee receiver supporting 433/860/915/960 MHz ISM bands with zero external components, in *IEEE International Solid-State Circuits Conference Digest of Technical Papers (ISSCC)*, 2014, pp. 164–165
39. N.M. Pletcher, S. Gambini, J. Rabaey, A 52 μW wake-up receiver with −72 dBm sensitivity using an uncertain-IF architecture. IEEE J. Solid-State Circuits **44**(1), 269–280 (2009)
40. F.X. Moncunill-Geniz, P. Pala-Schonwalder, O. Mas-Casals, A generic approach to the theory of superregenerative reception. IEEE Trans. Circuits Syst. Regul. Pap. **52**(1), 54–70 (2005)
41. J.-Y. Chen, M.P. Flynn, J.P. Hayes, A fully integrated auto-calibrated super-regenerative receiver in 0.13-μm CMOS. IEEE J. Solid-State Circuits **42**(9), 1976–1985 (2007)
42. Texas Instrument, CC2550. [Online]. Available: http://www.ti.com/product/cc2550?keyMatch=cc2550&tisearch=Search-EN
43. Nordic Semiconductor, nRF24E2. [Online]. Available: http://www.nordicsemi.com/eng/Products/2.4GHz-RF/nRF24E2/(language)/eng-GB
44. P.D. Bradley, An ultra low power, high performance Medical Implant Communication System (MICS) transceiver for implantable devices, in *IEEE Biomedical Circuits and Systems Conference*, 2006, pp. 158–161
45. J. Tan, W.-S. Liew, C.-H. Heng, Y. Lian, A 2.4 GHz ULP reconfigurable asymmetric transceiver for single-chip wireless neural recording IC. IEEE Trans. Biomed. Circuits Syst. **8**(4), 497–509 (2014)
46. J. Masuch, M. Delgado-Restituto, A 1.1-mW −81.4-dBm sensitivity CMOS transceiver for Bluetooth low energy. IEEE Trans. Microw. Theory Tech. **61**(4), 1660–1673 (2013)
47. J. Cheng, L. Xia, C. Ma, Y. Lian, X. Xu, C.P. Yue, Z. Hong, P.Y. Chiang, A near-threshold, multi-node, wireless body area sensor network powered by RF energy harvesting, in *Proceedings of the IEEE 2012 Custom Integrated Circuits Conference*, 2012, pp. 1–4
48. F. Zhang, K. Wang, Y. Miyahara, B. Otis, A 1.6 mW 300 mV-supply 2.4 GHz receiver with −94 dBm sensitivity for energy-harvesting applications, in *IEEE International Solid-State Circuits Conference Digest of Technical Papers*, 2013, pp. 456–457
49. T. Copani, S. Shashidharan, S. Chakraborty, M. Stevens, S. Kiaei, B. Bakkaloglu, A CMOS low-power transceiver with reconfigurable antenna interface for medical implant applications. IEEE Trans. Microw. Theory Tech. **59**(5), 1369–1378 (2011)

50. Y.-H. Liu, A. Ba, J.H. van den Heuvel, K. Philips, G. Dolmans, H. de Groot, 9.5 A 1.2 nJ/b 2.4 GHz receiver with a sliding-IF phase-to-digital converter for wireless personal/body-area networks, in *IEEE International Solid-State Circuits Conference Digest of Technical Papers (ISSCC)*, 2014, pp. 166–167
51. P. Choi, H.C. Park, S. Kim, S. Park, I. Nam, T.W. Kim, S. Park, S. Shin, M.S. Kim, K. Kang, Y. Ku, H. Choi, S.M. Park, K. Lee, An experimental coin-sized radio for extremely low-power WPAN (IEEE 802.15.4) application at 2.4 GHz. IEEE J. Solid-State Circuits **38**(12), 2258–2268 (2003)
52. M. Flatscher, M. Dielacher, T. Herndl, T. Lentsch, R. Matischek, J. Prainsack, W. Pribyl, H. Theuss, W. Weber, A bulk acoustic wave (BAW) based transceiver for an in-tire-pressure monitoring sensor node. IEEE J. Solid-State Circuits **45**(1), 167–177 (2010)
53. M. Vidojkovic, X. Huang, X. Wang, C. Zhou, A. Ba, M. Lont, Y.-H. Liu, P. Harpe, M. Ding, B. Busze, N. Kiyani, K. Kanda, S. Masui, K. Philips, H. de Groot, 9.7 A 0.33 nJ/b IEEE802.15.6/proprietary-MICS/ISM-band transceiver with scalable data-rate from 11 kb/s to 4.5 Mb/s for medical applications, in *IEEE International Solid-State Circuits Conference Digest of Technical Papers (ISSCC)*, 2014, pp. 170–171
54. W.-Z. Chen, T.-Y. Lu, W.-W. Ou, S.-T. Chou, S.-Y. Yang, A 2.4 GHz reference-less receiver for 1 Mbps QPSK demodulation. IEEE Trans. Circuits Syst. Regul. Pap. **59**(3), 505–514 (2012)
55. J.H. Jang, D.F. Berdy, J. Lee, D. Peroulis, B. Jung, A wireless sensor node for condition monitoring powered by a vibration energy harvester, in *IEEE Custom Integrated Circuits Conference (CICC)*, 2011, pp. 1–4
56. F.X. Moncunill-Geniz, P. Pala-Schonwalder, C. Dehollain, N. Joehl, M. Declercq, An 11-Mb/s 2.1-mW synchronous superregenerative receiver at 2.4 GHz. IEEE Trans. Microw. Theory Tech. **55**(6), 1355–1362 (2007)
57. H.-G. Park, J. Lee, J.-A. Jang, J.-H. Jang, D.-S. Lee, H. Kim, S.J. Kim, S.-G. Lee, K.-Y. Lee, An ultra-low-power super regeneration oscillator-based transceiver with 177 μW leakage-compensated PLL and automatic quench waveform generator. IEEE Trans. Microw. Theory Tech. **61**(9), 3381–3390 (2013)
58. D.C. Daly, P.P. Mercier, M. Bhardwaj, A.L. Stone, Z.N. Aldworth, T.L. Daniel, J. Voldman, J.G. Hildebrand, A.P. Chandrakasan, A pulsed UWB receiver SoC for insect motion control. IEEE J. Solid-State Circuits **45**(1), 153–166 (2010)

Channel Modeling for Wireless Body Area Networks

David B. Smith and Leif W. Hanlen

Abstract Wireless body area networks (BANs) are the latest generation of personal area networks (PANs) and describe radio networks of sensors, and/or actuators, placed in, on, around and some-times near the human body. BANs are motivated by the health-care application domain where reliable, long-term, operation is paramount. Hence understanding, and modeling, the body-area radio propagation channel is vital. In this chapter we describe channel models for wireless body area networks, in terms of operating scenarios—including on the human body, off the body, in the body, and body-to-body (or interfering); carrier frequencies from hundreds of MHz to several GHz; and bandwidth of operation, including narrowband and ultra-wideband. We describe particular challenges for accurate channel modeling such as the absence of wide-sense-stationarity in typical on-body narrowband BANs. We describe results following from a large amount of empirical data, and demonstrate that the BAN channel is dominated by shadowing with slowly-changing dynamics. Finally two particularly challenging scenarios for BAN operation are described: sleep-monitoring and also where there is a large number of co-located BANs.

Keywords Channel modeling • Radio propagation • Wireless body area networks

D.B. Smith (✉)
NICTA, 7 London Circuit, Canberra, ACT 2601, Australia

Australian National University (ANU), Canberra, ACT 0200, Australia
e-mail: David.Smith@nicta.com.au

L.W. Hanlen
NICTA, 7 London Circuit, Canberra, ACT 2601, Australia

Australian National University (ANU), Canberra, ACT 0200, Australia

University of Canberra, University Drive, Bruce, ACT 2617, Australia
e-mail: Leif.Hanlen@nicta.com.au

P.P. Mercier, A.P. Chandrakasan (eds.), *Ultra-Low-Power Short-Range Radios*,
Integrated Circuits and Systems, DOI 10.1007/978-3-319-14714-7_2

1 Introduction

Wireless body area networks (BANs) are radio networks of sensors and/or actuators, placed on, in, around and/or near the human body, and represent the latest generation of personal area networks. As such, BANs describe radio networks that will often employ ultra-low-power short-range radios. One of the principal application domains of BANs is for use in health-care, with other applications including consumer fitness, emergency services and consumer entertainment. Considering application in health-care, long-term, reliable operation at low-power is very important. We will show that reliable operation is a real challenge for BANs by considering typical characteristics of the radio channel. It is also then very important, so that system design can respond to these characteristics, to derive appropriate channel models for the BAN radio channel.

The main focus of this chapter will be the on-body radio channel, for communications from one location on a given subject's body to another location on the subjects body, which is envisaged as the most common BAN implementation. However there will be some focus on the off-body channel and the body-to-body channel. The body-to-body channel is important due to the anticipated prevalence of body area networks, where this interfering channel, with multiple co-located BANs, can dominate the on-body radio channel. It will be shown that there are various difficulties in channel modeling for BAN, which are particular to the BAN channel, underlining the importance of BAN reliability and life-time enhancing system design, such as relay-assisted communications, transmit power control and link adaptation. Important first and second-order statistics can be derived from extensive empirical campaigns, and alternate evaluations can be given directly from empirical data. The "everyday" BAN channel scenario presents a challenging environment for radio propagation and system design, but there are even more challenging environments in which BANs can operate, namely monitoring a person sleeping, and where there is a large number of coexisting BANs, which we will address.

2 Operational Scenarios for Wireless Body Area Network Channels

There are four scenarios for wireless body area network channels, namely

1. **On-body:** for radio communications from one part on the surface of the human body to another part on the surface of the human body;
2. **In-body:** for radio communications from inside the human body, typically to the body surface;
3. **Off-body:** for radio communications from the surface of the human body to a device closely located to the body, typically within 3 m of the body (or vice-versa, i.e., from off the body to on the body);

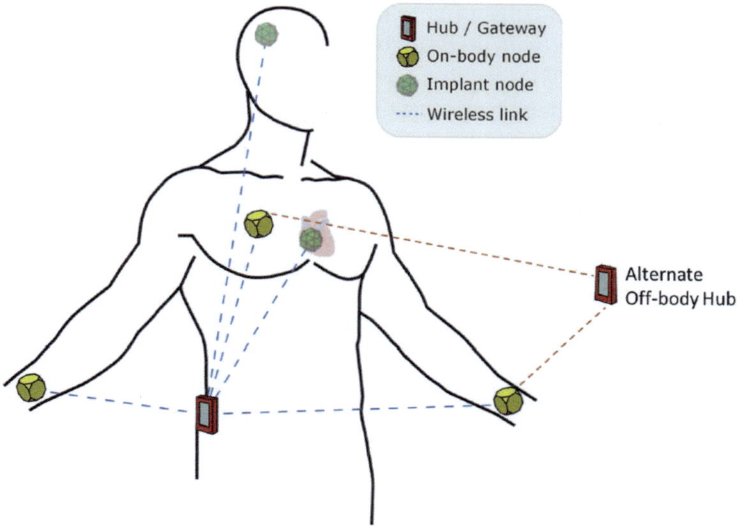

Fig. 1 BAN on a male subject, illustrating gateway (hub), sensors and in-body, on-body and off-body links, adapted from [44]

4. **Body-to-body or interfering:** for radio communications, target or interfering, from one subject's body to another subject's body.

A BAN on a subject, illustrating a gateway (hub), sensor nodes, on,-body in-body and off-body links, is shown in Fig. 1. The hub locations will be typically near the torso, either at the hips or on the chest; places where a subject could comfortably wear a device that is expected to be larger than a sensor node. These locations are also reasonably central on the human body.

We now describe the four scenarios in more detail, particularly with respect to challenges, operating environments and applications for each.

2.1 On-Body Channel

The on-body channel is the most prevalent channel for wireless body area networks and is the focus in this chapter. This channel will operate in various environments and will be dominated by slowly-varying dynamics from human-body movement and variations in shadowing by body parts. It presents significant difficulties to the radio systems designer, but there are also some benefits as follows:

- **Difficulties:** When operating with small low-power radios, long sensor/actuator radio lifetime is desired, thus requiring small power demands on the battery of the radio, as well as desired low electromagnetic radiation specific absorption rate (SAR) to the subject's body. This all leads to a desired transmit power significantly less than 0 dBm (or 1 mW) , −10 dBm (or 0.1 mW) may often be desirable. Further, as will be described later, at typical carrier frequencies of several hundreds of MHz up to a few gigahertz, communication on the human body provides a difficult radio channel, where instantaneous path losses can become very significant and typical (or median) path losses for a lot of on-body radio links are still (relatively) very large. Further the variations in signal strength are not uniform, from one time interval to the next, such that the channel is in general not wide-sense-stationarity.

- **Benefits:** However there are a few benefits/aids available to the radio system designer from the typical on-body radio channel, particularly with narrowband communications in everyday environments:

    1. The channel shows reciprocity, that is the radio channel for communications from position a. to position b. on the body, has the same channel profile as for communications b. to a.;
    2. The channel, for the majority of on-body BAN usage, is stable for at least hundreds of milliseconds (typically more than 0.5 s), enabling relatively accurate channel prediction across multiple communications frames, simply with the last channel gain sample, which can help transmit power control and resource allocation;
    3. Although the direct, sensor-to-hub, link may be in outage, the slowly varying on-body channel, and possible postures of the human body, means there will often be another dual-hop link between source and hub, through suitably located relay/s transmission paths, giving significant reliability benefit to radio communications;
    4. Although the overall information transfer over the whole on-body BAN may be large, for typical applications such as in health-care, high data rates for particular links may not be required (often in orders of tens of kilobits per second);
    5. Finally for narrowband BAN communications, although the on-body channel is slowly time-selective, it is frequency non-selective, with no resolvable multipath, and one channel tap, such that inter-symbol interference (ISI) does not need to be mitigated.[1]

2.2 In-Body Channel

The in-body channel will be, almost always, applied for medical applications, and mostly operate at lower carrier frequencies than the on-body channel. The main frequency of operation is most likely to be the medical implant communication system (MICS) band, which operates from 402–405 MHz. The in-body channel will

[1] However, we note that for typical IR-UWB, broadband, communications, IEEE 802.15.6 compliant, there are approximately ten resolvable channel taps.

also predominantly be from implants/devices, with miniature radios, to radios on the surface of, or just outside the body. Transmission from one radio in the body directly to a radio in another location inside the body will be highly uncommon. The in-body channel, apart from transmissions at tens of MHz, will suffer from significant attenuation for radiowaves propagating through the body, and will often depend on radio propagation from the nearest body surface to the implant radio device [32].

With respect to the mentioned challenging properties of the on-body channel, the in-body channel will be affected by similar challenges. However restrictions with respect to output Tx power, and reducing battery power consumption are even further magnified, as it is desirable for batteries inside the human body to have a lifetime of several years (frequent surgery is not desirable), as well as reducing radio-wave absorption inside the human body.

As the in-body communication channel includes various additional components (e.g., creeping waves) we shall not discuss this further in this chapter as it is significantly different to the other parts. We note that there is some good description of in-body communications in, e.g. [4, 32].

2.3 *Off-Body Channel*

The off-body channel is the radio channel the most similar to standard small cell and personal area networks radio communications. However transmission from one part of the human body to a gateway/hub radio at a small distance from the human body will also often be dominated by shadowing, similar as for the on-body channel. It is also slowly time selective and a one-tap channel—but it can reasonably be expected that it is more wide-sense-stationary than the on-body channel, and also median path losses will often be lower, even though often over a greater distance than on-body links. In applications such as health-care, suitable placement of the radio device/s off the body may be particularly important to maximize the typical channel gains from desired off-body transmission, or to enhance the on-body communications, where one or more relays is placed off the human body. Also the off-body channel may have less energy-constrained relays than the on-body radio channel. All the other benefits for radio systems design for the on-body channel also apply for the off-body channel, such as reciprocity—but the data rates may sometimes be larger than for the on-body channel.

2.4 *Body-to-Body (or Interference) Channel*

In most wireless body area networks, it is unlikely that one network will be spread over multiple human bodies, apart from obvious exceptions for uses such as in the military and emergency services. But the body-to-body radio channel characteristics are still very important, as in many cases for BAN operation there will be some, or significant, mobility, which coupled with the large anticipated large take-up of BANs, implies there will often be multiple people wearing BANs closely located, requiring coexistence without coordination between BANs. Thus, understanding radio propagation from one BAN to the sensor, relay or hub, of another BAN becomes very important.

This interfering channel will often demonstrate lower path losses than for an on-body Tx/Rx radio link-of-interest, due to on-body shadowing, and a lack of shadowing from the body-to-body interfering channel. Further the body-to-body channel does not demonstrate free-space path loss, and is strictly not distance dependent, unless a slowly varying shadowing factor is added to a distance-based path loss description with a larger path loss exponent than free space. The dynamics of the on-body channel, and body-to-body channel, are also similar to each other in that they are slowly time selective and frequency non-selective when considering narrowband communications.

The operation of BANs can also be significantly enhanced, when co-located with other BANs experiencing body-to-body interference, by both transmit power control and relay-assisted communications. In fact these two techniques may be particularly important to achieve performance benchmarks for on-body BANs to coexist with other BANs.

3 Technical Requirements for IEEE 802.15.6 BANs

There are various technical requirements, or, more precisely, guidelines for BANs from the IEEE 802.15.6 [47]. These broadly represent how BANs should operate and significantly influence key parameters for channel modeling.

- BANs should be scalable up to 256 nodes.
- A BAN link should support bit-rates between 10 kb/s and 10 Mb/s.
- The packet error rate (PER) should be $\leq$10 % for a 256 octet payload (i.e., 256$\times$8 bits of data) for the 95 % best-performing links according to PER (i.e., at a given signal-to-noise ratio, those 5 % of channels that give the worst PER performance should not be used to determine whether this PER guideline is met).
- Maximum radiated Tx power should be 0 dBm (or 1 mW), and all devices should be able to transmit at -10 dBm (or 0.1 mW).[2] This automatically meets specific-absorption-rate (SAR) guideline of the FCC of 1.6 W/kg in 1 g of body tissue [13] (which equates to a max Tx radiated power of 1.6 mW).
- Nodes should be able to be added and removed (insertion/de-insertion) to/from the network in less than 3 s.
- Reliability, latency (delay) and jitter (variation of one-way transmission delay) should be supported for those BAN applications that need them. Latency in

[2]Please note this maximum Tx power is a requirement in the standard.

medical applications should be less than 125 ms, and should be less than 250 ms in non-medical applications. Jitter should be less than 50 ms.

- Power saving mechanisms (such as duty cycling) should be provided.
- The physical layer should support co-located operation of at least ten randomly distributed BANs (i.e. up to 2,560 nodes) in a $6 \times 6 \times 6m^3$ volume.
- In-body BAN and on-body BAN should coexist in and around the body.

4 Narrowband and UWB Radio Channels for BANs

BANs can use narrowband communications or UWB communications, classifications in terms of carrier frequencies and bandwidths are given in Table 1. We exclude mm-wave communications, such as at 60 GHz carrier frequency, as there is no BAN specification for this, and with very large path losses around the body at these frequencies, reliable communications is very difficult. We also exclude optical wireless and human-body communications (using body conduction), as typical radios do not use these techniques.

4.1 BAN Propagation Scenarios

There are two physical layer radio propagation methods defined by the IEEE 802.15.6 BAN standard [23],

1. Narrowband communications: The use of narrowband in healthcare has been described extensively, e.g., [17, 24]. Narrowband communications is better suited to most healthcare applications due to its lower carrier frequencies that suffer less attenuation from the human body and due to better established electromagnetic compatibility. Its smaller bandwidth (1 MHz or less) also means that multipath is unlikely to cause significant inter-symbol-interference (ISI) [36].

Table 1 Frequency bands and channel bandwidths (BW) for the two BAN radio propagation methods: Narrowband, Ultra-wideband [23]

Narrowband communications				UWB communications	
Frequency band	Bandwidth	Frequency band (MHz)	Bandwidth	Frequency band (GHz)	Bandwidth (MHz)
402–405 MHz	300 kHz	420–450	300 kHz	3.2–4.7	499
863–870 MHz	400 kHz	902–928	500 kHz	6.2–10.2	499
950–956 MHz	400 kHz	2,360–2,400	1 MHz		
2,400–2,483.5 MHz	1 MHz				

For any of the methods, IEEE 802.15.6 compliant devices must operate in one of the associated bands

2. Ultra-wideband (UWB) communications: Frequency-modulated FM-UWB and impulse-radio IR-UWB are both supported by the standard, with IR-UWB being better suited to BAN, because for IR-UWB noncoherent receivers can be implemented very efficiently and promises low power consumption to meet stringent constraints on battery autonomy [26]. One particularly suitable application of UWB in BAN is in consumer electronics as UWB offers higher throughput due to its larger bandwidth; each UWB channel has a bandwidth of 499 MHz in IEEE 802.15.6 [23].

5 Suitable Small-Scale First Order Statistics of BAN Channels

First-order small-scale statistical modeling of narrowband channels, has been performed by fitting statistical distributions that are commonly used to describe fading (Rayleigh, normal, lognormal, Ricean, Nakagami-m, Weibull, gamma) to measured channel gain (channel gain is the inverse of path loss) data, e.g., [8, 17, 35, 39], and, some unusual (e.g., kappa-mu ($\kappa - \mu$)) distributions [9]. Statistical modeling of channel gain has mostly been performed indoors e.g., [17, 35]. A chart that summarizes the distributions considered, from [44], is given in Fig. 2.

In general, lognormal, gamma and Weibull are most-often found to be a best-fit.[3] Whilst Nakagami-m is often attempted as a fit, it has a smaller success rate; and Ricean has considerably smaller success rate than Nakagami-m. Further, it is very clear from Fig. 2 that the Rayleigh distribution is a poor fit for almost every scenario and environment for which it is attempted. Conversely, for any distribution, an author has invariably found at least one scenario that fit.

The lognormal, gamma and Weibull distributions are specified as follows:

- Lognormal

$$f(x|\mu_l, \sigma_l) = \frac{1}{x\sigma_l\sqrt{2\pi}} \exp\left\{\frac{-(\ln(x) - \mu_l)^2}{2\sigma_l^2}\right\}, \tag{1}$$

  where $\ln(\cdot)$ is the natural logarithm.
- Gamma

$$f(x|a, b) = \frac{1}{b^a \Gamma(a)} x^{a-1} \exp\left\{-\frac{x}{b}\right\}, \tag{2}$$

  where $\Gamma(\cdot)$ is the Gamma function.

[3] According to the ratio of the first two bars for each of these in Fig. 2 (only considering those distributions tested ten or more times.)

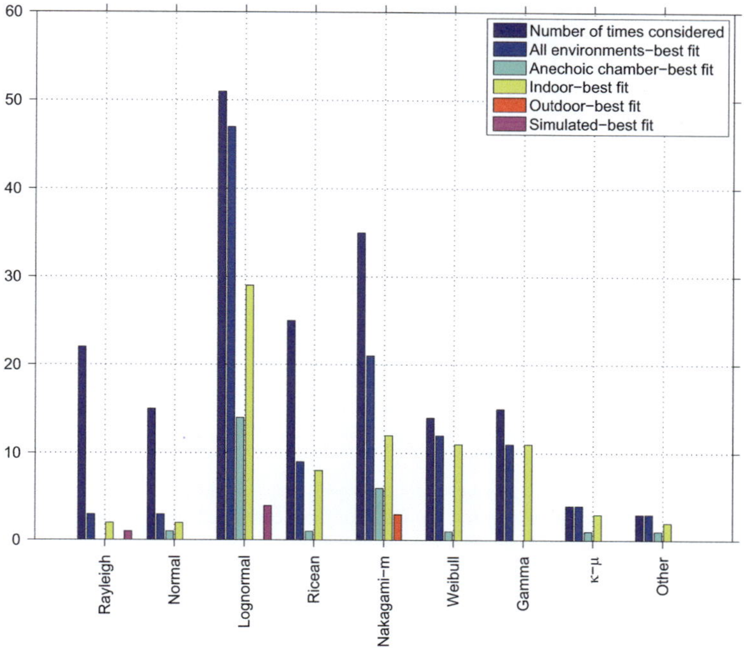

Fig. 2 Distributions, number of times considered, and times found to be best fit for all environments and per-environment from [44]."Other" includes one log-logistic and one $\eta-\mu$ distribution, as well as one distribution representation by empirical histogram representations of channel gain data

- Weibull

$$f(x|a_w, b_w) = b_w a_w^{-b} x^{b_w - 1} \exp\left\{-(x/a_w)^{b_w}\right\}. \tag{3}$$

We now give three example narrowband scenarios where the lognormal, Weibull, and gamma distributions are good fits for measured fading statistics of channel gain data, all scenarios' data is open-access [43].

5.1 *Experimental Narrowband Measurement Campaigns*

1. *On-Body:* We captured hundreds of hours of on-body channel gain data for "everyday" mixed activity of ten different adult subjects, using a range of transceiver Tx/Receiver(Rx) locations. The everyday mixed activity included indoor office work, at-home general activity, driving in a car and jogging outdoors, as well as transitions between each activity. Small wearable radios as described in [19], were used to capture the data. The radios transmit 540 kHz

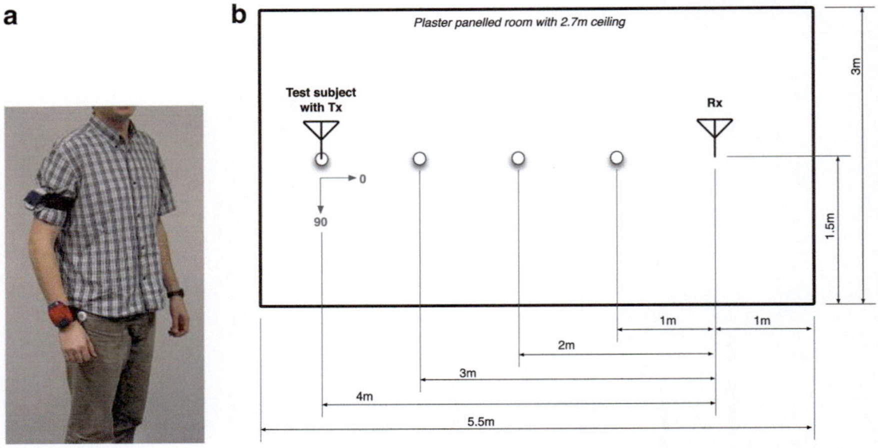

Fig. 3 On-body and off-body experiment scenarios [39]. (**a**) Subject wearing two wearable radios, Rx at *right wrist*, Tx at *upper right arm*. (**b**) Off-Body experimental environment. An angle of 0° corresponds to the test subject facing the receive antenna

bandwidth signals at a carrier frequency of 2,360 MHz, with the Rx radio sampling digital channel gain at 200 Hz. Each subject wore between 3 and 20 of these radios, some which operated as Rx, some as Tx, and some as both Tx and Rx. A subject wearing two of these wearable radios is shown in Fig. 3a. The measured data for each Tx/Rx link was normalized by the mean path loss for that link, and the data for all links was agglomerated, i.e., combined into one large set of channel gain samples. A typical channel gain profile from a subset of the complete open-access "everyday" data [43], is shown in Fig. 4a. The empirical probability density function (pdf) histogram of the complete normalized agglomerate data is shown in Fig. 4b, with various distribution fits overlayed. It is clear from Fig. 4b that the gamma and Weibull distributions provide excellent fits. In fact, gamma fading is a slightly better fit than Weibull according to a negative log-likelihood criterion of the parameter estimates. The very poor fits of Rayleigh and Normal distributions are also obvious in Fig. 4b. The gamma distribution fits to the 10 main Tx/Rx links channel gains, and overall fits, are given in Table 2.

2. *On-Body:* We chose a set dynamic activity, with a male adult subject running on the spot; a single Tx to Rx link from back to the chest; and a bandwidth of 10 MHz. Complex channel gain data was sampled over 2,048 data bits every 2.5 ms, for a 10 s period. Here the lognormal distribution is the best fit to normalized channel gain, as shown in Fig. 5a. Interestingly, the gamma distribution also provides a good fit. Once again the very poor fit of the Rayleigh distribution is obvious from Fig. 5a.
3. *Off-Body* channel measurements were made using a commercial wearable antenna at carrier frequencies, 427, 820 and 2,360 MHz, for 10 MHz bandwidth and 100 kHz bandwidth, with a male adult test-subject walking on the spot

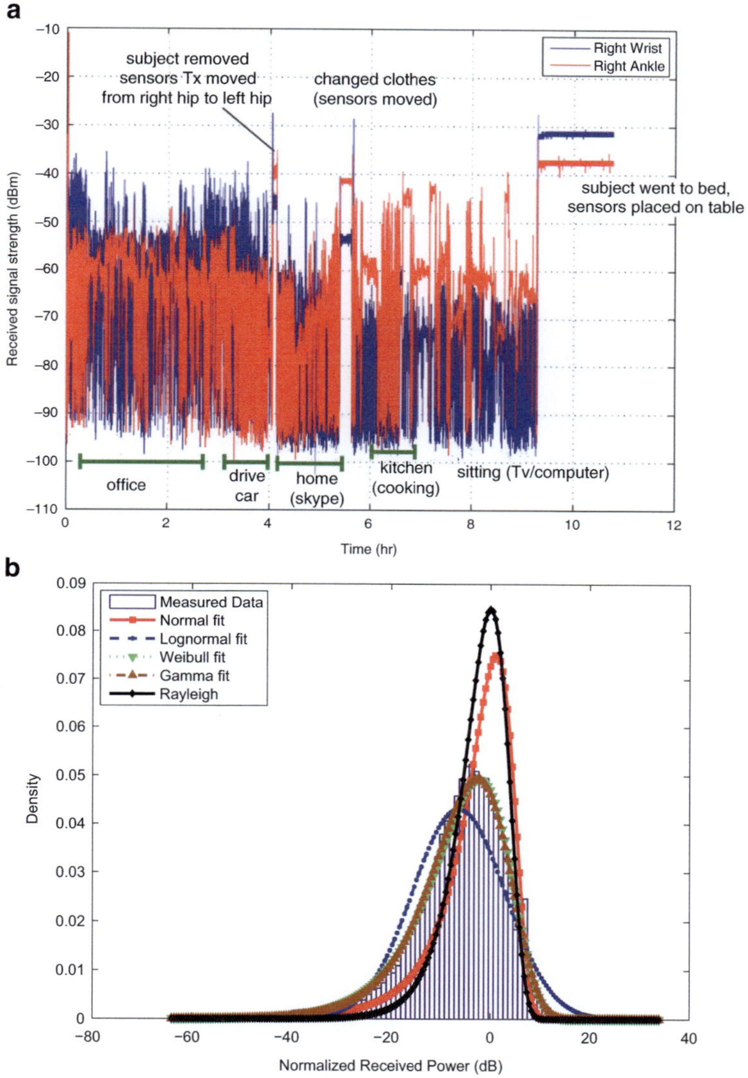

Fig. 4 Typical "everyday" channel gain profiles for one subject and agglomerate PDF over ten subjects. (**a**) Typical channel gain profile over 9 h for "everyday" activity from [41]. (**b**) PDF on-body channel gain agglomerate from everyday activity of ten subjects, 540 kHz bandwidth at 2,360 MHz [39]

Table 2 First-order statistics fits to everyday activity channel sounder channel-gain data, where the data has been normalized to mean of each link data-set [46]

Radio tx–Rx link	Mean channel gain (dB)	Median channel gain (dB)	Gamma distribution
Left hip–Chest	−50.1	−62.5	$a = 1.59, b = 0.486$
Left hip–Left wrist	−53.8	−61.7	$a = 1.44, b = 0.472$
Left hip–Right wrist	−60.2	−69.1	$a = 1.82, b = 0.431$
Left hip–Right ankle	−61	−69.7	$a = 2.16, b = 0.38$
Left hip–Head	−64.3	−71.2	$a = 2.17, b = 0.372$
Chest–Left wrist	−58.4	−62.5	$a = 1.82, b = 0.412$
Chest–Right wrist	−66.3	−70	$a = 1.86, b = 0.419$
Chest–Right ankle	−69.4	−77.5	$a = 2.63, b = 0.317$
Chest–Left ankle	−78.8	−82	$a = 2.9, b = 0.29$
Left hip–Right hip	−53.2	−83.5	$a = 2.31, b = 0.348$
Overall	−56.4	−69.8	$a = 1.81, b = 0.43$
O'all no normalization	–	–	LN ($\mu_l = -7.74, \sigma_l = 1.1$)

The gamma distribution (2), is the best fitting distribution type for this data, apart from overall sets, non-normalized, which is lognormal (LN), (1)

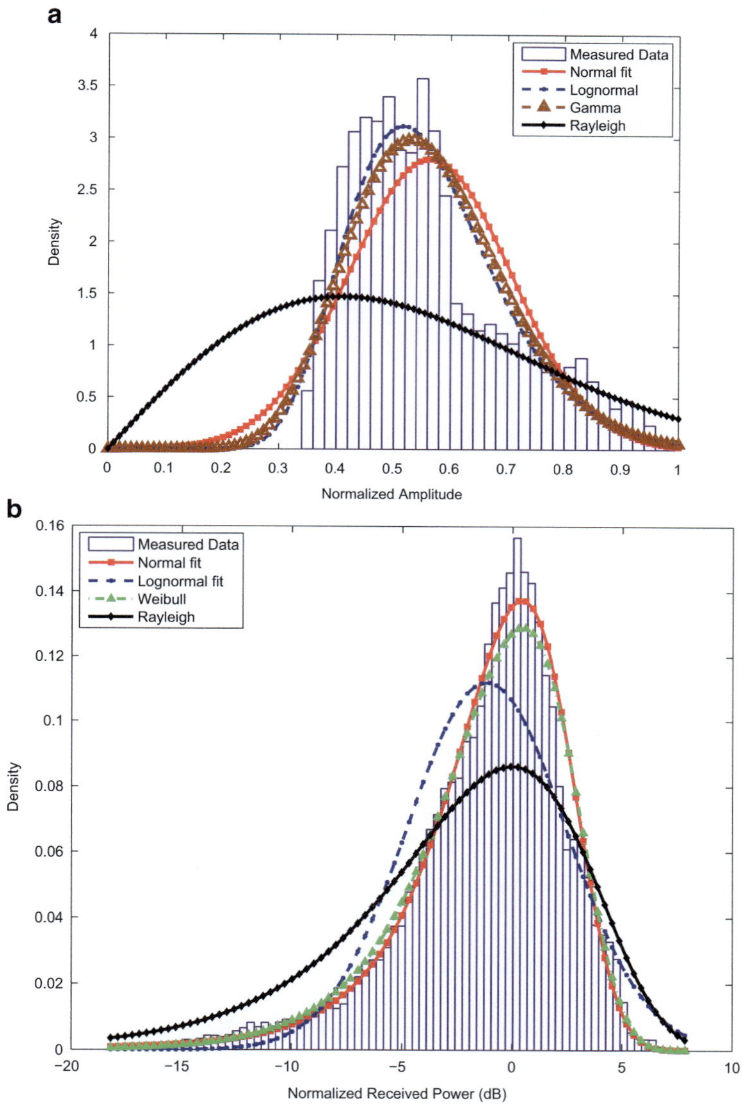

Fig. 5 On-body and off-body probability density functions (PDFs) with running and walking activity respectively from [39]. (**a**) PDF back to chest, running, 10 MHz bandwidth at 2,360 MHz. (**b**) PDF off-body agglomerate of subject walking, 10 MHz bandwidth, at 820 MHz

Table 3 Agglomerate scenarios, bandwidths and the best fitting model with parameters (in brackets) to those off-body scenarios at 427 MHz, 820 MHz and 2.36 GHz [39]

Action	Carrier frequency (MHz)	Bandwidth	Fading distribution
Moving	820	10 MHz	Weibull,($a_w = 1.05, b_w = 3.04$)
Moving	2, 360	10 MHz	Nakagami-m,($m = 1.6, \omega = 1$)
Moving	427	100 kHz	Weibull,($a_w = 1.02, b_w = 2.25$)
Moving	2, 360	100 kHz	Weibull,($a_w = 1.01, b_w = 2.15$)
Standing	820	10 MHz	Lognormal,($\mu_l = -0.000839, \sigma_l = 0.0289$)
Standing	2, 360	10 MHz	Gamma,($a = 384, b = 0.0026$)
Standing	427	100 kHz	Gamma,($a = 44, b = 0.0224$)
Standing	2, 360	100 kHz	Normal,($\mu = 0.987, \sigma = 0.161$)

for 5 s. Measurements were taken with a vector signal analyzer (VSA) with the test subject placed in four different locations in a room, with set-up in Fig. 3b. The horizontal distance between the test subject and Rx was either 1, 2, 3 or 4 m at each location. At each location measurements were taken with the subject facing in four directions: 0°, 90°, 180° and 270°, with 0° when the subject faced the Rx and 90° when he moved 90° clockwise from the 0° position. In Fig. 5b the best-fitting distribution to channel gain is Weibull for the scenario of the subject Walking, 10 MHz bandwidth, at 820 MHz carrier frequency, considering all distances and directions [39]. The lognormal distribution also provides a good fit in Fig. 5b. A summary table of best fitting distributions for all scenarios is given in Table 3.

5.2 *First-Order UWB BAN Channel Modeling*

The lognormal distribution is by far the most commonly found best fit for UWB BAN channels, this lognormal fit, and measurements campaigns used in this characterization, can be found in, e.g., [11, 15, 31]. BAN channels, particularly those with large bandwidths, contain a large number of factors that contribute to the attenuation of the transmitted signal; these include diffraction, reflection, energy absorption, antenna losses, etc..., which are additive in the log-domain [15]. The addition of multiple lognormally distributed paths results in another lognormal distribution.[4] When compared to narrowband BAN, there is also more large-scale fading with UWB from larger path losses due to its higher carrier frequencies [46].

[4]The negative effects of multipath are more common in UWB as there is increased inter-symbol interference (ISI) with its higher sampling rates.

5.3 Difficulty Choosing the Best Channel Model

It is clear from the variety of models and modeling techniques the choice of a most-suitable or best BAN model is difficult. There is also a bigger issue that we will address in the following section, that of non wide-sense-stationarity.

In terms of choosing the best statistical model, particularly according to small-scale first-order statistics, an argument has been made for separate characterization of individual links (i.e., from a Tx at a particular position to an Rx at another position on the body) in various literature [11, 35]. In fact, the best-fitting statistical distributions have often been specified with their relevant parameter estimates, according to particular positions on the body. However, in some studies, a fit to normalized agglomerate data from many on-body links, where each link's channel gain data is normalized by the mean path loss, i.e., mean-removed, is made for the whole-body with fewer parameters [39, 41]. This is often preferable, because channel dynamics are more important than the static attenuation represented by the mean path loss for any individual link. Whilst a parameterized "model" might give better fit by specifying the precise location of the sensor nodes, such a model is useless to a sensor node designer: would they use different radios for each part of the body? Would a consumer be told: "this sensor only goes on your arm, this one only works on your ankle"? A good model fit in such a setup is meaningless.

5.4 Body-to-Body Ban Interference Modeling

The following summary of body-to-body BAN modeling, follows from description in [10], and further details can be found in [10]. In [22] the body-to-body radio channel was investigated for carrier frequencies of 2.45 and 5.8 GHz with two subjects. Channel gains followed a gamma distribution with mean and variance values following a power law in terms of distance between two BANs; almost independent of carrier frequency but dependent upon on-body antenna position and body orientation. Small-scale Ricean fading was found with the Ricean K-parameter depending mainly upon on-body antenna position rather than Tx-Rx separation, while large-scale gamma fading was found at a constant distance, with large-scale lognormal fading when the distance changed randomly.

Investigations on UWB body-to-body communications have been described in [33]. Measured data was obtained in an anechoic chamber for two subjects standing at various distances with different body orientations and showed that the path loss was strongly related to the placement of the devices on the body as well as to the relative position of the human bodies.

In [20] it was shown, for an indoors environment, that the interference channel gain is dominated mostly by subject movements and not the distance between BANs. The results showed that the signal-to-interference-ratio could be very low, with greater interfering channel gain, than for target on-body signal, because of significant shadowing from the human body. It was also shown that on-body links and interfering links are uncorrelated.

6 Important Second-Order Statistics for BANs

The BAN channel is significantly influenced by the movement of the wearer of the BAN (whether moving very slowly or quickly). Considering such BAN dynamics, second-order statistics are also important for characterizing BAN channels, both on-body and off-body. As also described in [44], the following summarizes key second-order statistical characterizations for BAN radio propagation:

1. Delay spread and the power delay profile of BAN channels can be used to determine the number of channel taps and hence the presence of inter-symbol interference (ISI). Significant multiple resolvable signal paths; i.e., significant multiple channel taps, and hence ISI, only occurs in the UWB BAN channel [16, 29]. This is different to the narrowband BAN channel, with bandwidths up to 10 MHz, which can be well-approximated by a single-tap channel [36]. This is an obvious result as the amount of ISI in a channel increases with its bandwidth. The 499 MHz UWB channel bandwidth specified for IEEE 802.15.6 [23] is approximately 50 times that of the peak narrowband channel bandwidth. Measurements using 500 MHz bandwidth IR-UWB report that more than 10 channel taps can be resolved [14].
2. Average fade duration—i.e., the average time the received signal strength is below any given level—can be used to determine the amount of time for which successful packet transmission on a given Tx/Rx link may not be possible. Hence it is an important parameter for BAN communications. The level crossing rate (LCR)—i.e., the average rate at which the signal strength crosses from above to below any given signal level (particularly at the mean path loss [38])—can be used to infer the rate of fading. The LCR can be used to determine the Doppler spread, which is approximately 1 Hz in "everyday" BAN channels [44], but can be above 4 Hz with someone running [37]. It has been determined that both average fade duration and level crossing rate are highly dependent on channel dynamics, as they depend on the rate and amount of body movement [38, 39]. In many typical BAN channels the average fade duration is 300 ms or more [38], significantly larger than the 250 ms latency requirement for many BAN applications [25] (as outlined in Sect. 3).
3. Autocorrelation of time-varying channel gain, which can be used to determine coherence time [37], for any BAN link can determine for how much time successful packet transmission is possible, as with average fade duration. Thus autocorrelation drives the design of packet lengths, as well as driving the placement of pilots for channel estimation, making it an important parameter for BAN communications. It is also important for power control based on channel prediction [40]. Longer coherence times, of up to 1 s for the 'everyday' mixed activity for on-body narrowband BAN channel [39, 40], allow for successful transmit power control over the duration of multiple BAN superframes (even when a superframe is hundreds of milliseconds in length). With continuous movement, the channel coherence time can drop to between 70 and 25 ms [37], indicating much smaller time for successful packet transmission.

4. Cross-correlation is of some importance and has been investigated in [11, 45]. It is important because BAN sensors may be densely placed on the body, and the quality of one gateway-to-sensor link could be used to determine the quality of the same gateway to another proximate sensor link via the cross-correlation of their signal strengths. However, we have found that with a medium density of 10 on-body sensors such spatial cross-correlation coefficients are 0.5 or lower.[5] This may not be sufficient given that spatial cross-correlation is generally considered to be significant for values of 0.7 or greater.

7 Significant Issues in Wireless BAN Channel Modeling

There are several issues, or challenges presented, in determining suitable channel models for wireless BANs, which will be outlined here.

7.1 Statistical Fits: User Beware

The narrowband on-body BAN channel, is not wide-sense-stationary (WSS) outside timeframes of 500 ms or less [6], unlike networks such as mobile cellular communications and wireless LANs. This implies that any channel model, no matter how seemingly accurate, will provide a limited representation of the channel with respect to accuracy in terms of statistics of any order, across time, for all time.[6] This implies that resource allocation, based on long-term statistical analysis, may not be a practical mechanism for narrowband BAN. Although lack of wide-sense stationarity has only been shown, thus far, for narrowband BAN, it can be reasonably be expected to also be present for UWB BAN. The fact that BAN radio channels are not wide-sense-stationary, calls into question the statistical fits in the open literature where WSS is implicitly assumed.

Amongst statistical fits, the Rayleigh distribution is a very poor choice for BAN fading statistics. Although the Rayleigh distribution is a good fit when various multipath in the radio channel are additive in the linear domain. Thus, in contrast to many other radio networks, the combinations of multipath that occur in the BAN are not additive in the linear domain—these effects are additive in the log-domain, as indicated by the good fit of the lognormal distribution; and the small-scale fading is also often dominated by shadowing, as indicated by the good-fit for gamma

[5] This is corroborated by results for 5 on-body sensors in [11].

[6] As wide-sense-stationarity is generally considered to be both necessary and sufficient for nth-order statistical channel characterization across time.

fading [1]. Although the Rayleigh model is a consistently poor fit in most cases, [12] notes "we can approximate the fading statistic with a Rayleigh distribution," suggesting that a Rayleigh model might be useful in some cases.

Unfortunately, most authors provide only their goodness-of-fit result, based on their particular measurement and comparison criteria: one cannot retrospectively test if the measurements might support a new model choice, nor can one test the impact of an invalid stationarity assumption. This means that the models in the literature cannot necessarily be relied upon. The lack of reliable representation reinforces the need for large datasets, such as the "open-access" dataset [43], capturing many hundreds of hours of BAN link data, to test the appropriateness and validity of various radio system designs using deterministic modeling with respect to reliable empirical data, rather than statistical modeling. It also reinforces that traditional approaches to system design are not applicable to BANs and that the presumptions implicit in standard radio communications must be validated in BANs.

7.2 *Issue: Path Loss for BAN Channels Is Not Well-Characterized by Propagation Distance*

Some of the efforts for large-scale statistical modeling have been to model expected path loss in terms of distance for both narrowband and ultra-wideband propagation, and hence determine path loss exponents as a function of the carrier frequency, e.g., [4, 7, 15, 29]. A wide variation of path loss exponents for both narrowband and UWB, even within similar environments, have been reported, e.g., [3, 5, 15, 29]. Such variation suggests that the distance-based path loss modeling approach is poor; path loss exponents in the UWB bands for indoors measurements, have been reported from below 2 (better than free space) [29] to above 7 [15], and even up to 10 [3]. When measured in anechoic chambers, 2.4 GHz narrowband path loss exponents have been reported from below 3 [4], but have also been reported to be above 6 [48]. It is clear that path loss exponents are very much measurement campaign and environment dependent—this is more severe than simply "indoor" or "outdoor" and seems to indicate that the specifics of the building would be needed before distance-based path loss could be used reliably.

A distance-based path loss model, which ignores sensor placement and movement, produces a misleading model of the received signal strength for a BAN link. This is seen by measured path losses for set activities (standing, walking, and running) given in Table 4, see [28]. Two points are immediately clear:

1. the "distance" between the hip and wrist/ankle is very different for standing still vs running;
2. the path-loss is dominated by the whether-or-not of the path that includes the human body: the direct distance back-to-chest is much less than hip-to-ankle, yet the path loss is lower for the longer distance—because the path to the ankle is predominantly free-space, while back-to-chest is shadowed.

Table 4 Average path loss (dB), set activities, at 2.36 GHz [28].

	Receiver at right hip						Receiver at chest		
	Transmitter at						Transmitter at		
Action	Chest	Right wrist	Left wrist	Right ankle	Left ankle	Back	Right wrist	Right ankle	Back
Standing	65.3	44.5	74.7	60.9	70.7	75.3	70.5	66.3	73.0
Walking	59.1	47.3	59.8	53.9	58.5	67.4	64.9	62.4	72.0
Running	55.9	36.3	52.5	55.0	59.0	68.5	57.4	63.3	71.7

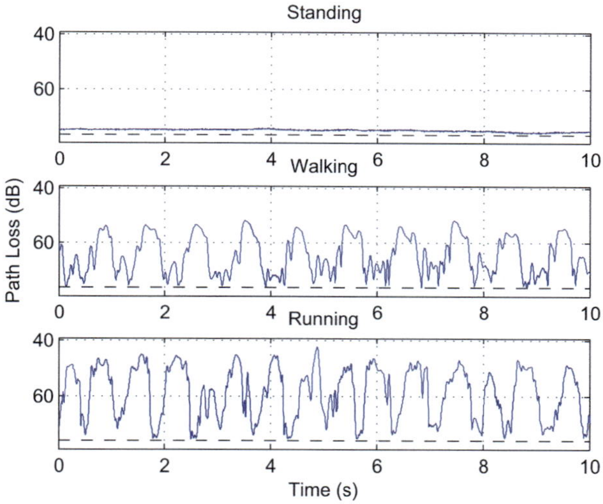

Fig. 6 Left wrist to right hip path loss v. time, 10 MHz bandwidth at 2.36 GHz [28], subject standing, walking and running

7.3 *Issue: It Is Not Appropriate to Categorize On-Body BAN Links as Either Non-Line-of-Sight (NLOS) or Line-of-Sight (LOS)*

Consider a left wrist to right hip link. Figure 6 shows a transition between NLOS and partially obstructed LOS as the subject movements increasing from standing to running. The subject is exposing and blocking the LOS link between the wrist and hip with their torso. Hence, characterizing a link as either LOS or NLOS is not meaningful for this link. The number of signal states is more than just LOS and NLOS, particularly with respect to dynamics, with moving body parts and changes in posture. It is more appropriate to capture the rate of movement and statistically characterize the path loss for this link.

8 Alternative Model Evaluations

Much of the previous statistical description has been based on relative comparisons of different statistical models. Here we provide an absolute measure, following from the description in [34, 41], to evaluate accuracy of channel models.

8.1 New Goodness-of-Fit Criterion to Characterize BAN Channel

In order to choose the best characterization of data we propose a goodness of fit function [34, 41]. This function, which represents a generalization of various criteria for model selection,[7] for a model with p parameters $\boldsymbol{\theta} = \{\theta_1, \ldots, \theta_p\}$ applied to data $\mathbf{x}$ with n samples is:

$$\mathscr{G}\{\boldsymbol{\theta}, \mathbf{x}\} \triangleq \mathscr{E}\{\boldsymbol{\theta}, \mathbf{x}\} + \mathscr{C}\{\boldsymbol{\theta}, \mathbf{x}\}, \tag{4}$$

where $\mathscr{E}\{\cdot\}$ is an increasing function of error between model and data, and $\mathscr{C}\{\cdot\}$ is a monotonically increasing function of number of parameters, for a given number of samples. Here goodness-of-fit improves as $\mathscr{G}\{\cdot\} \to 0$.

The Akaike-information-criterion (AIC) [2] (which has previously been used to determine best BAN model selection, e.g., [15, 39]) can be represented according to the framework of (4),

$$\mathscr{G}_{\mathrm{AIC}}\{\boldsymbol{\theta}, \mathbf{x}\} = \underbrace{\left[-2\ln\left(L(\hat{\boldsymbol{\theta}}|\mathbf{x})\right)\right]}_{\mathscr{E}_{\mathrm{AIC}}\{\boldsymbol{\theta},\mathbf{x}\}} + \underbrace{\left[2p + \frac{2p(p+1)}{(n-p-1)}\right]}_{\mathscr{C}_{\mathrm{AIC}}\{\boldsymbol{\theta},\mathbf{x}\}}, \tag{5}$$

where $\mathscr{G}.\{\cdot\}$ implies goodness and $\ln(L(\hat{\boldsymbol{\theta}}|\mathbf{x}))$ is the maximized log-likelihood, based upon the maximum-likelihood estimate of model parameters $\boldsymbol{\theta}$, given the data $\mathbf{x}$.

For BANs with measurements across many Tx/Rx links the AIC approach suffers from the problem that it only provides an ordering of models. For different data sets with different parameterizations it is meaningless to compare AIC values. The goodness-of-fit form (4) can be used to develop a natural reference point, which is the joint empirical histograms of the many-link channel gain data sets. That is, given M data sets, we choose B histogram bins, and for each set $m = \{1, \ldots, M\}$ find the histogram $H_m(b)$ with $b = \{1, \ldots, B\}$. This 'model' has $P = M \times B$ free parameters.

[7]Hence this function is not limited to propagation data.

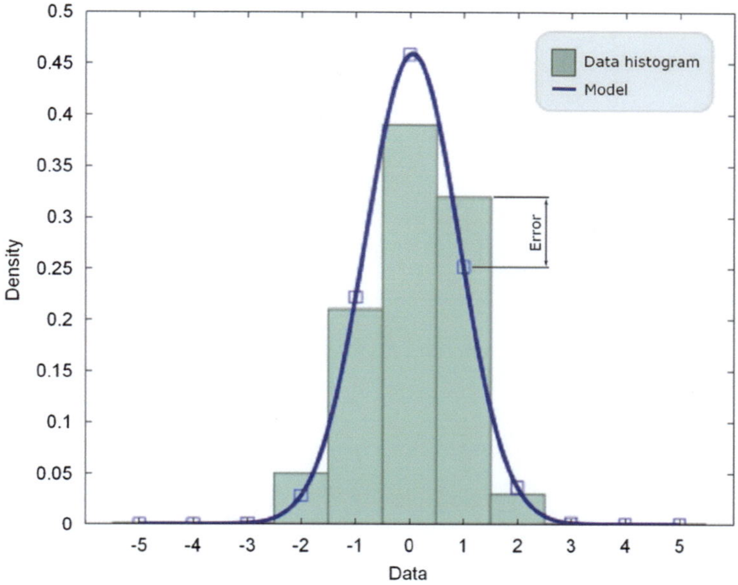

Fig. 7 Example error calculation between fitted distribution (model) and reference histogram [44]

Criteria for our systematic goodness of fit include:

- The comparison of any model against a reference histogram, with a given number of bins, is a metric. This metric is defined as the sum of errors (squared) between the model pdf and the reference histogram, evaluated at the histogram bin centres. The application of the metric is illustrated in Fig. 7, with histogram and fitted model.
- The number of parameters is given by the number of model options M over all sets, as well as the number of free-parameters, p_m, per option, $m = \{1, \ldots, M\}$.

This is formulated as follows. Consider M 'empirical models' comprising univariate empirical histograms. Each histogram, H_m, $m = \{1 \ldots, M\}$, comprises a set of values $H_m(\beta_b)$. Consider M 'continuous models', with density functions $F_m(\mathbf{x})$, that may be evaluated at histogram points β_b. An absolute goodness-of-fit $\mathcal{G}$ follows as

$$\mathcal{G} \triangleq \underbrace{\frac{1}{MB} \sum_{m,b} \left| H_m(\beta_b) - F_m(\beta_b) \right|^2}_{\mathcal{E}} + \underbrace{\log_2 \left(\sum_{m=1}^{M} p_m \right)}_{\mathcal{C}}, \tag{6}$$

and the base-2 logarithm, $\log_2(\cdot)$ above, follows complexity suggestions of [21]. Note that in (6) we assume (very) large n, which implies that the complexity is predominantly due to the number of parameters similar to the AIC approach in (5), and hence we ignore the number of samples n.

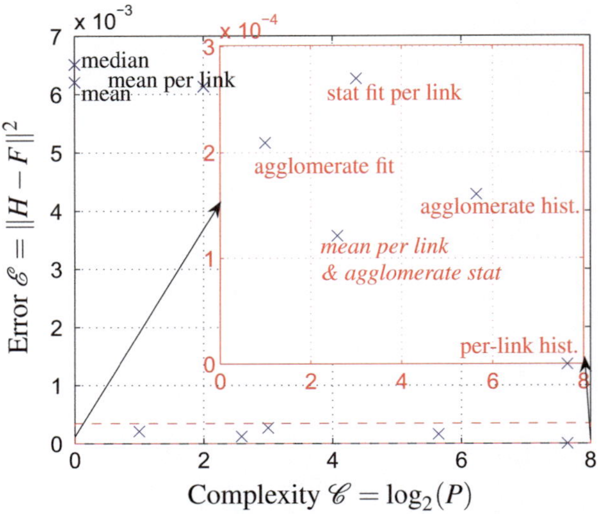

Fig. 8 Error versus complexity, first-order statistics, for the everyday data [41]. Note *insets* are the zoomed-in lower portions of graph

8.1.1 Evaluation of First-Order Statistics by Absolute Measure

With a subset of the complete open-access dataset in [43], the everyday activities of an adult male subject (height 1.84 m) over a period of 9 h are captured. There are $M = 4$ links (left/right-hip $\rightarrow$ right wrist; left/right hip $\rightarrow$ right-ankle) and the data contains over 2.9 million samples per link, sampled at 200 Hz. Figure 8 shows error $\mathscr{E}$ vs complexity $\mathscr{C}$ for various model options for the everyday data. Equivalent goodness $\mathscr{G}$ is given by $\mathscr{E} + \mathscr{C} =$ constant and 'better' models will appear closer to the origin.

In all cases the lognormal distribution was the best-fit. The number of parameters for the *mean per-link & agglomerate stat* is $P = M + 2$, since there are M means, and 2 free parameters. In terms of goodness $\mathscr{G}$ and as a trade-off between error $\mathscr{E}$ and complexity $\mathscr{C}$, Fig. 8 shows that one of either: (a) a mean-per-link with a lognormal statistic (1) fitted to agglomerate data with mean-removed from each link (parameters $\mu_l = -1.02, \sigma_l = 0.87$); or (b) a lognormal fit to agglomerate data (parameters $\mu_l = -7.66, \sigma_l = 1.02$); is the preferable model. Option (a) is preferable in terms of $\mathscr{E}$, and (b) is preferable in terms of $\mathscr{C}$.

8.1.2 Evaluation of Second-Order Statistics by Absolute Measure

It is important to note that in various earlier radio propagation literature, direct statistical characterization of second-order statistics of level crossing intervals[8] and

[8]Level-crossing interval is the inverse of level crossing rate.

fade durations has been performed, in, e.g., [27, 30]. Such an approach has been adopted in some BAN propagation characterization, in, e.g., [38, 46]. With direct characterization of second-order statistics we apply the same measure of (6) to ∼150 h of the "everyday" on-body link dataset in [43]. We show some results for comparing different direct statistical characterization techniques for fade durations and level crossing intervals in Fig. 9.

Figure 9a, b show that the empirical histogram for all data sets gives zero error but excessive complexity, $P = MT$. Similarly, a combined histogram is also complex, $P = T$, and has moderate error. The error caused by using simple agglomerate mean level crossing interval, in Fig. 9a or simple agglomerate average fade duration at median channel gain in Fig. 9b, is very large. Similarly a set of mean level crossing intervals or average fade durations per link also has large error. For Fig. 9a

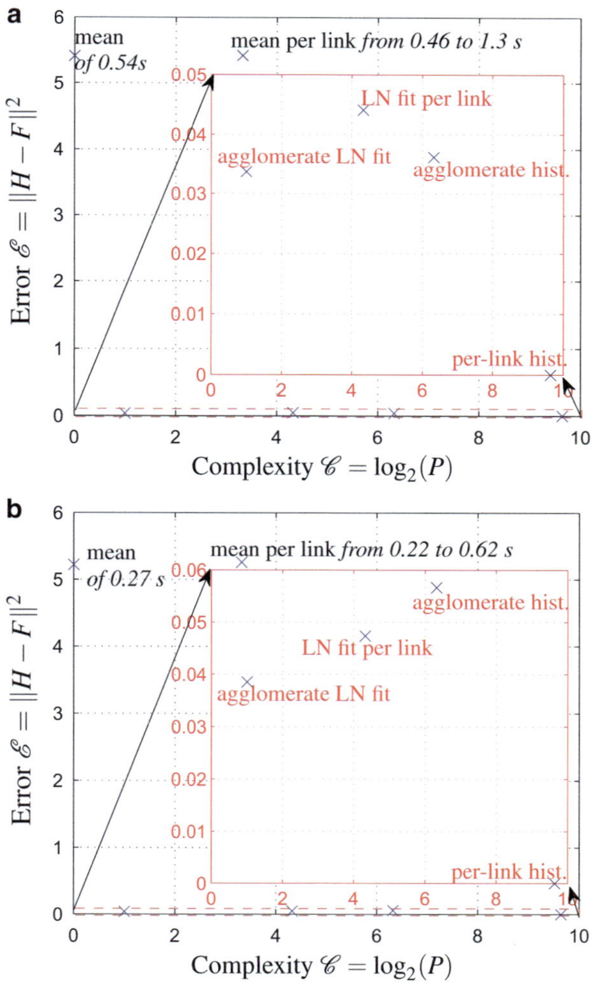

Fig. 9 Error v. complexity, second-order statistics [34]; LN-lognormal. (**a**) Error $\mathscr{E}$ v. complexity $\mathscr{C}$ for models of level crossing intervals with respect to median channel gains. Mean intervals range specified. (**b**) Error $\mathscr{E}$ v. complexity $\mathscr{C}$ for models of fade duration data with respect to median channel gains. Mean durations range specified

Table 5 Best lognormal (LN) agglomerate fits, parameters: μ_l is log-mean, σ_l is log-standard deviation

Statistic	LN parameters at h_{md}	LN parameters at h_m
Level crossing interval (s)	$\mu_l = -2.71, \sigma_l = 1.62$	$\mu_l = -2.78, \sigma_l = 1.68$
Fade duration (s)	$\mu_l = -3.81, \sigma_l = 1.65$	$\mu_l = -3.82, \sigma_l = 1.82$
Non-fade duration (s)	$\mu_l = -3.81, \sigma_l = 1.67$	$\mu_l = -3.93, \sigma_l = 1.56$

Statistics, in seconds (s), captured at median and mean channel gains, h_{md} and h_m respectively [34]

for level crossing intervals, and Fig. 9b for fade durations, goodness $\mathscr{G}$ is clearly optimized across all links simply with a 2 parameter lognormal fit.

The best lognormal agglomerate fits, with respect to both mean and median channel gains (where channel gain is the inverse of path loss), for level crossing intervals, fade duration and non-fade duration data, measured in seconds, are summarized in Table 5. It can be observed that for respective statistics, whether fade duration, non-fade duration or level crossing interval, that the best lognormal fit is very similar (according to both parameters of log-mean and log-standard deviation) whether with respect to mean or median channel gains—even though mean channel gain is typically several-dB larger than the median gain.

9 Particularly Difficult Scenarios for BAN Operation

Although BANs may be used in any scenario, they are motivated from a healthcare viewpoint. As such, significant work is needed to ensure that the BAN is functional when a subject sleeps, and when a subject is in close proximity to others. Both scenarios are unusual and difficult from a wireless communication standpoint.

9.1 BAN Channels for Sleep-Monitoring

We demonstrate effective performance measures and show that transmit-receive (Tx-Rx) links are often in outages for periods of minutes over a range of receive sensitivities [42]. The outages are in excess of latency requirements for many medical BAN applications [25, 47], with a packet error rate greater than 10 % at a very optimistic Rx sensitivity of −100 dBm, 100 dB below transmit power. The sleeping experiment set-up with the particular links is outlined in Fig. 10.

The on-body channel gain profiles for the left-wrist to the hip (back), and the off-body time series for the hip (front) to the radio next to the bed (head) are given in Fig. 11a, *There is clearly channel temporal stability with long periods of little movement while subjects are sleeping. The channel provides unreliable communications due to very low channel gain.* The empirical outage probability

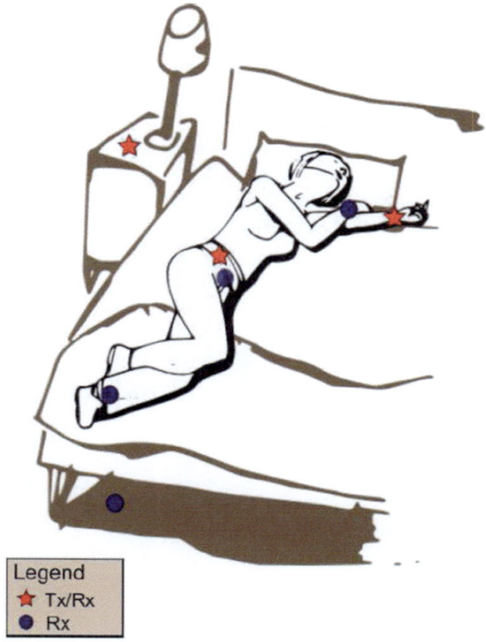

Fig. 10 Illustration of the sleeping experiment set-up [42]

for the on-body and off-body channel is shown in Fig. 11b. The best case outage probability is more than 10 % for both on-body (13.5 %) and off-body (10.9 %) channels. Figure 11b illustrates that the packet error rate (PER) for a BAN radio will be at least 10 % for a standard one-hop star topology with a person sleeping—which demonstrates the need for relays, and potential two-hop links.

The sleeping channel is also best characterized with gamma fading. For the on-body sleeping channel the shape parameter $a = 1.60$, and the scale parameter $b = 0.480$; and for the off-body channel $a = 3.54$ and $b = 0.254$ [42], with median path losses of 80 dB for both these channels.

In [42] it is also shown that, e.g., *a receiver with a sensitivity of −88 dBm, or 88 dB below transmit power of 0 dBm, will experience outages of larger than 1,000 s 5 % of the time*. Further, in terms of BAN latency requirements for medical applications at 88 dB below transmit power for example, outages of larger than a typical latency requirement of 125 ms [25, 47], occur more than 22 % of the time.

9.2 Large Numbers of Co-located BANs

Up to 10 BANs must be capable of coexisting (operating properly) within a $6 \times 6 \times 6\,\mathrm{m}^3$ cube. For example, if a group of subjects enters an elevator. BANs do not have a global coordination mechanism, hence understanding, and mitigating the interference of multiple co-located BANs, the body-to-body channel, becomes very important. Further it provides a particularly challenging scenario for the operation

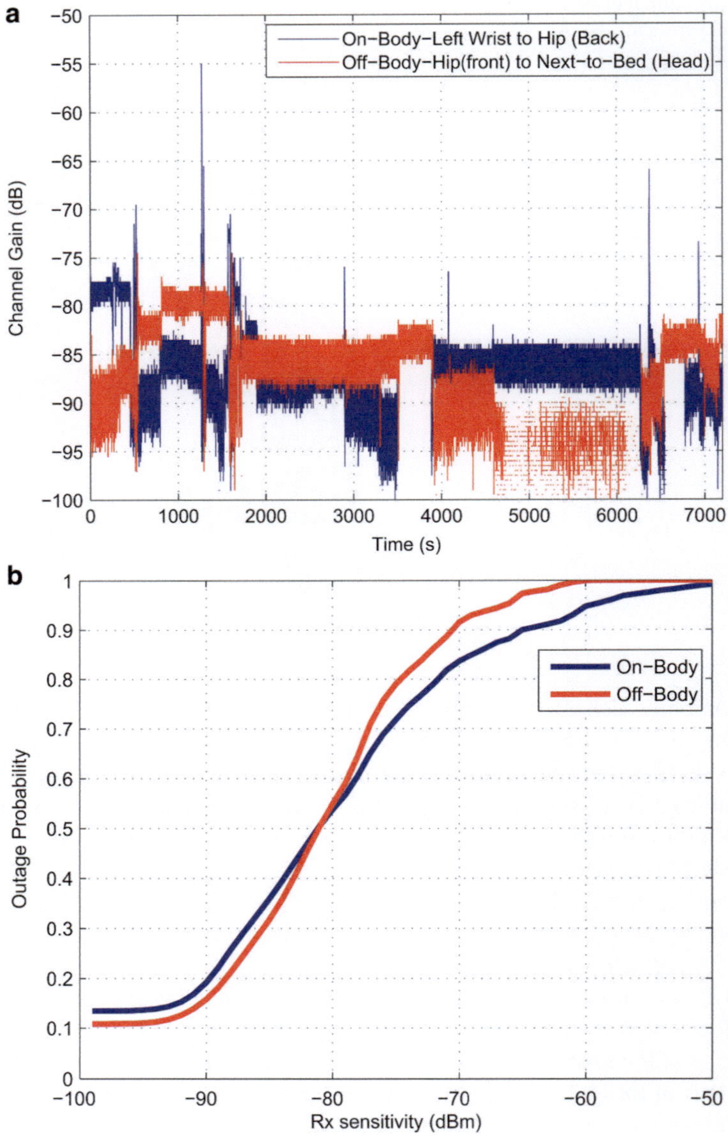

Fig. 11 Characterization of the impact that a subject sleeping has on channel outage [42]. (**a**) Typical on-body and off-body channel gain time series profile for the BAN sleep-monitoring channel. (**b**) Outage probability as a function of receive sensitivity, with Tx power of 0 dBm

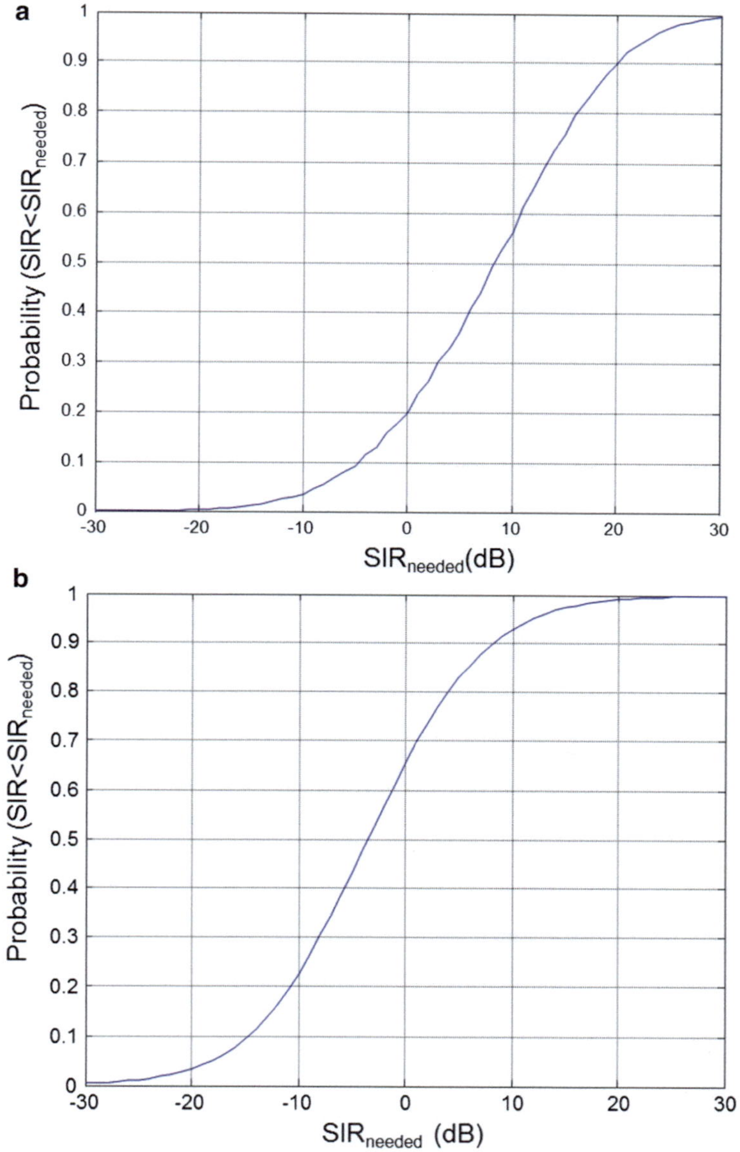

Fig. 12 CDF for required (needed) SIR to achieve outage probability value given in y-axis for co-located BANs [18]. (**a**) SIR outage for BAN with 1 co-channel interferer. (**b**) SIR outage for BAN with 9 co-channel interferers

of multiple BANs. To illustrate this in Fig. 12, we show from extensive interference measurements, that to ensure 10 % outage with 9 cochannel interferers, e.g., Tx/Rx links from 10 BANs operating in the same time-division-multiple-access (TDMA) time slot, the SIR that needs to be tolerated is −15 dB, Fig. 12b, with one co-channel interferer this is −5 dB, Fig. 12a, [18]. This is very difficult, and underlines the need for interference mitigation techniques including significant duty cycling to ensure best operation, and demonstrates that re-transmits may often be necessary when significant numbers of BANs are co-located.

10 Conclusion

In this chapter we have investigated channel modeling for wireless body area networks (BANs). We have shown that the BAN radio channel is particularly different from other typical radio channels—and in consideration of the stated technical requirements for BANs as they employ ultra-low-power short-range radios, there are many challenges presented to the radio system designer, including large path losses and non wide-sense-stationarity over any significant length of time. But, as described, there are mitigating benefits, including BAN channel temporal stability and channel reciprocity. We have highlighted the importance of mitigating interference with large numbers of co-located, non globally-coordinated, BANs, as well as other difficult channels for BAN operation, such as sleep monitoring. Also emphasized has been the importance of long-term radio channel measurements, to use as the basis for radio design, particularly considering non wide-sense-stationarity. In terms of first-order statistics, lognormal, and sometimes gamma or Weibull distributed fading characterizations have been shown to be most prevalent, but Rayleigh fading *is not a good characterization.* Finally an alternative means of evaluation, using an absolute histogram representation with respect to measurements, has been described that is very helpful in deciding the best characterization of the BAN channel, for both first and second-order statistics.

References

1. A. Abdi, M. Kaveh, On the utility of gamma pdf in modeling shadow fading (slow fading), in *IEEE 49th Vehicular Technology Conference*, vol. 3 (1999), pp. 2308–2312 doi:10.1109/VETEC.1999.778479
2. H. Akaike, A new look at the statistical model identification. IEEE Trans. Autom. Control **19**(6), 716–723 (1974)
3. A. Alomainy, Y. Hao, Radio channel models for UWB body-centric networks with compact planar antenna, in *IEEE Antennas and Propagation Society International Symposium* (IEEE, Albuerquerque, NM, 2006), pp. 2173–2176
4. A. Alomainy, Y. Hao, Modeling and characterization of biotelemetric radio channel from ingested implants considering organ contents. IEEE Trans. Antennas Propag. **57**(4), 999–1005 (2009). doi:10.1109/TAP.2009.2014531

5. A. Alomainy, Y. Hao, Y. Yuan, Y. Liu, Modelling and characterisation of radio propagation from wireless implants at different frequencies, in *The 9th European Conference on Wireless Technology*, Manchester, 2006, pp. 119–122. doi:10.1109/ECWT.2006.280449
6. V. Chaganti, L. Hanlen, D. Smith, Are narrowband wireless on-body networks wide-sense stationary? IEEE Trans. Wirel. Commun. **13**(5), 2432–2442 (2014). doi:10.1109/TWC.2014.031914.130303
7. X. Chen, X. Lu, D. Jin, L. Su, L. Zeng, Channel modeling of UWB-based wireless body area networks, in *IEEE International Conference on Communications (ICC)*, Kyoto, 2011, pp. 1–5. doi:10.1109/icc.2011.5962687
8. S.L. Cotton, W.G. Scanlon, A statistical analysis of indoor multipath fading for a narrowband wireless body area network, in *IEEE Personal, Indoor and Mobile Radio Communications Symposium, PIMRC 2006*, Helsinki, 2006, pp. 1–5
9. S.L. Cotton, W.G. Scanlon, Higher-order statistics for the κ-μ distribution. Electron. Lett. **43**(22), 1215–1217 (2007)
10. S.L. Cotton, R. D'Errico, C. Oestges, A review of radio channel models for body centric communications. Radio Sci. **49**(6), 371–388 (2014)
11. R. D'Errico, L. Ouvry, Delay dispersion of the on-body dynamic channel, in *Proceedings of the Fourth European Conference on Antennas and Propagation (EuCAP)*, Barcelona, 2010, pp. 1–5
12. R. D'Errico, L. Ouvry, A statistical model for on-body dynamic channels. Int. J. Wirel. Inf. Netw. **17**, 92–104 (2010)
13. Federal Communications Commission (FCC) Guidelines. (1997) [Online] http://www.fcc.gov/oet/rfsafety/sar.html#sec1
14. A. Fort, Body area communications: channel characterization and ultra-wideband system-level approach for low power, Ph.D thesis, Vrije Universiteit Brussel, Brussels, 2007
15. A. Fort, C. Desset, P. De Doncker, P. Wambacq, L. Van Biesen, An ultra-wideband body area propagation channel model—from statistics to implementation. IEEE Trans. Microw. Theory Tech. **54**(4), 1820–1826 (2006)
16. A. Fort, J. Ryckaert, C. Desset, P. De Doncker, P. Wambacq, L. Van Biesen, Ultra-wideband channel model for communication around the human body. IEEE J. Sel. Areas Commun. **24**(4), 927–933 (2006)
17. A. Fort, C. Desset, P. Wambacq, L. Biesen, Indoor body-area channel model for narrowband communications. IET Microw. Antennas Propag. **1**(6), 1197–1203 (2007)
18. L. Hanlen, D. Miniutti, D. Smith, A. Zhang, D. Rodda, B. Gilbert, A. Boulis, Network-to-network interference measurements ID: 802.15-09-0520-01-0006. IEEE Submission (2009)
19. L.W. Hanlen, V.G. Chaganti, B. Gilbert, D. Rodda, T. Lamahewa, D.B. Smith, Open-source testbed for body area networks: 200 sample/sec, 12 hrs continuous measurement, in *IEEE International Symposium on Personal, Indoor and Mobile Radio Communications (PIMRC)*, Istanbul, 2010, pp. 66–71
20. L.W. Hanlen, D. Miniutti, D.B. Smith, D. Rodda, B. Gilbert, Co-Channel interference in body area networks with indoor measurements at 2.4 GHz: distance-to-interferer is a poor estimate of received interference power. Int. J. Wirel. Inf. Netw. **17**, 113–125 (2010)
21. R.V.L. Hartley, Transmission of information. Bell Syst. Tech. J. **7**(3), 535–563 (1928)
22. Z.H. Hu, Y. Nechayev, P. Hall, Measurements and statistical analysis of the transmission channel between two wireless body area networks at 2.45 GHz and 5.8 GHz, in *ICECom, 2010 Conference Proceedings* (IEEE, 2010), pp. 1–4
23. IEEE standard for local and metropolitan area networks part 15.6: wireless body area networks. IEEE Std 802156-2012 (2012), pp. 1–271. doi:10.1109/IEEESTD.2012.6161600
24. M. Kim, J.I. Takada, Statistical model for 4.5-GHz narrowband on-body propagation channel with specific actions. IEEE Antennas Wirel. Propag. Lett. **8**, 1250–1254 (2009). doi:10.1109/LAWP.2009.2036570
25. D. Lewis, 802.15.6 call for applications - response summary ID: 802.15-08-0407-05. IEEE Submission (2008)

26. H. Luecken, T. Zasowski, C. Steiner, F. Troesch, A. Wittneben, Location-aware adaptation and precoding for low complexity IR-UWB receivers, in *IEEE International Conference on Ultra-Wideband, 2008. ICUWB 2008*, vol. 3, Hannover, 2008, pp. 31–34. doi:10.1109/ICUWB.2008.4653409
27. T. Mimaki, H. Sato, M. Tanabe, A study on the multi-peak properties of the level-crossing intervals of a random process. Signal Process. **7**(3), 251–265 (1984)
28. D. Miniutti, L.W. Hanlen, D.B. Smith, J.A. Zhang, D. Lewis, D. Rodda, B. Gilbert, Narrowband channel characterization for body area networks ID: 802.15.08.0421. IEEE Submission (2008)
29. A.F. Molisch, D. Cassioli, C.C. Chong, S. Emami, A. Fort, B. Kannan, J. Karedal, J. Kunisch, A comprehensive standardized model for ultrawideband propagation channels. IEEE Trans. Antennas Propag. **54**(11), 3151–3166 (2006)
30. S. Rice, Distribution of the duration of fades in radio transmission: Gaussian noise model. Bell Syst. Tech. J. **37**(3), 581–635 (1958)
31. C. Roblin, J.M. Laheurte, R. D'Errico, A. Gati, D. Lautru, T. Alvès, H. Terchoune, F. Bouttout, Antenna design and channel modeling in the BAN context part i: antennas. Ann. Telecommun. **66**, 139–155 (2011)
32. K. Sayrafian-Pour, W.B. Yang, J. Hagedorn, J. Terrill, K.Y. Yazdandoost, K. Hamaguchi, Channel models for medical implant communication. Int. J. Wirel. Inf. Netw. **17**(3–4), 105–112 (2010)
33. T.S. See, J. Hee, C. Ong, L. Ong, Z.N. Chen, Inter-body channel model for UWB communications, in *Proceedings of the Third European Conference on Antennas and Propagation (EuCAP) 2009*, Berlin, 2009, pp. 3519–3522
34. D.B. Smith, L.W. Hanlen, The body area network channel model: a new look at second-order statistics, in *Proceedings of the Eighth European Conference on Antennas and Propagation (EuCAP)*, The Hague, 2014, pp. 4351–4354
35. D.B. Smith, L.W. Hanlen, D. Miniutti, J.A. Zhang, D. Rodda, B. Gilbert, Statistical characterization of the dynamic narrowband body area channel, in *International symposium on applied sciences in biomedical and communication (ISABEL)*, Aalborg, 2008, pp. 1–5
36. D. Smith, D. Miniutti, L. Hanlen, A. Zhang, D. Lewis, D. Rodda, B. Gilbert, Power delay profiles for dynamic narrowband body area network channels id:15-09-0187-01-0006. IEEE Submission (2009)
37. D. Smith, J. Zhang, L. Hanlen, D. Miniutti, D. Rodda, B. Gilbert, Temporal correlation of dynamic on-body area radio channel. Electron. Lett. **45**(24), 1212–1213 (2009). doi:10.1049/el.2009.2057
38. D. Smith, D. Miniutti, L.W. Hanlen, D. Rodda, B. Gilbert, Dynamic narrowband body area communications: link-margin based performance analysis and second-order temporal statistics, in *IEEE Wireless Communications and Networking Conference (WCNC)*, Sydney, 2010, pp. 1–6
39. D. Smith, L. Hanlen, J. Zhang, D. Miniutti, D. Rodda, B. Gilbert, First- and second-order statistical characterizations of the dynamic body area propagation channel of various bandwidths. Ann. Telecommun. **66**, 187–203 (2011)
40. D. Smith, T. Lamahewa, L. Hanlen, F. Miniutti, Simple prediction-based power control for the on-body area communications channel, in *IEEE International Conference on Communications (ICC)* (2011), pp. 1–5. doi:10.1109/icc.2011.5963355
41. D.B. Smith, L.W. Hanlen, T.A. Lamahewa, A new look at the body area network channel model, in *Proceedings of the Fifth European Conference on Antennas and Propagation (EuCAP)*, Rome, 2011, pp. 2987–2991
42. D.B. Smith, D. Miniutti, L.W. Hanlen, Characterization of the body-area propagation channel for monitoring a subject sleeping. IEEE Trans. Antennas Propag. **59**(11), 4388–4392 (2011). doi:10.1109/TAP.2011.2164209
43. D. Smith, L. Hanlen, D. Rodda, B. Gilbert, J. Dong, V. Chaganti, Body Area Network Radio Channel Measurement Set (2012). [Online] http://opennicta.com/datasets

44. D. Smith, D. Miniutti, T. Lamahewa, L. Hanlen, Propagation models for body-area networks: a survey and new outlook. IEEE Antennas Propag. Mag. **55**(5), 97–117 (2013). doi:10.1109/MAP.2013.6735479
45. X.D. Yang, Q. Abbasi, A. Alomainy, Y. Hao, Spatial correlation analysis of on-body radio channels considering statistical significance. IEEE Antennas IEEE Antennas Wirel. Propag. Lett. **10**, 780–783 (2011). doi:10.1109/LAWP.2011.2163378
46. K. Yazdandoost, K. Sayrafian-Pour, TG6 channel model ID: 802.15-08-0780-12-0006. IEEE Submission (2010)
47. B. Zhen, M. Patel, S. Lee, E. Won, Body area network (BAN) technical requirements ID:15-08-0037-01-0006. IEEE Submission (2008)
48. B. Zhen, K. Takizawa, T. Aoyagi, R. Kohno, A body surface coordinator for implanted biosensor networks, in *IEEE International Conference on Communications, ICC '09*, Dresden, 2009, pp. 1–5. doi:10.1109/ICC.2009.5198579

Circuit Techniques for Ultra-Low Power Radios

Jagdish Pandey and Brian Otis

Abstract As low power radio circuits are enabling new technology avenues such as health monitoring, efforts to tackle the design challenge of making extremely reliable yet lost cost, low power CMOS radios have gained prominence. Despite advancements in battery technology and energy harvesting, there remains a wide gap between the available and desired performance. Innovative system architectures and circuits need to be explored to bridge this gap. Over the last few years, there have been numerous sub-mW integrated system offerings that trade off performance and reliability to achieve low power consumption. At the same time, an ever growing list of applications has led to a number of commercial products that guarantee robust operation with power consumption in 10′s of mWs. However, a vast number of applications demand sub-mW power consumption or complete energy autonomy while demanding robust operation and high performance over a highly variable environment. Implantable systems are a prime example of this. In this chapter, we will begin with an overview of system considerations, application driven challenges, and proceed to discuss the existing design approaches to make the reader aware of the practical limitations of many of these techniques. We will then identify the fundamental design challenges and explore fresh angles to approach the problem.

Keywords Implantable systems • Wearable technologies • Low power RF circuits • Energy harvesting • Energy autonomy • Sensor networks

1 Introduction

Both industry and academic interest in the development of fully autonomous sensors networks, in particular the implantable or wearable health monitoring devices constituting a body area network, has seen a rapid growth in recent years. Tire pressure monitoring system (TPMS), smart buildings with intelligent regulation of

J. Pandey (✉)
Qualcomm Inc., San Diego, CA, USA
e-mail: jnpandey@uw.edu; jnp1981@gmail.com

B. Otis
Google Inc, Mountain View, CA, USA
e-mail: botis@uw.edu

P.P. Mercier, A.P. Chandrakasan (eds.), *Ultra-Low-Power Short-Range Radios*, Integrated Circuits and Systems, DOI 10.1007/978-3-319-14714-7_3

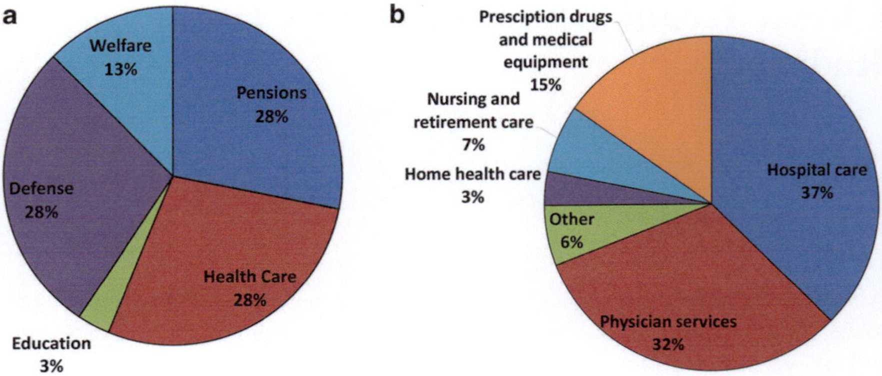

Fig. 1 (**a**) US government spending for financial year 2013. More then a quarter of the total federal expenditure is on healthcare [27]. (**b**) The U.S. national healthcare expenditure for year 2011 [26]

energy consumption and industrial monitoring are some other emerging applications facing a common challenge: a near or complete energy autonomy [21]. A number of these networks, most notably the Body Area Networks, comprise highly miniaturized, inexpensive sensor nodes where both the transmit and the receive ends are severely energy constrained, often harvesting energy from ambient sources. In such scenarios, the performance burden needs to be shared equally between the transmitter and the receiver while simultaneously achieving low energy per transceived bit. However, it has been extremely difficult to demonstrate a fully autonomous, robust, symmetric wireless links supporting reasonably high data rates. In this chapter, we will explore design techniques both at the level of system architecture and circuits to realize the same.

Tackling the design challenge of ultra-low power radio is at the heart of realizing viable body area networks (BAN). These networks have perhaps the most direct impact of human lives, scientific research and national economies. For example, the United States spends the highest percentage of its GDP on healthcare (>16 %) among all nations [26]. In year 2013, the federal expenditure on healthcare constituted 28 % of its total expenditure (Fig. 1a). The cost breakdown of this spending is shown in Fig. 1b. Nearly 70 % of this cost is incurred on hospital care and clinician services. Body Area Networks consisting of highly miniaturized, low cost sensor nodes can reduce this cost by allowing home monitoring of patient health. Apart from reducing the cost of health care, a continuous monitoring of vital signs can also improve the quality of diagnosis and treatment.

Figure 2 shows the leading cause of deaths worldwide according to World Health Organization report [28]. Coronary heart disease and stroke are not only the leading causes of deaths but also are alarmingly on the rise over the last decade. Deaths due to diabetes and hypertension have seen the highest increase in mortality. A constant monitoring of heart rate, blood pressure and blood glucose level can go a long way in saving lives. It is perhaps not an exaggeration to say that the development of

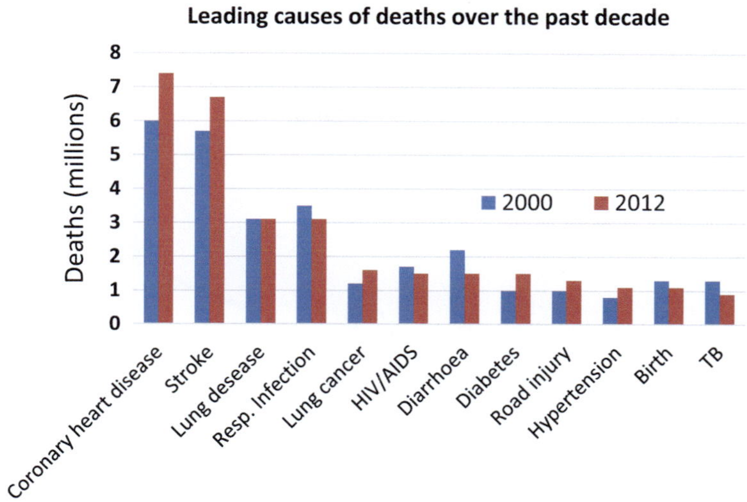

Fig. 2 Leading causes of death worldwide according World Health Organization (WHO) report, 2013 [28]

Table 1 Various scenarios for body area networks

Scenario	Description	Frequency band	Channel model
S1	Implant to implant	402–405 MHz	CM1
S2	Implant to body surface	402–405 MHz	CM2
S3	Implant to external	402–405 MHz	CM2
S4	Body surface to body surface (LOS)	13.5, 50, 400, 600, 900 MHz, 2.4, 3.1–10.6 GHz	CM3
S5	Body surface to body surface (NLOS)	13.5, 50, 400, 600, 900 MHz, 2.4, 3.1–10.6 GHz	CM3
S6	Body surface to external (LOS)	900 MHz, 2.4, 3.1–10.6 GHz	CM4
S7	Body surface to external (NLOS)	900 MHz, 2.4, 3.1–10.6 GHz	CM4

reliable, low cost CMOS wireless sensors will perhaps have the greatest impact on human lives yet!

The IEEE 802.15.6 standard for the Body Area Network has envisaged it to be a peer-to-peer network consisting of both embedded (body surface and implant) and body worn devices communicating with each other together with a hand-held device or a computer using UWB/Bluetooth wireless backhaul [36]. Table 1 and Figure 3 various scenarios for the body area networks. Except for the scenarios S6 and S7, 400 MHz MICS band is suitable for all scenarios. In this chapter, we will illustrate the design challenges and methods for 402–405 MHz MICS band for body-worn devices. In addition, we will focus on the narrow-band PHY for the reason that the architectures and circuit techniques explored in this work can

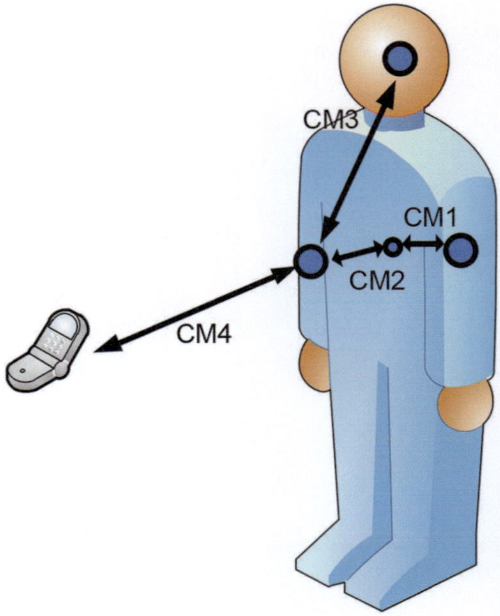

Fig. 3 The U.S. national healthcare expenditure for year 2011 [36]

easily be extended to other frequency bands as well as other applications such as intelligent homes, industrial monitoring and implants. Note that wide-band and human-body communication systems are discussed in chapters "Pulsed Ultra-Wideband Transceivers" and "Human Body Communication Transceiver for Energy Efficient BAN" respectively.

1.1 Energy Considerations

In order to fully comprehend the challenge of designing these wireless sensors, let us understand the constraints posed by the energy sources and performance requirements. Table 2 shows the energy density of common energy harvesters. Although the power density varies a lot across energy sources, available power in volume and application constrained systems can be as little as $100\,\mu W/cm^3$ [18]. To make things more difficult, energy harvesting transducers and thin film batteries typically exhibit high source impedances limiting their peak current. This can pose challenge for burst mode of operation, a popular technique to lower average power consumption using duty cycling. A detailed discussion on energy harvesting can be found in the chapter "Energy Harvesting Opportunities for Low-Power Radios".

The story is not very different for battery powered wearable systems. Figure 4 show the performance of single cell, high capacity Zinc-Air microbatteries. Although the energy density remain roughly constant as the size shrinks, internal resistance dramatically rises eventually limiting battery's ability to supply high peak currents. Additionally, high current drain and pulsed loads can further degrade the

Table 2 Energy density of common energy scavengers [30]

Energy	Power density and performance
Acoustic noise	0.003 μW/cm^3 @75 Db 0.96 μW/cm^3 @100 Db
Temperature variation	10μW/cm^2
Ambient radio frequency	1μW/cm^2
Ambient light	100 mW/cm^2 (direct sun) 100 μW/cm^2 (illuminated office)
Thermoelectric	60 μW/cm^2
Vibration (micro generator)	4 μW/cm^3 (human motion–Hz) 800 μW/cm^3 (machines–kHz)
Vibrations (Piezoelectric)	200 μW/cm^3
Airflow	1 μW/cm^2
Push buttons	50 μJ/N
Shoe inserts	330μW/cm^2
Hand generators	30 W/kg
Heel strike	7 W/cm^2

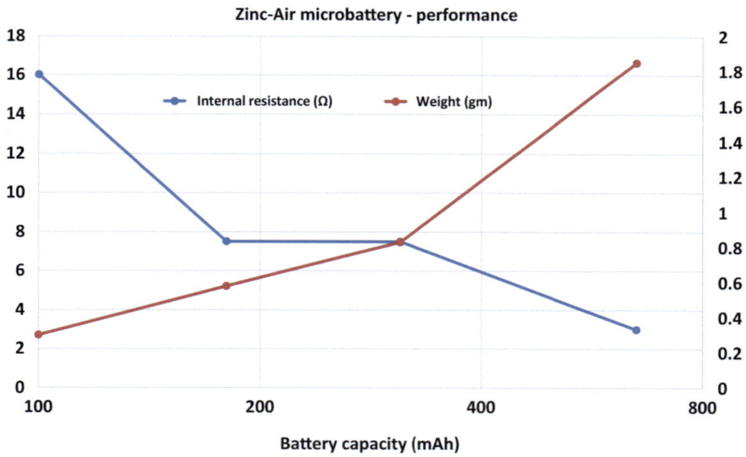

Fig. 4 Zinc-Air microbattery performance: internal resistance increases dramatically as weight/volume drops [31]

batter capacity [32]. Thin film batteries offer superior volume densities but their capacities are a fraction of single cell microbatteries [33].

Table 3 presents peak and average power constraints imposed by the battery capacity and an expected lifetime of 10 years. For typical data rates of 1 kbps and 1 % duty cycle, 100 μA of current and 1 nJ/bit of energy efficiency is desired.

Table 3 Energy budget analysis for weight/volume constrained microsystems powered by a single cell battery for 10 year life span

Battery capacity	100 mAh
Average load current	1 μA (10 years life)
Average load current at 1 % duty cycle	100 μA
Data rate @source	1 kbps
Data rate @chip at 1 % duty cycle	100 kbps
Required energy efficiency	1 nJ/bit/10 years

1.2 RF Link Budget Analysis

After becoming familiar with the energy constraints, let us look at the RF transceiver which tends to dominate the power consumption of the entire system. As mentioned before, we will use the example of IEEE 802.15.6 standard to derive the specifications, but the exercise can be easily repeated for other cases. The 400 MHz band has the advantage of fewer jammers leading to relaxed linearity in turn facilitating lower power consumption. The current users of the band are weather balloons. Maximum effective isotropic radiated power is −16 dBm and channel width is 300 kHz. For miniaturized systems operating at sub-GHz carrier frequencies, antenna losses can be high leading to reduced link budget to work with. However following many advantages acrue due to application environment that work in favor of 400 MHz band:

1. Low BW wastage due to TX and RX crystal errors
2. Lower path loss
3. Less crowded spectrum
4. Higher diffraction around edges leading to better coverage and penetration
5. Lower tissue absorption

Literature reports highly miniaturized implantable antenna for MICS band with −16 dBi gain [12]. This makes RX sensitivity and TX power efficiency extremely difficult requirements given the power budget. Table 4 presents a detailed break-up of the RF link budget. Assuming both transmit and receiver blocks energy constrained with equal performance burden, implants have only 53 dB of link budget to work with. If the communication is from an implant to a wearable, the improved efficiency of receive antenna can increase this budget by 10 dB.

1.3 Existing Solutions

Figure 5 depicts the existing personal area network solutions. Their power consumption typically goes down linearly with date-rates. Body area networks and similar systems demand data-rates comparable to that of ZigBee but power consumption

Table 4 RF link budget for a typical medical transceiver operating in 400 MHz MICS band

Term	Power/gain	Comment
P_{TX}	−10 dBm	From the chip (−16dBm maximum EIRP)
G_{TX}	−16 dB	2.5 % antenna efficiency
G_{RX}	−16 dB	RX antenna loss
$N_{RX,in}$	−120 dBm	300 kHz wide channel
SNR	10 dB	0.1 % BER
RX NF	15 dB	Reasonable target for ULP RX
$P_{RX,in}$	−95 dBm	Required power after the receive antenna (sensitivity)
Path loss	53 dB	Maximum tolerable path loss

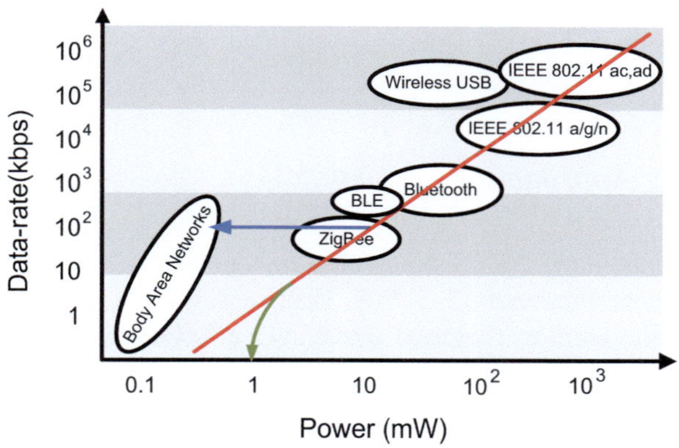

Fig. 5 Existing wireless PAN solutions. Body area networks and other wireless sensors demand data-rates comparable to ZigBee but power consumption an order of magnitude lower

an order of magnitude lower. These systems, specially the health monitoring devices, need extremely reliable performance while operating in a very uncertain environment (due to varying body postures and neighboring objects as discussed in the chapter "Channel Modeling for Wireless Body Area Networks"). As we will understand in future sections that beyond a point, simply lowering down data-rates does not mean a commensurate reduction in power consumption which tends to hit a lower limit. Reducing power dissipation beyond this limit while delivering good performance calls for a fresh look at the problem and identifying the fundamental issues. In addition, existing personal area networks (PAN) do not meet the medical regulations such as proximity to human tissue.

Lastly, as BAN nodes employ aggressive duty cycling to reduce average power while maintaining low latency, synchronization poses a tough challenge. Development of power efficient asynchronous MACs, accompanying low-power wake-up radio and a fast start-up to reduce transient energy become imperative. The average power overhead in synchronization can be significant. While evaluating the energy efficiency of a wireless link, wake-up current and transient energy need to be taken into account.

2 Ultra-Low Power TX Design

Figure 6 shows a 0.3 g, 7.6×8.7 mm^2 microsystem including a fully integrated biosignal interface and MICS band transmitter, antenna and a battery with 70 h of continuous operation from a single hearing aid cell. The transmitter consumes nearly 1 mW of power. The battery weighs 0.17 g and dominates the overall system weight. It becomes apparent that TX power consumption ($\approx$1 mW) needs to be minimized to lower system weight and volume. Let us take a brief look at the existing approaches to solve this problem.

2.1 Existing Approaches

A number of integrated MICS band ultra-low power (ULP) transmitters have appeared in recent literature [3–5, 10]. Energy per bit is the most commonly metric used in literature to represent transmitter efficiency. Figure 7 shows some representative cases with different design techniques to lower transmit power. Let us critically examine the advantages and disadvantages of each case.

1. A common approach to reducing TX power consumption is to shift the performance burden on to the receiver which is then designed to be more power hungry and high performance. While suitable for some applications, the approach does

Fig. 6 A 0.3 g biosignal recording and transmitting platform called Bumblebee. The battery weight (0.17 g) dominates the total system weight [13]

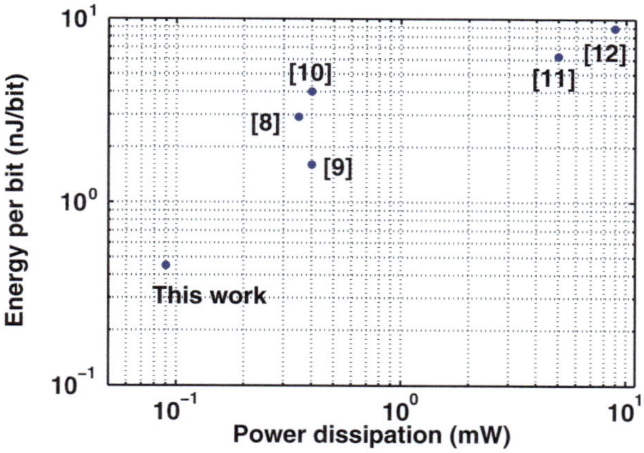

Fig. 7 State-of-the-art in MICS band transmitters

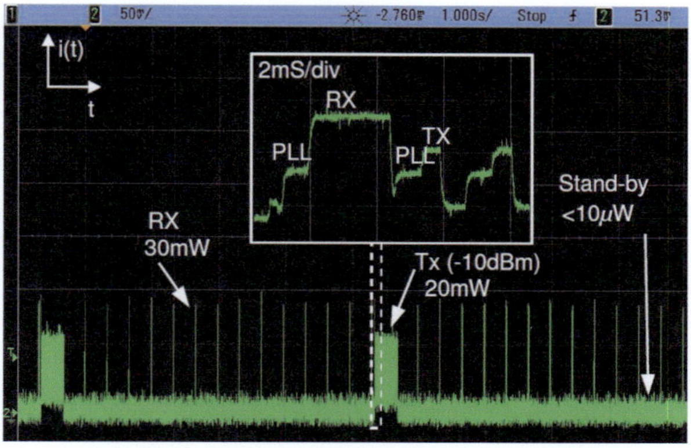

Fig. 8 Asynchronous MAC protocols, low-latency and short packets typically lead to synthesizer locking time being comparable to the transmitter active time. In addition, the average wake-up radio power can be significant

not work for autonomous sensors in a peer-to-peer network where radio link needs to be symmetric in terms of energy consumption. That is, the power and complexity burdens cannot be shifted from transmitter to receiver and vice versa. These designs can boast of very low energy per bit for the transmitter but their energy per *transceived* bit is very high!

2. Ultra-low power radio systems typically employ very aggressive duty cycling to conserve power. In such cases, the finite locking time of carrier generation loop places the upper limit on the duty cycling, and the start-up energy overhead begins to dominate the energy used in communication, significantly increasing energy per bit [7]. Figure 8 shows the measured time domain power consumption of a commercial off-the-shelf transceiver wherein the PLL lock time exceeds the communication time period when the packet size becomes relatively small. Additionally, FCC regulations in MICS band currently specify a Listen-Before-Talk (LBT) protocol to avoid channel collision. However, as the number of devices operating in this band increases, the time spent in finding a clear

channel may significantly increase, leading to higher drain on the battery. In the wake of this, FCC and other regulatory agencies have specified a Low-Power Low-Duty Cycle (LP-LDC) protocol for interference mitigation wherein extremely agile duty cycling (0.1 %) at very low-output power (1 μW) can be used without significant carrier-to-interference degradation from other such transmitter operating even in the same channel [19]. In such scenarios, the LO settling time should be made extremely small. Start-up energy is usually not accounted for when reporting energy per bit.

3. To reduce power dissipation in carrier generation that typically dominates the total power budget in ULP transmitters, an open loop oscillator is used [3, 4]. The necessary frequency stability is obtained using a frequency correction/calibration loop running at the receiver. Others have adopted an RFID-style passive tag to perform medical sensing [23, 24]. This again requires a power hungry RFID receiver.

In summary, all the low-power transmitters described above shift the performance burden to a complex receiver resulting in high energy per transceived bit and a highly asymmetrical link unsuitable for peer-to-peer applications. Finally, an important characteristic of a portable transmitter is a high global efficiency: the ratio of power delivered to the antenna to the total transmitter power consumption. In systems such as GSM phones where the typical power delivered to the antenna exceeds a Watt, global transmitter efficiency approaches the efficiency of the power amplifier since carrier generation and baseband power overheads become a negligible fraction of the total power consumption [2]. However, in ultra-low power transmitters, the average power delivered to the antenna is very small; MICS compatible systems, for example, must transmit no greater than 25 μW of power outside the human body [29]. In such systems, the carrier generation, not the PA efficiency, is the dominant source of power dissipation leading to very poor global efficiencies (<6.5 %) [3–5, 10]. This calls for change in design paradigm from PA centric TX design to LO centric TX design. Not only the LO needs to be ultra low power, its settling time needs to be minimized while offering robustness performance.

2.2 MICS TX Specifications

1. The maximum isotropic transmitted power is 25 μW (−16 dBm) to prevent interference to the incumbent users of the band (meteorological aid systems such as weather balloons.)
2. The maximum emission bandwidth is 300 kHz (for narrowband PHY).
3. Adjacent channel power ratio (ACPR) better than −26 dB.
4. Spurious emissions within the MICS band and 250 kHz or above and below the MICS band must be 20 dB down relative to the maximum level of the modulated carrier. For other frequency bands in 216–960 MHz range, the spurious emission measured at 3 m distance needs to be lower than −50 dBm.

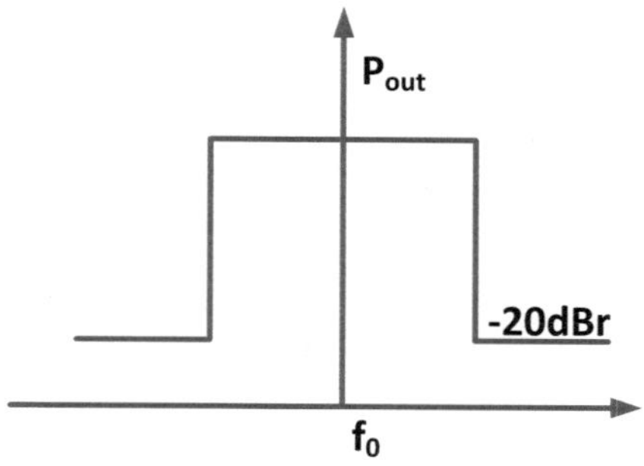

Fig. 9 MICS TX spectral mask

Figure 9 shows the spectrum mask for the MICS complaint transmitter for a channel at the edge of the band where tighter specifications apply. The 433 MHz ISM band in the US and Europe can also be used for medical telemetry with FCC imposing less than −16 dBm of transmit power in this band. In addition to this, FCC has reserved the use of 174–216 MHz and 470–668 MHz bands for biotelemetry applications [34].

2.3 *Low Power Circuit Techniques for ULP TX Design*

1. Due to high frequency of operation, RF components typically dominate the total power budget. A design that tries to lower down the number of RF circuit blocks has a chance to lower the overall power consumption. Combining functions and current reuse are common techniques to achieve this.
2. Try to eliminate the slow and area hungry phase locked loops without resorting to free running local oscillator. Even for systems that have a narrow temperature range, need periodic calibration to bring BER under control. Injection locking to generate LO locked to a stable reference is an attractive option.
3. Frequency multiplication (FM) can be utilized to synthesize LO at a lower frequency. However PLLs running at lower frequencies are slower unless more power hungry fractional synthesizers are utilized. FM and injection locking together can provide a low power alternative to RF carrier generation.
4. Harness process scaling! Designs that are more digital in nature will benefit from Moore's law.
5. Constant envelope modulation schemes allow for high-efficiency non-linear PA.
6. Exploit low data-rates to trade spectral efficiency with lower power consumption. Quadrature modulation schemes though more spectrally efficient tend to consume more power since they need quadrature LOs and quadrature mixers.
7. In PLL-less designs, data modulation can be done at baseband which can also open the possibility of removing RF mixer.

In this section, we will design a sub-100 μW MICS/433 MHz ISM band transmitter, realizing a significant improvement in state-of-the-art. We will study the technique of cascaded multi-phase injection locking and frequency multiplication to realize a high figure-of-merit local oscillator locked to a stable crystal reference.

Using this technique, the new transmitter architecture will eliminate the slow phase/delay locked loops used in carrier generation, exhibiting a fast settling time of less than 250 ns. This permits agile duty cycling of the transmitter to conserve energy. Finally, while delivering 20 μW output power, it achieves a high global efficiency of 22 %. A highly digital transmitter architecture leads to a very small active die area of 0.04 mm^2 in 0.13 μm CMOS. Figure 7 places this work in context of the published results on MICS band transmitters.

2.4 Proposed TX Architecture

Conventional ULP transmitters perform frequency synthesis and data modulation at the carrier frequency leading to poor global efficiency and high power consumption (Fig. 10a). The proposed transmitter achieves very low power dissipation by performing these operations at a much reduced frequency and employing an edge-combiner merged into the power amplifier (PA) (Fig. 10b). In order to obtain the equally spaced edges necessary for the edge-combiner, a delay chain or a ring oscillator locked in a PLL/DLL is typically needed. The additional components in the loop such as charge-pump and loop filter present significant area/power overhead, while the settling time of the loop constrains the maximum possible duty cycling of the transmitter [16]. The use of frequency multiplication by a relatively large factor can reduce the carrier generation frequency to permit fundamental harmonic injection-locking by the crystal oscillator. However, any attempt to directly injection-lock the multi-phase low-frequency ring oscillator using the single phase crystal reference will introduce significant mismatch in the delayed waveforms of the RO. We address this issue by using cascaded injection-locking (Fig. 11). The design of a low-power high-efficiency edge-combiner is the second key challenge. Authors in Chien et al. [6] have reported an edge-combiner that uses differential MOS transistor pairs to commutate current. We propose the use of digitally driven MOS transistor switches to increase its power efficiency. The proposed overall architecture is an example of digitally-assisted RF design that benefits from CMOS and power supply scaling. The key building blocks of our

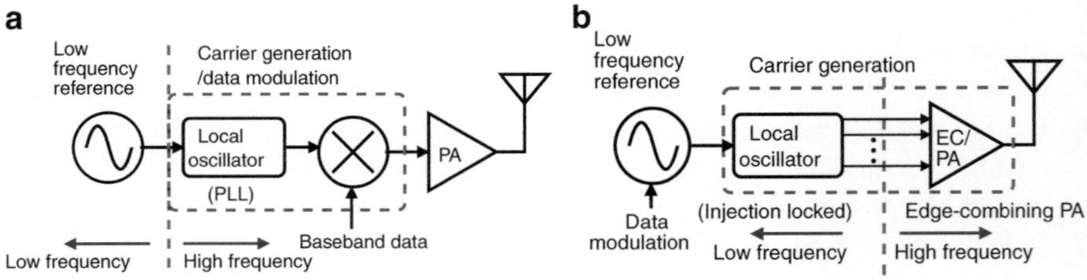

Fig. 10 (**a**) A conventional transmitter architecture. (**b**) Proposed injection-locked frequency multiplying transmitter architecture

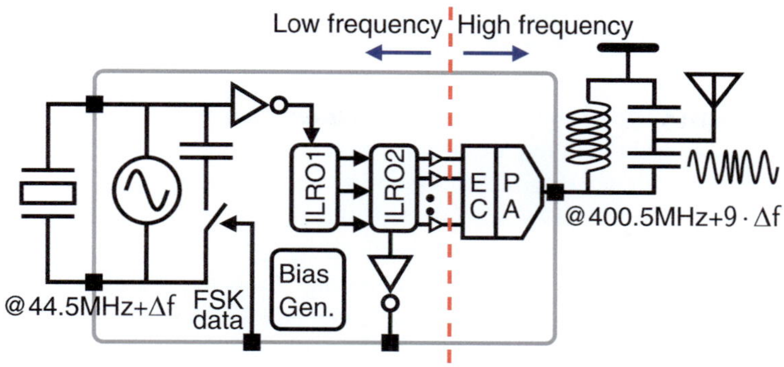

Fig. 11 Detailed block diagram of the proposed transmitter architecture

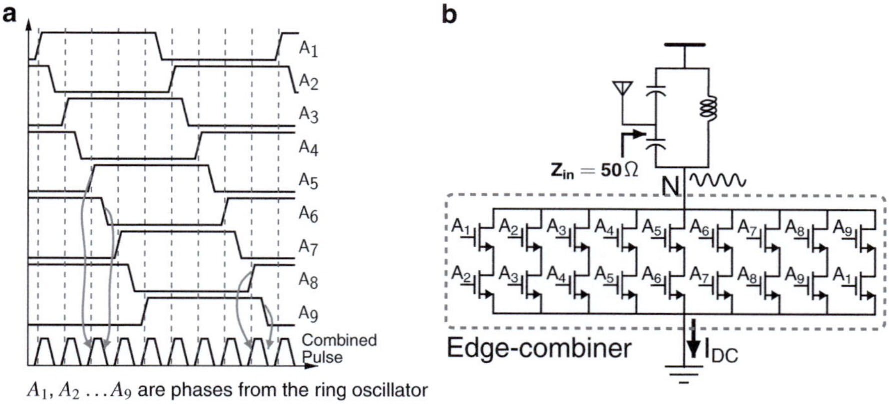

Fig. 12 (**a**) The principle of edge-combining. (**b**) Schematic of the edge-combiner showing frequency multiplication by 9

proposed architecture—the edge-combining PA, the injection-locked oscillator, and data modulation—are described below.

2.5 *Edge-Combiner*

The low-power frequency multiplier is based on the principle of edge-combining. Let $A_1, A_2 \ldots A_N$ be the waveforms from a digital ring oscillator running at frequency f_{RO}. The waveform $\Sigma(A_1A_2 + A_2A_3 \ldots A_NA_1)$ is a square wave of frequency Nf_{RO} where N is both the factor of multiplication and the number of stages in the ring oscillator. For our nine-stage ring oscillator, the waveforms $A_1, A_2 \ldots A_9$ are spaced apart by a period of $T/18$, where T is the time period of the reference input at 44.5 MHz (Fig. 12a). We use MOS transistor switches to perform an AND operation and sum the switched currents to realize an OR operation (Fig. 12b). Equivalently, 18-transistor switch network acts like a single composite switch operated at 400.5 MHz.

We take advantage of the low output power requirement of the MICS standard (and the body area network requirements in general) to combine the PA functionality with the edge-combiner (Fig. 12b). Assuming abrupt switching of the composite switch, square current pulses are injected into the LC tank which filters out of harmonics. The amplitude at node N is given by $\frac{2}{\pi} I_{\mathrm{DC}} R_p$ where I_{DC} is the DC current into the edge-combiner switch matrix and R_p is the impedance of the LC-tank at resonance. R_p is transformed using a tapped-capacitor matching network to match low impedance antenna (e.g. 50 Ω). This high-Q load at the edge-combiner output also attenuates the out-of-band spurs resulting from mismatch in the low-frequency ring oscillator.

2.6 Injection-Locked LO Design

The 9× frequency multiplication allows the ring oscillator to run at 45 MHz, which can thus be directly locked to a crystal oscillator. Instead of using a phase-locked loop, we injection-lock the low-frequency ring oscillator to an on-chip crystal reference, thereby eliminating the longer settling times, stability issues, and associated loop filter components. Under locked conditions, the injection-locked system behaves as a first order PLL with unconditional stability and fast settling [1]. The equivalent loop-bandwidth is the one-sided locking range of the oscillator. Due to their poor quality factors, ring oscillators display wide-locking range and therefore permit fast locking. The fast lock-time (on the order of 100 ns) of the LO allows aggressive duty cycling of the transmitter to further save power. Figure 13 shows the schematic of the three-phase two-stage injection-locked oscillator. The first stage of injection locking by the crystal oscillator ensures the correct frequency and low phase noise. However, the single-phase injection introduces asymmetry in otherwise equally spaced phases (A, B, and C) of the ring oscillator. This asymmetry will lead to large reference spurs in the frequency multiplied output. The second stage of injection-locking attenuates this phase imbalance by using three-stage symmetrical injection in a nine-stage ring oscillator. As shown in [9], cascaded multiphase injection-locked oscillators can be used to correct the phase and amplitude mismatches. The phase mismatches in phases $A_1, A_2, \ldots A_9$ are approximately an order of magnitude smaller than those in A, B and C (Fig. 14). Multiphase-injection locking also increases the locking bandwidth, ensuring reliable operation across PVT variation.

Figure 15 shows the simulated PA output with a settling time less than 100 ns.

Figure 16 shows the layout of the ring oscillator and the PA transistors. The PA switching transistors and stage-1 ring oscillator inverters are interleaved with stage-2 ring oscillator inverters. This results in a highly symmetric and compact layout.

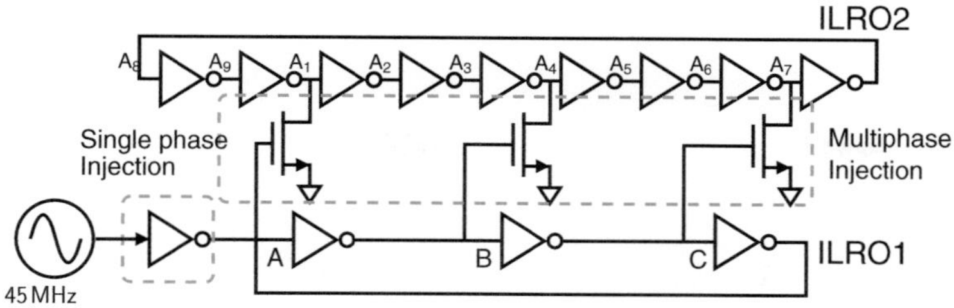

Fig. 13 Schematic of the two-stage multi-phase injection-locked ring oscillator

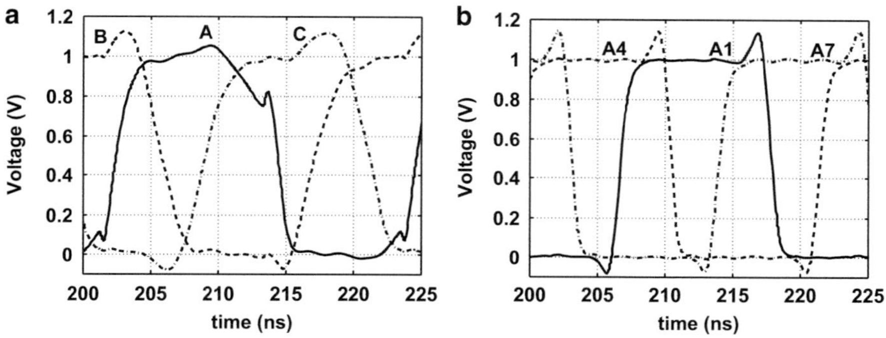

Fig. 14 (**a**) Simulated voltage waveforms of the three stage ring oscillator (ILRO1) and (**b**) nodes A_1, A_4 and A_7 of the nine-stage oscillator (ILRO2) showing improvement in phase mismatch due to 2-stage injection locking

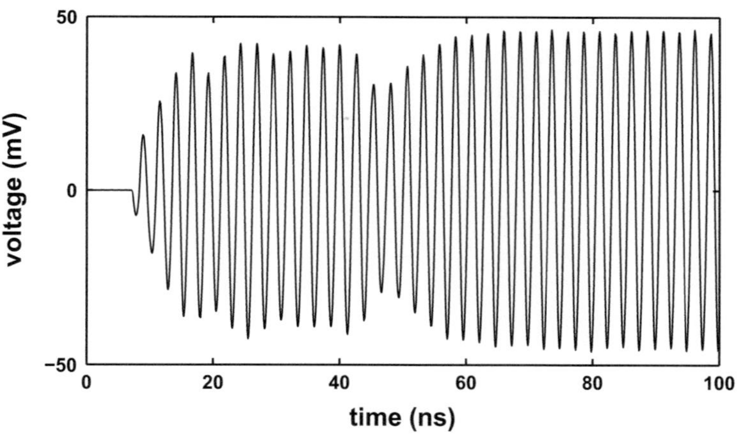

Fig. 15 Simulated transmitter output showing a <100 ns settling time of the two stage injection-locked ring oscillator

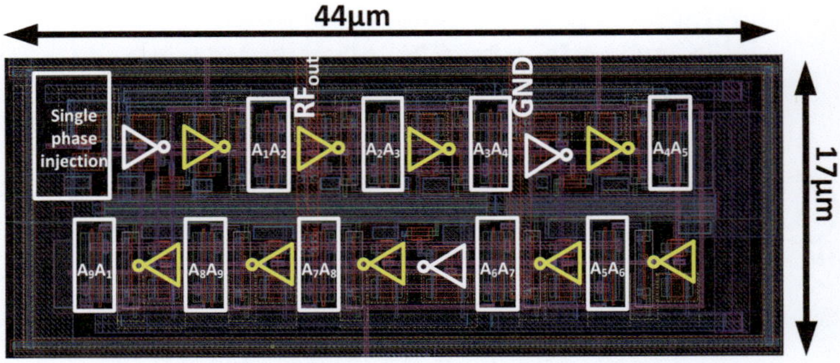

Fig. 16 Layout of the multi-phase, multi-stage, injection-locked ring oscillator. Inverters in *white* correspond to the three-stage ring oscillator and those in *yellow* constitute the nine stage oscillator. The PA switches are also interwoven and the entire layout of the transmitter except for the crystal oscillator is only 44 μm×17 μm

2.7 Data Modulation

On-chip FSK modulation is accomplished by pulling the quartz reference clock. The resulting frequency deviation is multiplied by 9×. Figure 17 shows the schematic of the Pierce oscillator and the equivalent circuit with load-pulling capacitance ΔC. The frequency of oscillation is given by $f = f_0(1 + p)$, where $f_0 = 1/2\pi\sqrt{L_s C_s}$ and p is the fractional frequency deviation given by Vittoz et al. [20]

$$\frac{\Delta f}{f_0} = \frac{C_s}{2(C_L + C_0 + \Delta C)} \tag{1}$$

where $C_L = \frac{C}{2} + C_p$ and C_p represents the parasitic capacitance of the circuit.

By modulating ΔC, we can pull the crystal frequency by a few hundred ppm. Frequency deviation is limited by the presence of shunt capacitance C_0, parasitic capacitance C_p, and the oscillator start-up requirements. For a 45 MHz crystal, we obtained a total of 20 kHz (440 ppm) of frequency pulling resulting in a Δf of 180 kHz at the antenna.

2.8 Measurement Results

This section presents the measured results from our transmitter prototype, implemented in 0.13 μm CMOS, based on the architecture described above. The entire system is integrated except for the crystal and the matching network. Figure 18 shows the frequency multiplied output with a −17 dBm output power. The carrier-to-spur ratio (CSR) is better than 44 dB indicating excellent matching in phases $A_1, A_2, \ldots A_9$. The unwanted radiation outside the 402–405 MHz MICS band must be less than 200 μV/m field strength at 3 m distance or an EIRP of −49.2 dBm [29].

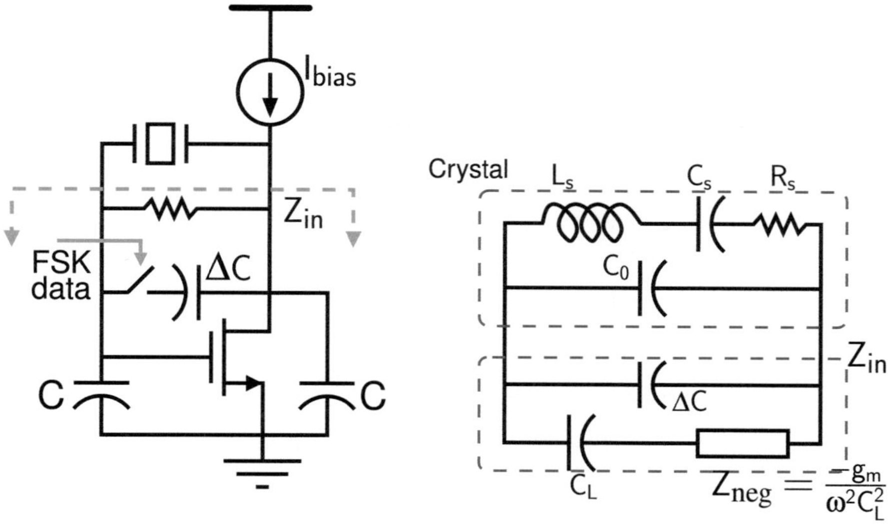

Fig. 17 Pierce oscillator used for crystal pulling and its equivalent circuit

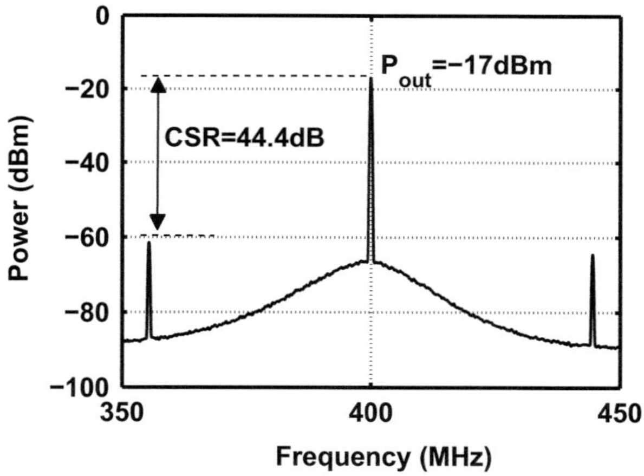

Fig. 18 A measured carrier-to-spur ratio of 44.4 dB was achieved with $P_{out} > -17$ dBm

The measured reference spurs for our transmitter on either side of the 400 MHz carrier frequency are less than −61 dBm.

Figure 19a shows the overlaid spectrum of the free running and the injection-locked ring oscillator. As shown in Fig. 19b, the close-in phase noise of the injection-locked ring oscillator is vastly improved. The overall locking range of the 45 MHz free-running ring oscillator assembly (ILRO1 and ILRO2) extends from 32–52 MHz (44 % locking bandwidth). This allows operation in the 433 MHz ISM band using a 48 MHz crystal. The FSK modulated waveforms of the locked oscillator and the frequency multiplied output are captured in Fig. 20. An FSK modulated

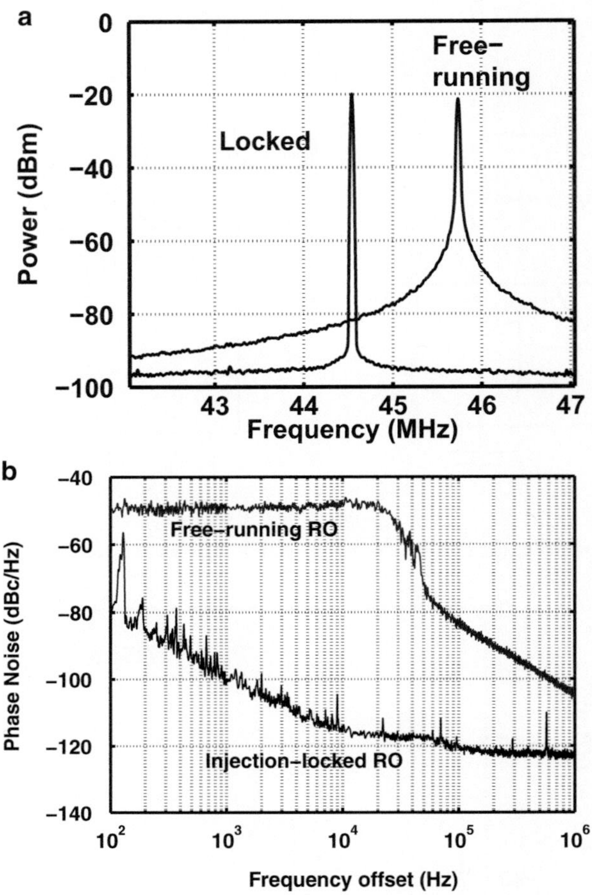

Fig. 19 (**a**) The spectrum and (**b**) the measured phase noise of the free running and the injection-locked ring oscillator

pseudo-random data sequence was successfully detected using a commercial off-the-shelf receiver at a data rate of 200 kbps.

Since the LO is locked to a stable crystal reference, its frequency drift over 15–55 °C temperature range (of interest to wearable systems and implants) is only ±2 ppm (Fig. 21).

Figure 22 presents the phase noise of the injection-locked ring oscillator and the frequency multiplied output. At a given offset, the phase noise of the frequency multiplied output is higher than that of the 44.5 MHz ring oscillator by $20\log_{10}(9) \approx 19$ dB. At 300 kHz offset, the frequency-multiplied output achieves a phase noise of −104 dBc/Hz. The oscillator figure-of-merit (FoM) is given as

$$\mathrm{FoM(dB)} = -\mathscr{L}(\Delta\omega) + 20\log\left(\frac{f_0}{\Delta f}\right) - 10\log(\mathrm{P(mW)})$$

For our 400.5 MHz frequency multiplied oscillator, the FoM is 203 dB.

a

b

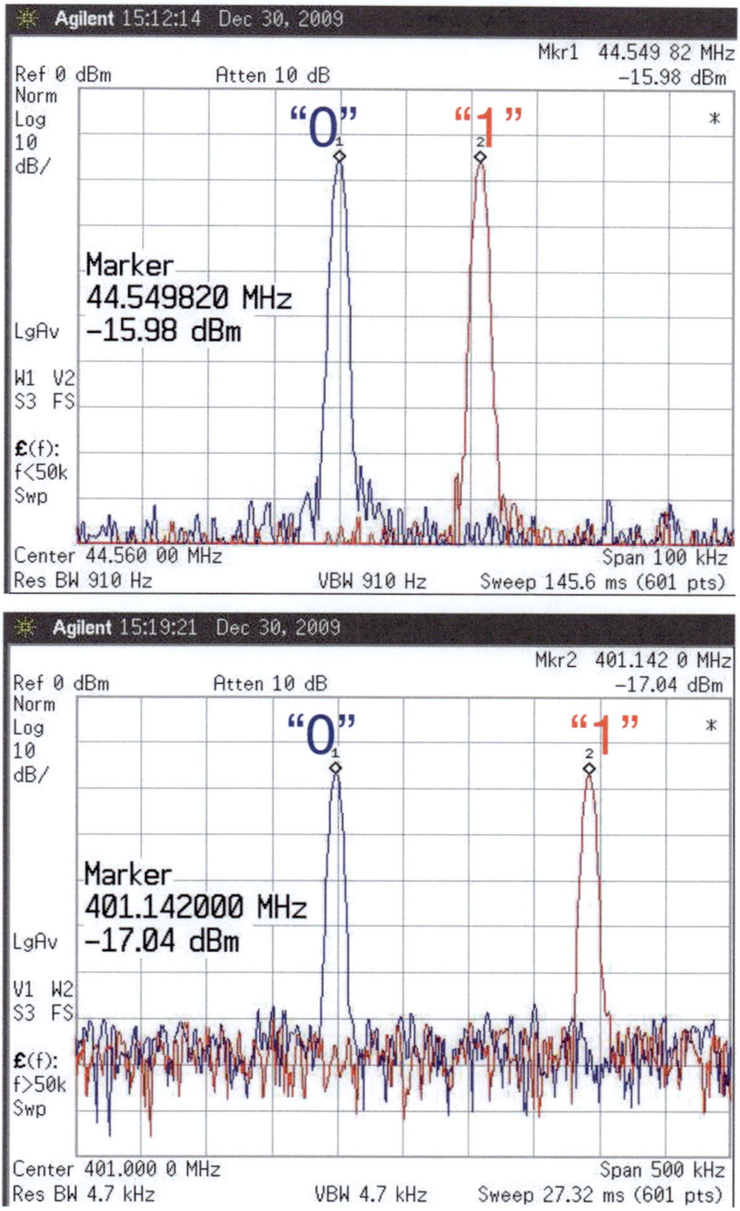

Fig. 20 (**a**) FSK modulated 44.5 MHz injection-locked ring oscillator. $\Delta f_{RO} = 20\,\text{kHz}$. (**b**) FSK modulated frequency multiplied output at 400.5 MHz. $\Delta f_{FM} = 180\,\text{kHz}$

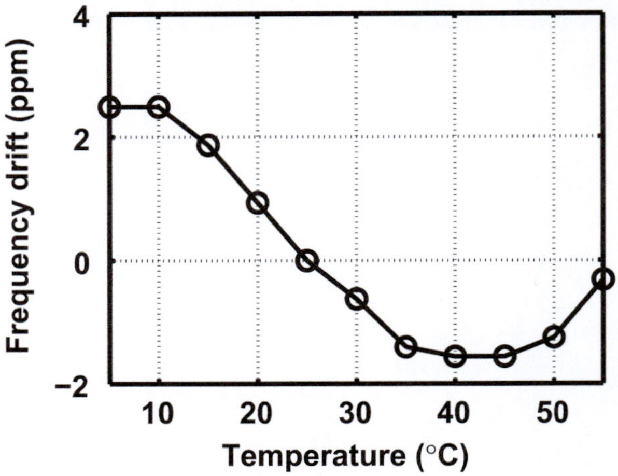

Fig. 21 Fractional frequency drift of the RF output

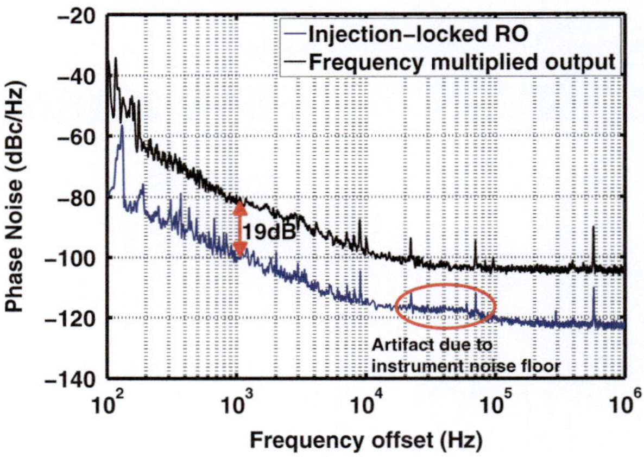

Fig. 22 Measured phase noise of the injection-locked LO and the frequency multiplied output using an Agilent E4446A spectrum analyzer

Using our proposed techniques of high efficiency frequency multiplication and multi-phase cascaded injection locking, we were able to reduce the total power dissipation in carrier generation and data modulation to less than 24 μW, including the crystal oscillator. The measured PA drain efficiency is higher than 30 % for an output power greater than 20 μW (Fig. 23). It can deliver a maximum of −11 dBm of output power.

Figure 24 presents the measured input return loss for the transmitter. Excellent matching with a measured $|S_{11}| < -20$ dB was achieved.

Figure 25 shows the chip micrograph of the die, implemented in 0.13 μm CMOS. Due to the highly digital architecture of the proposed transmitter, and the absence

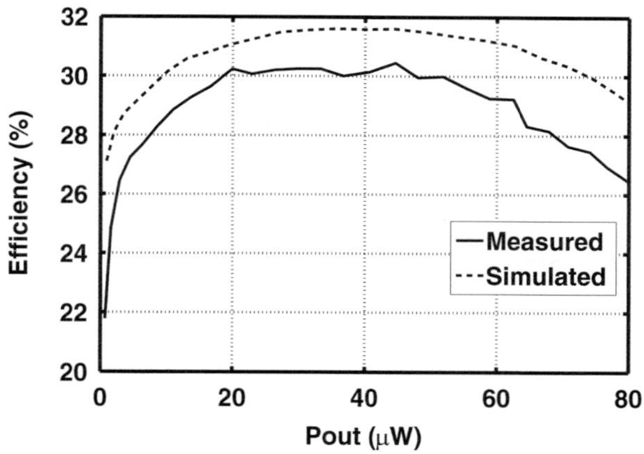

Fig. 23 Measured and simulated PA drain efficiency

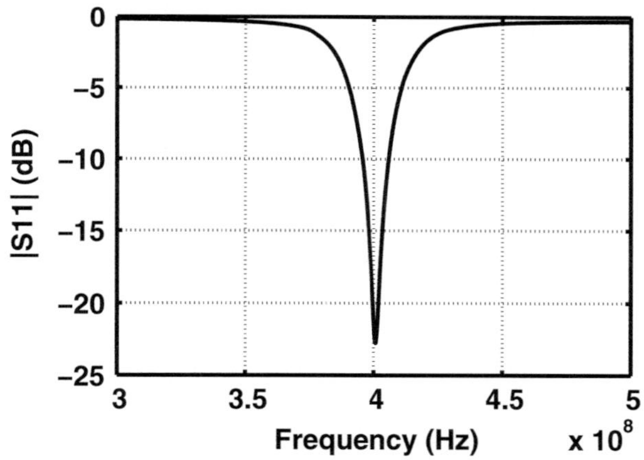

Fig. 24 Measured input return loss showing $|S_{11}| < -10\,\text{dB}$

of the synthesizer loop along with its large loop filter capacitors, the active area is less than $200\,\mu\text{m} \times 200\,\mu\text{m}$.

Figure 26 shows the transmitter start-up with an already settled crystal oscillator. Figure 27 presents the eye diagram using Tektronix RSA 3408A spectrum analyzer at 200 kbps data rate using square wave input showing sufficiently wide eye opening. Figure 28 shows the transmit data and received data and clock for a 10-bit psuedo-random bit sequence at 200 kbps with BER $< 0.1\,\%$ using TI CC1101 receiver [35] at 5 m distance. The received signal strength for the receiver was −85dBm.

Table 5 captures the summary of the latest reported transmitters for the MICS band, along with our proposed work. Our transmitter has a global efficiency of 22 % and energy efficiency of 450 pJ/bit which is a 3× improvement in the state-of-the-art.

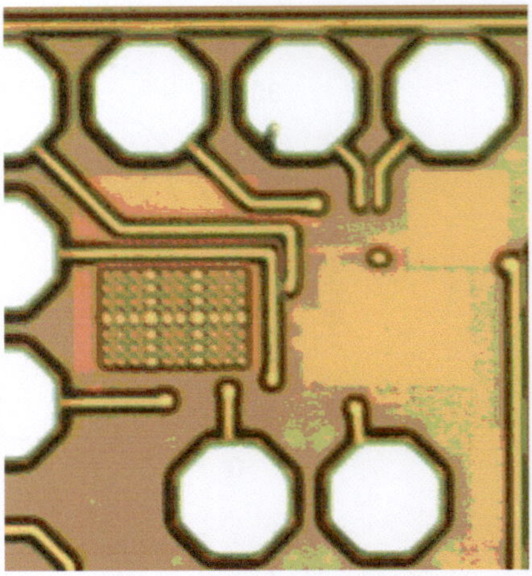

Fig. 25 The chip micrograph of the transmitter. The active area of the transmitter is $\approx 200\,\mu\text{m} \times 200\,\mu\text{m}$

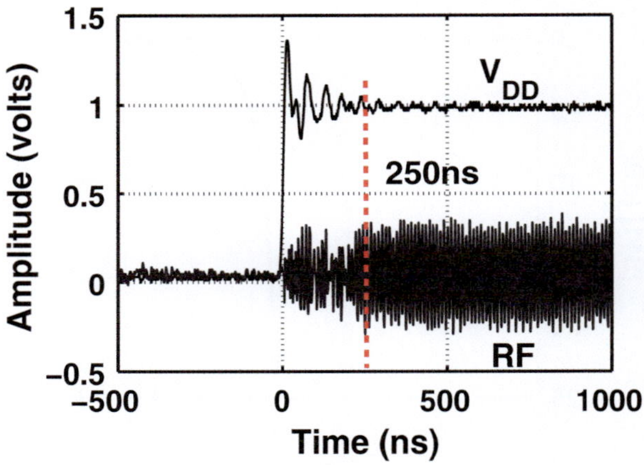

Fig. 26 A settling time of <250 ns was measured

2.9 Channel Selection

One of the main drawbacks of the above design is lack of channel selection due to lack of a traditional frequency synthesizer. However Wang et al. [22] have reported channelization techniques that obviate the needs for a PLL.

3 Ultra-Low Power RX Design

In this section, we will design an adaptive ultra-low power (ULP) RX aimed at fully or semi-autonomous peer-to-peer wireless links. We will take advantage of fewer jammers in MICS band and trade relaxed linearity specifications with lower power consumption.

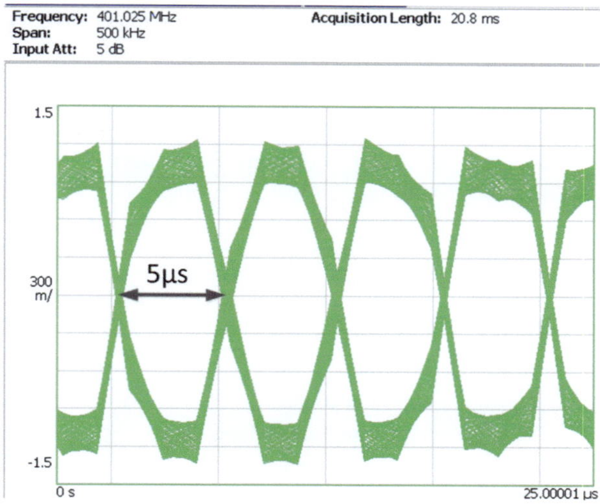

Fig. 27 Measured eye diagram using 200 kbps square wave data

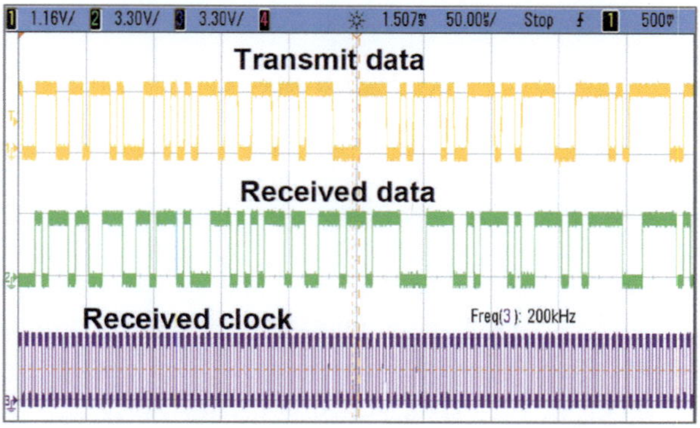

Fig. 28 Transmit and received pseudo-random bit sequence at 200 kbps

Adaptive performance for ULP receivers is an extremely desirable trait since they run on a shoe-string power budget. Hence the receiver should be able to efficiently trade performance with power dissipation when received signal strength is high.

In the next few pages, we will study the design of a 120 μW MICS/ISM band receiver with −90 dBm sensitivity at a data-rate of 200 kb/s with BER $< 0.1\,\%$. The receiver incorporates a 44 μW low-power (LP) mode that achieves −70 dBm sensitivity at 200 kb/s and 0.1 % BER. At a lower data-rate of 20 kb/s, the LP mode can be used as a wake-up receiver with increased sensitivity of −75 dBm with 1 % BER and 38 μW power consumption.

Table 5 Performance summary of the ULP transmitters and the proposed transmitter

	[4]	[3]	[16]	Bradley	[10]	This work
Power diss.	350 μW	400 μW	400 μW	>10 mW	8.9 mW	90 μW
Data-rate	120 kbps	250 kbps	100 kbps	800 kbps	1 Mbps	200 kbps
P_{out}	NA	−16 dBm	−16 dBm	−4 to −17 dBm	−12 dBm	−17 dBm
Energy/bit	2.9 nJ/bit	1.4 nJ/bit	4 nJ/bit	>12.5 nJ/bit	8.9 nJ/bit	0.45 nJ/bit
Process	90 nm	130 nm	130 nm	180 nm	180 nm	130 nm
Modulation	MSK	BFSK	BFSK	BFSK	BFSK	BFSK

Table 6 Channel model of various scenarios for the 400 MHz medical implants band

Scenario	Description	Operating distance	Loss (3σ)
S1	Implant to implant	100 mm	78 dB
S2	Implant to body surface	100 mm	83 dB
S3	Implant to external	100 mm	84 dB
S4	Body surface to body surface (LOS)	20 mm	54 dB
S5	Body surface to body surface (NLOS)	2 m	59 dB

3.1 Receiver Specifications

In order to derive the system and block level specifications for RX, we need to understand the link budget and path loss constraints imposed by the target application. For medical wearables and implants, the IEEE 802.15.6 standard has provided the recommendation for the losses in 400 MHz MICS band for each of the BAN scenarios based on extensive studies on more than 300 male body parts (Table 6).

1. To select a channel for communication, the receiver must monitor the channel for a period of 10 ms with a sensitivity of no less than

$$10\log_{10} B - 150\,\text{dBm/Hz} + G(\text{dBi})$$

   where B is the emission bandwidth and G is the antenna gain. For an isotropic antenna, this results in a sensitivity of −95 dBm for the receiver.
2. For data rates of 75 kbps and 150 kbps, the receiver sensitivity should be −95 dBm and −92dBm respectively.
3. Adjacent channel rejection of 17 dB.

We target a BER of 0.1 %. For the orthogonal FSK modulation, this requires an SNR of 10 dB. The target receiver sensitivity is −95 dBm. The noise figure, NF, of the front end is given by [17],

$$P_{in,min} = -174\,\text{dBm/Hz} + 10\log_{10}(B) + NF + \text{SNR}_{\text{min}}$$

For B= 300 kHz, the required NF $<$ 14.2 dB. Allowing some margin, we will target a single-side band NF of 12 dB.

For communication range, assuming line-of-sight communication,

$$P_{RX} = P_{TX} + G_{TX} - L_{FS} - L_M + G_{RX}$$

where: P_{RX} = received power (dBm)
P_{TX} = transmitter output power (dBm)
G_{TX} = transmitter antenna gain (dBi)
L_{FS} = free space loss or path loss (dB)
L_M = miscellaneous losses (fading margin, cable or trace losses etc) (dB) G_{RX} = receiver antenna gain (dBi)
The free space path loss is given as,

$$L_{FS} = 10\log_{10}\left(\frac{4\pi df}{c}\right)^2 = 20\log_{10}(d) + 20\log_{10}(f) - 147.56\,\text{dB}$$

where d is distance in meters and f is in Hz.

For applications in home-monitoring and operating theater scenarios, fading must be taken into account while calculating the range of operation. Johansson [8] reports 18 dB of fading loss in an experimental set-up of hospital environment. Assuming isotropic antennae, an additional 10 dB link margin (L_M =28 dB), $P_{RX} = -95$ dBm and $P_{TX} = -20$ dBm, the range d can be calculated to be 12 m which is sufficient for the application. For body worn derives using a miniaturized antenna with a link budget of 55–60 dB, cases S_4 and S_5 in Table 6 can be targeted.

To derive linearity specifications, we target a 40 dB spurious free dynamic range (SFDR) using which we derive the linearity specification for the receiver [17].

$$\text{SFDR} = \frac{2(P_{\text{IIP3}} - F)}{3} - \text{SNR}_{\text{min}}$$

where F is the receiver noise floor given by -174 dBm/Hz+NF+10log$_{10}$(B), which evaluates to -107 dBm. This leads to P_{IIP3} of -32 dBm.

Table 7 summarizes the target specifications for the receiver front-end.

Next we determine local oscillator phase noise using the information on the interferer. Phase noise can be estimated using the following equation [17]

$$PN(\text{dBc/Hz}) = P_{sig} - P_{int} - 10\log_{10}(B) - SNR_{min} \quad (2)$$

where P_{sig} and P_{int} denote the signal and interferer levels respectively. We target a signal-to-noise ratio of 20 dB due to the next channel (interferer) so that the small signal noise figure dominates the carrier to noise ratio at demodulator output. Assuming adjacent channel jammer to be 30 dB higher than the received signal, $P_{sig} - P_{int}$ is -30 dB. Using (2), the receiver LO phase noise should be no worse

Table 7 MICS-band receiver front-end specifications for BER of less than 0.1 %

Description	Target specification
Frequency band	MICS(402–405 MHz)
Sensitivity	−95 dBm
Noise figure	12 dB
Linearity (IIP3)	−32 dBm
Power dissipation	≈ 100 μW
Spurious free dynamic range	40 dB
LO Phase noise	−105 dBc/Hz @ 300 kHz
External components	Crystal and matching network

Table 8 Summary of the latest reported receivers

	Power	Data-rate	Sensitivity	Energy/bit	Process	Architecture
Bohorquez [4]	400 μW	120 kbps	−93 dBm	3.3 nJ/bit	90 nm	Super-regenerative
Bradley [5]	>10 mW	200 kbps	−96 dBm	50 nJ/bit	180 nm	Low-IF
Bae [3]	490 μW	250 kbps	−98 dBm	2 nJ/bit	180 nm	Super-regenerative
Liu [11]	910 μW	156 kbps	−80 dBm	5.8 nJ/bit	180 nm	Super-regenerative
Porret [15]	1 mW	24 kbps	−95 dBm	40 nJ/bit	0.5 μm	Zero-IF

than −105 dBc/Hz at 300 kHz offset. This is an extremely tough requirement for a low power LO.

3.2 Existing Approaches

Table 8 presents the state-of-the-art receivers operating in the MICS band. In traditional receiver architectures, the LO generation circuitry and the LNA typically consume the bulk of the power consumption [5, 15]. As a result, the super-regenerative architecture has been a popular choice to achieve high sensitivity at lower power dissipation [3, 4, 11]. However its selectivity is extremely poor. Also, due to free running nature of the oscillator, channel selection is difficult and the VCO needs frequent calibration for frequency stability This introduces asymmetry in the radio link and hence renders it unsuitable for autonomous body area network applications. In this section, we will explore RX architectures that overcome this limitation by creating a high-frequency"virtual LO" from a low-frequency stable quartz reference.

3.3 Low Power Circuit Techniques for ULP RX Design

The following ideas and methods could be considered to lower power consumption of a receiver. Many of these techniques are application specific.

1. LNA and LO dominate the power budget. For low LO power consumption, avoid low, area and power hungry PLLs (just like in case of ULP TX). Injection locking and frequency multiplication should be explored to generate LO locked to stable crystal reference.
2. Quadrature demodulators need quadrature LOs. Take advantage of low data rates and explore demodulator designs that can obviate the need for quadrature LO.
3. Due to lack of strong jammers, LNA g_m transistors can be biased in weak/moderate inversion region to get better transconductor efficiency.
4. Since linearity specifications are relaxed, high-Q passives can be used in the matching network to provide matching gain. This helps reduce system noise figure without burning current.
5. Save power with adaptive performance. Transceiver power dissipation and performance should be adaptive to save power when the link distance is short.
6. Bias current reuse within a block (to gain same performance for the lower current, for example) and/or between different blocks (if permissible, to get more functions out of the same current)
7. A more digital architecture can leverage the power of process scaling to lower power consumption in lower technology nodes.

Figure 29 shows the block diagram of the proposed ultra-low-power FSK receiver. We utilize a low-IF (1.5 MHz IF) architecture to reduce the influence of 1/f noise on the noise figure of the receiver. We perform 9× frequency multiplication in the mixer to allow LO operation at 44.5 MHz. Carrier generation at a fraction of the RF frequency allows extremely low-power operation. Using multistage, multiphase injection locking, we generate nine equally spaced phases from the 44.5 MHz ring oscillator locked to a quartz oscillator at the same frequency with a settling time of <100 ns permitting aggressive duty-cycling of the receiver [14]. The 9× subharmonic mixer utilizes these nine phases to effectively switch the RF current at 400.5 MHz, creating a virtual LO. The proposed receiver system is fully integrated

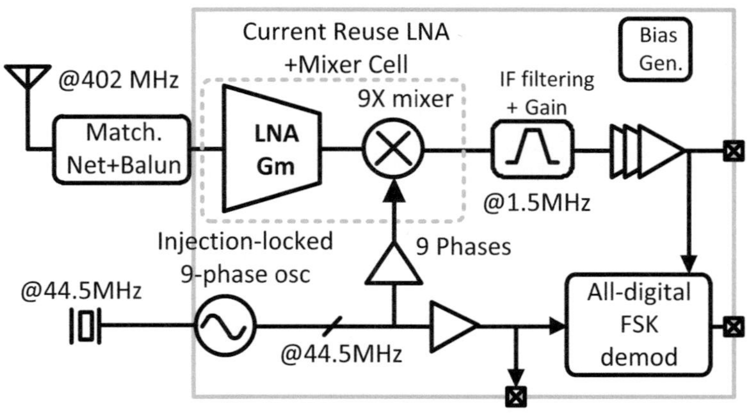

Fig. 29 Block diagram of the proposed ultra-low power receiver

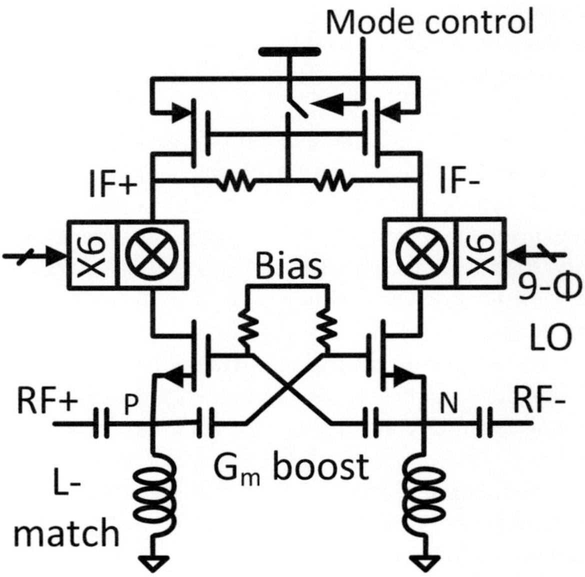

Fig. 30 Schematic of current-sharing receiver front-end

except for a crystal, matching network, and input balun. The chip represents a >3× improvement over previously-published MICS band receivers both in terms of energy/bit and total power consumption [3, 4].

Figure 30 shows the receiver front-end schematic. We use a g_m-boosted common-gate LNA [25] with stacked subharmonic mixer to save power. The g_m transistors are biased in subthreshold region to maximize their transconductance efficiency. The high-Q input matching network provides voltage gain and helps reduce the system noise figure while matching the high impedance LNA input (due to very low bias current) to the 50 Ω source. The voltage gain from the RF input to the IF output of the front-end is

$$A_v = 2 \cdot \frac{2}{\pi} \cdot \sqrt{1 + Q^2} \cdot g_m \cdot R_{\mathrm{IF}}$$

where R_{IF} is the IF load impedance and the factor "2" accounts for the g_m-boosting. Assuming input matching and a transconductance efficiency (g_m/I_d) of η, the voltage gain of the front-end is approximately given as

$$A_v = \frac{2}{\pi} \cdot Q \cdot \eta \cdot I_d \cdot R_{\mathrm{IF}} \tag{3}$$

Figure 31 shows the equivalent circuit for the input of the LNA+Mixer block. The impedance at node P is limited by the inductor R_p. The Q of the matching

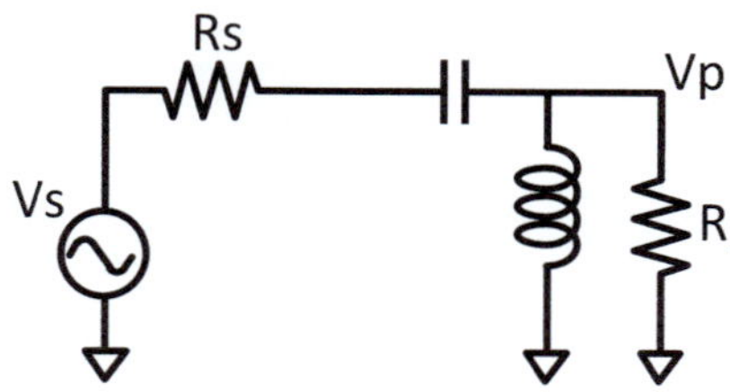

Fig. 31 Equivalent circuit of the input matching network. The impedance at node P is limited by the inductor R_p

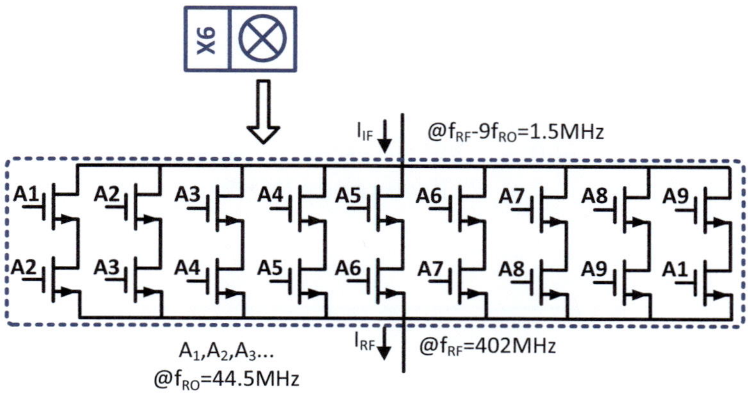

Fig. 32 Schematic of the edge-combining sub-harmonic mixer

network is given by

$$Q = \sqrt{\frac{R_p}{R_s} - 1}$$

For our off-chip matching network with 39 nH inductor, the R_p is $\approx 2.5\,\text{k}\Omega$. For the $50\,\Omega$ source impedance, this results in a Q of $\approx$7. For this design, we used $\eta = 20$, $I_d = 75\,\mu\text{A}$ and $R_{\text{IF}} = 5\,\text{k}\Omega$. Using (3), the receiver front-end voltage gain A_v is 30.5 dB.

The schematic of the sub-harmonic mixer is shown in Fig. 32. We use a differential RF path in the mixer to reject LO feedthrough in the IF signal present due to random mismatches in the nine-phase LO. The 4-stage IF amplifiers provide 80 dB of voltage gain and are AC-coupled to reject DC offsets.

Figure 33 shows the input matching. The input return loss is less than −10 dB.

Figure 34 shows the voltage conversion gain of the receiver front-end. Noise figure (NF) of the front-end is shown in Fig. 35. At a 1.5 MHz IF, the receiver front-end in the main mode exhibits >30 dB of voltage conversion gain and a 13 dB NF. In LP mode, the DC current though the mixer and LNA is disabled, effectively converting the front-end to a passive mixer. The low conversion gain (5 dB) (Fig. 34) and higher NF (18 dB) (Fig. 35) in the front-end lead to approximately 20 dB higher system NF and 20 dB reduced sensitivity in this mode.

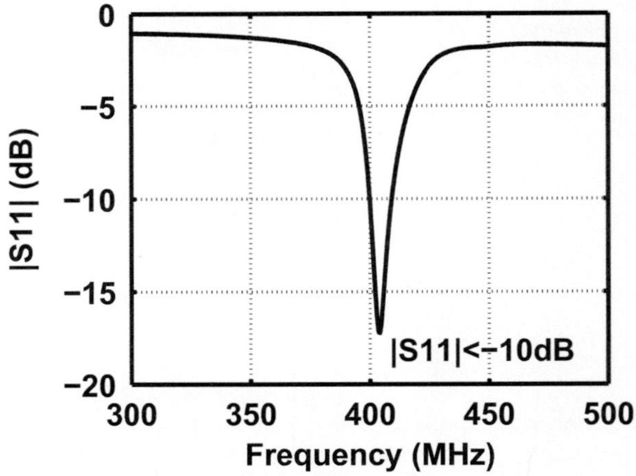

Fig. 33 Input return loss of the receiver front-end. $|S_{11}| < -10\,\text{dB}$ is observed

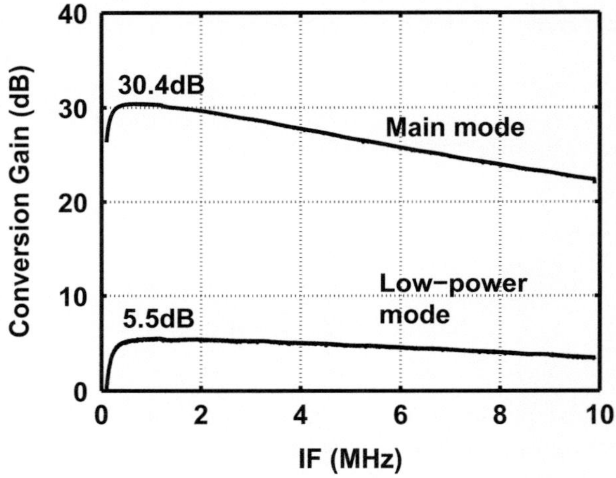

Fig. 34 Voltage conversion gain of the front-end

Front-end linearity in main mode and low-power modes are plotted vs. the IF frequency in Figs. 36 and 37 respectively. In the main mode, the front-end achieves an P_{IIP3} of −23 dBm. 1-dB compression points of both the main and the LP modes are shown in Fig. 38.

Figure 39 presents the schematic and measured results of the 44.5 MHz injection-locked ring oscillator. The use of frequency multiplication by a large factor (9×) reduces the LO power and permits direct locking to a crystal reference, completely avoiding the need for a PLL/DLL. However, direct single-phase injection into a nine-stage oscillator disturbs its phase symmetry, leading to increased 44.5 MHz LO feedthrough to the IF. We solve this problem by first locking a three-stage

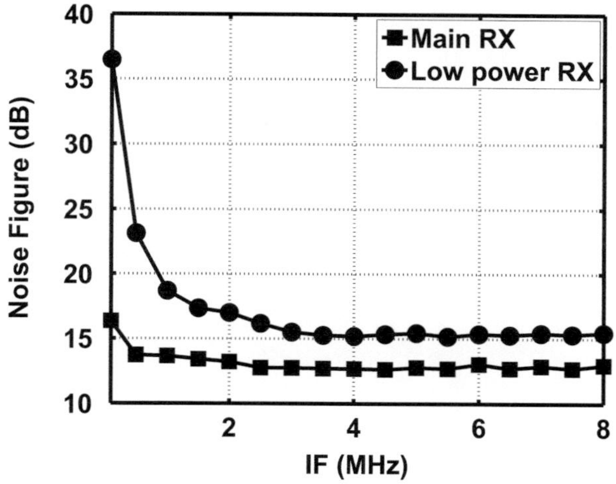

Fig. 35 Noise figure of the main and the LP mode receiver

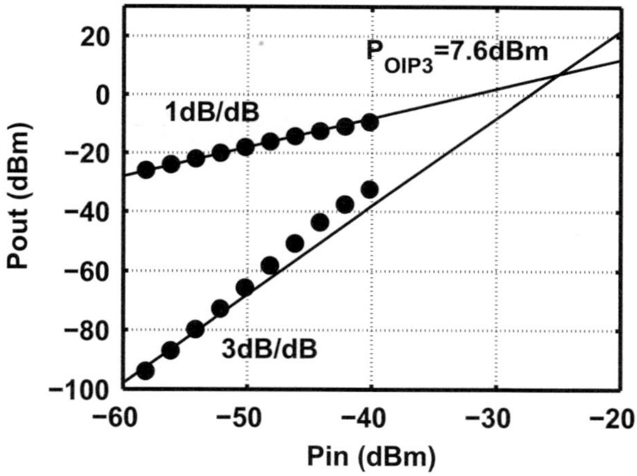

Fig. 36 IIP3 of the main mode receiver

ring oscillator (ILRO1). These mismatched edges from ILRO1 are symmetrically injected into a nine-stage oscillator (ILRO2) that reduces the phase mismatch. The injection-locked oscillator can be modeled as a first order PLL with the one-sided lock range determining the equivalent bandwidth [1]. The proposed LO architecture eliminates the phase/delay-locked-loops saving associated loop filter area and power while avoiding stability issues and a long settling time. The measured lock-range of the cascaded oscillator stages is 32–52 MHz with a settling time < 100 ns. The spectra and phase noise of the free-running and locked oscillators are shown in Figs. 39, 40 and 41 respectively. Figure 42 shows the measured phase noise of the 400.5 MHz "virtual" LO using Agilent 5052B signal source analyzer. At 300 kHz

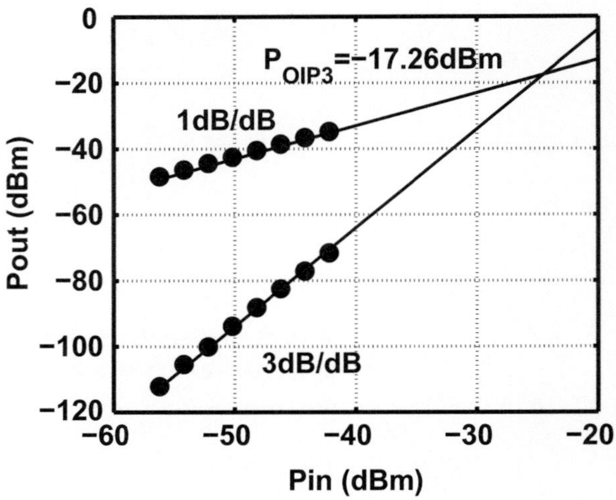

Fig. 37 IIP3 of the low-power mode receiver

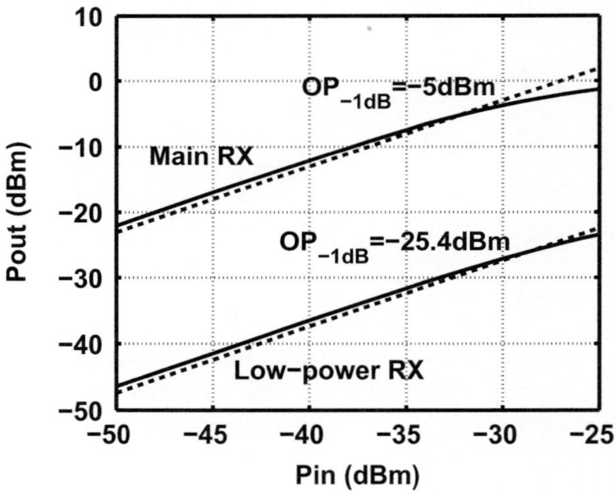

Fig. 38 1-dB compression point of the main and the low-power mode

offset, its phase noise is -104 dBc/Hz. The total power dissipation of the crystal and both ring oscillators is 22 μW.

Figure 43 explains the operation of our all-digital FSK demodulator. The output of the crystal oscillator (44.5 MHz) is fed into the "measurement" counter (counter2) in the FSK demodulator as the reference clock. The IF output is fed into the "window" counter (counter1), which operates for N_1 cycles and gates the measurement counter. This counter measures the number of periods of the reference clock in each measurement window. The time of measurement, Tmeas,

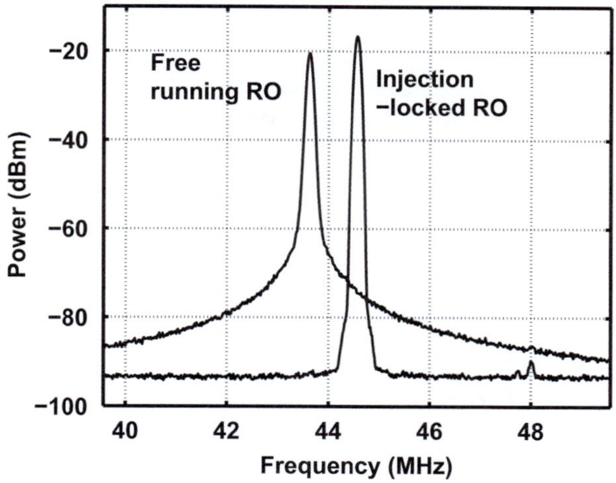

Fig. 39 Spectrum of the free-running and injection-locked LO

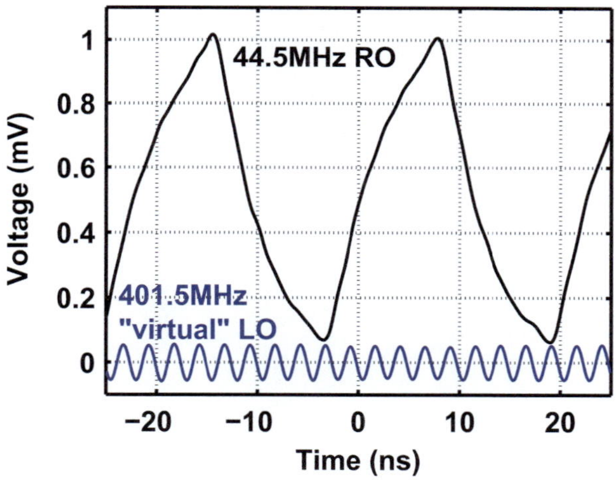

Fig. 40 Time domain waveforms of the 44.5 MHz ring oscillator and 401.5 MHz 'virtual' LO

is a multiple of the unknown IF frequency and can be controlled by the predefined value N_1. $T_{\text{meas}} = N_1/f_{\text{if}}$. Also, $T_{\text{meas}} = N_2/f_{\text{ref}}$. It follows that the measurement counter output N_2 contains the unknown IF frequency. $N_2 = N_1 \cdot f_{\text{ref}}/f_{\text{if}}$. A change in the IF frequency will change Tmeas and N_2. The FSK-modulated IF signal can be demodulated according to the changes in N_2. The resolution of this frequency-detection scheme is determined by f_{ref}, T_{meas}, and the FSK deviation. The asynchronous nature of sampling reduces the error margin for the comparator and limits the maximum data-rate to 100 kb/s when the input SNR drops to 10 dB. The

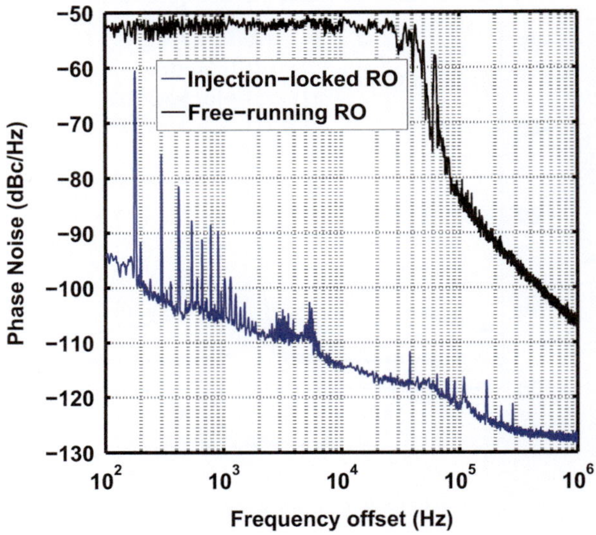

Fig. 41 Phase noise of the 44.5 MHz injection-locked ring oscillator and the free-running ring oscillator

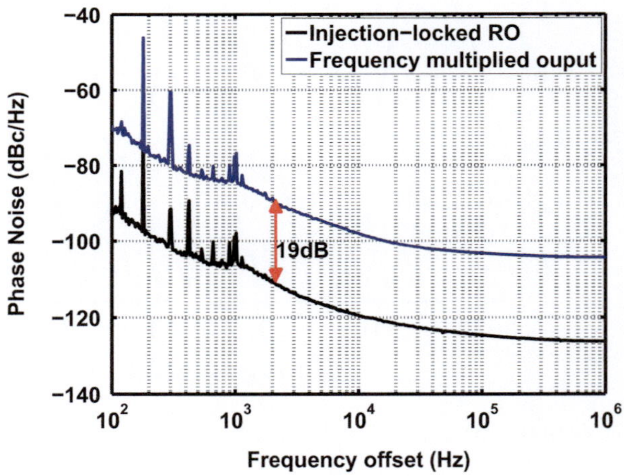

Fig. 42 Measured phase noise of the injection-locked LO and the frequency multiplied output using an Agilent 5052B signal source analyzer

measured output of the demodulated FSK signal with the input pseudo-random data is also shown in Fig. 44.

Table 9 presents the performance comparison with the existing work. Power breakdown of the system is given in Table 10. Our main mode receiver achieves a data-rate of 200 kb/s at 600 pJ/bit with BER < 0.1 %. At the expense of 20 dB loss in sensitivity, the LP mode receiver achieves 220 pJ/bit. For MICS-band transmitters

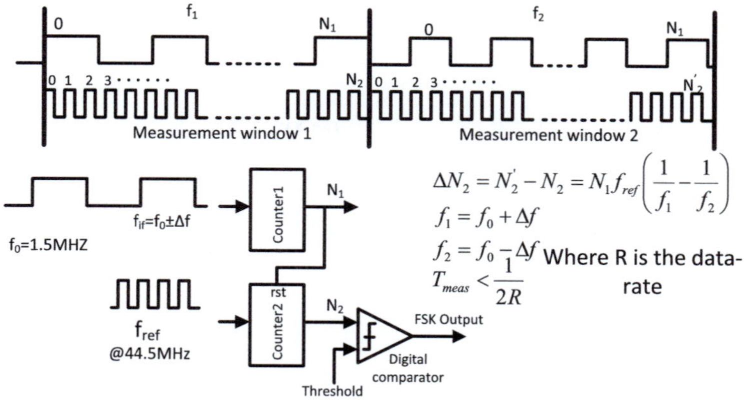

$$\Delta N_2 = N_2' - N_2 = N_1 f_{ref}\left(\frac{1}{f_1} - \frac{1}{f_2}\right)$$

$$f_1 = f_0 + \Delta f$$

$$f_2 = f_0 - \Delta f$$

$$T_{meas} < \frac{1}{2R}$$

Fig. 43 Schematic of the all-digital FSK demodulator

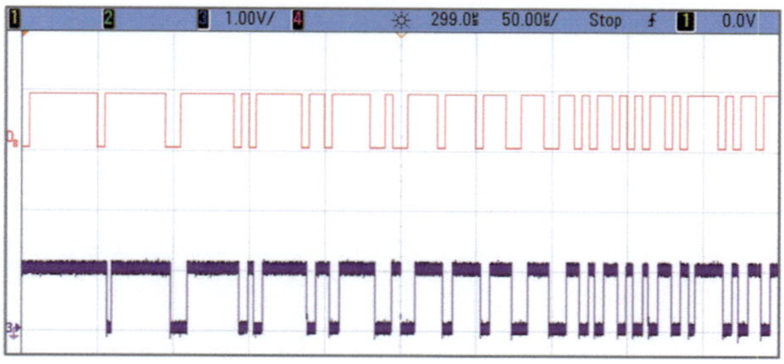

Fig. 44 Input pseudo-random data and the resulting demodulator output for the a 200 kbps data-rate and −90 dBm input

with a −16 dBm output, the LP mode can permit wireless operation over 2 m of distance. The main mode can be turned on in case of additional losses such as body tissue attenuation or fading losses in a hospital scenario. The front-end and IF limiting amplifiers consume 75 μW and 11 μW from a 1 V supply, respectively. The total die area of the receiver is 0.5 mm^2 in a 130 nm CMOS process (Fig. 45). With our previously discussed sub-100 μW FSK transmitter [14], this receiver will enable a fully autonomous symmetric wireless link in a peer-to-peer network with an energy per transceived bit of 1 nJ/bit.

4 Conclusion

Ultra-low power transmitters and receivers hold the key to the viability of peer-to-peer fully autonomous wireless sensing network harvesting power from common surroundings. Applications such as Body Area Networks hold promise to greatly improve lives by both bringing down the cost of healthcare and improving the quality of health monitoring. Although the body worn devices, in general, can

Table 9 Performance comparison with the state-of-the-art in MICS/ISM band receivers

Performance metric	[4]	[5]	[3]	[11]	This work Main mode	This work LP mode
Power	400 μW	>10 mW	490 μW	910 μW	120 μW	44 μW
Date-rate	120 kbps	200 kbps	250 kbps	156 kbps	200 kbps	100 kbps
Sensitivity	−93 dBm	−96 dBm	−98 dBm	−80 dBm	−90 dBm	−70 dBm
Energy/bit	10 nJ/bit	50 nJ/bit	2 nJ/bit	5.8 nJ/bit	0.6 nJ/bit	0.22 nJ/bit
Process	90 nm	180 nm	180 nm	180 nm	130 nm	130 nm
Architecture	Super-reg.	Low-IF	Low-IF	Super-reg.	Low-IF	Low-IF
Modulation	OOK	FSK	FSK	OOK	FSK	FSK

Table 10 Power consumption breakdown of the receiver in both main and low-power modes

Circuit block	Power dissipation Main mode	Power dissipation LP mode
Front-end (LNA+mixer)	75 μW	0
LO+crystal osc.	22 μW	22 μW
IF amps + limiters	11 μW	11 μW
Demodulator	12 μW	12 μW
Total	120 μW	44 μW

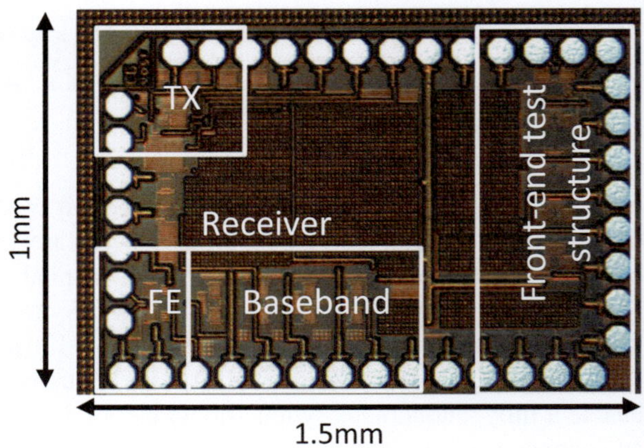

Fig. 45 Chip micrograph of the receiver

afford to have a small battery, its weight and the accompanying increase in the form factor and cost encourage greater reliance on harvested energy. In addition to low available power, these harvested energy sources generally exhibit relatively large source impedances that limit the peak power available from such sources.

Smaller signal dynamic range and low transmit power imply that the receiver RF amplifier and transmitter PA are no longer the major sources of power dissipation. The carrier generation block instead dominates the total power consumption. This unique situation renders traditional transmitter and receiver architectures power hungry and less energy efficient. Attempts at introducing a frequency multiplier in order to run the local oscillator at a fraction of the carrier frequency typically meet two challenges: (1) The frequency multiplier running at RF frequency is itself quite power hungry; and (2) Although desirable, it is difficult to perform frequency multiplication by a large factor without paying a power penalty and sacrificing performance at the same time. Use of free-running oscillators to circumvent the power problem implies: (1) Frequency calibrations; and (2) Both the transmitter and receiver cannot have a free running oscillator. At least one of them must be locked to a stable crystal reference and must run a frequency correction loop to correct for the long term frequency drift of the other end of the wireless link. This essentially makes the radio link asymmetric.

In this chapter, we have introduced a power-efficient frequency multiplier (FM) based on edge-combining. The proposed multiplier supports large factors of frequency multiplication and is digital in nature that benefits from process scaling. In addition, we have incorporated the FM block as a sub-harmonic mixer on the receiver side and as a PA on the transmit side without increasing the RF components leading to overall reduction in the power dissipation. We have also demonstrated a new injection-locked ring oscillator topology that provides large number of symmetric edges for frequency multiplication while locked to a stable crystal reference. This eliminates the need of a PLL/DLL loop that brings with it the stability issues and additional area/power overheads in form of a PFD, charge-pump, loop filter and dividers.

The TX and RX architectures presented in this chapter incorporate circuit elements that directly benefit from process scaling. Finally, although the techniques presented here are demonstrated for the 400 MHz MICS band, they are easily applicable in other frequency bands and standards such as 2.4 GHz ZigBee to explore the possibility of reducing the power consumption.

References

1. R. Adler, A study of locking phenomena in oscillators. Proc. IEEE **61**(10), 1380–1385 (1973)
2. I. Aoki, S. Kee, R. Magoon, R. Aparicio, F. Bohn, J. Zachan, G. Hatcher, D. McClymont, A. Hajimiri, A fully-integrated quad-band GSM/GPRS CMOS power amplifier. IEEE J. Solid-State Circuits **43**(12), 2747–2758 (2008)
3. J. Bae, N. Cho, H.-J. Yoo, A 490 μW fully MICS compatible FSK transceiver for implantable devices, in *IEEE symposium on VLSI Circuits,* June 2009
4. J. Bohorquez, A. Chandrakasan, J. Dawson, A 350 μW CMOS MSK transmitter and 400 μW OOK super-regenerative receiver for medical implant communications. IEEE J. Solid-State Circuits **44**(4), 1248–1259 (2009)
5. P. Bradley, An ultra low power, high performance medical implant communication system (MICS) transceiver for implantable devices, in *IEEE Biomedical Circuits and Systems (BioCAS)*, 2006
6. G. Chien, P.R. Gray, A 900 MHz local oscillator using a DLL-based frequency multiplier technique for PCS applications. IEEE J. Solid-State Circuits **35**(12), 1996–1999 (2000)

7. S. Cho, A.P. Chandrakasan, Energy efficient protocols for low duty cycle wireless microsensor networks, in *IEEE International Conference on Acoustics, Speech, and Signal Processing*, 2001, pp. 2041–2044
8. A.J. Johansson, Performance of a radio link between a base station and a medical implant utilising the MICS standard, in *IEEE International Conference on Engineering in Medicine and Biology Society*, 1–5 Sept 2004, vol. 1, pp. 2113–2116
9. P. Kinget, R. Melville, D. Long, V. Gopinathan, An injection-locking scheme for precision quadrature generation. IEEE J. Solid-State Circuits **37**(7), 845–851 (2002)
10. K.-C. Liao, P.-S. Huang, W.-H. Chiu, T.-H. Lin, A 400 MHz/900 MHz/2.4 GHz multi-band FSK transmitter in 0.18 μm CMOS, in *IEEE Asian Solid-State Circuits Conference*, 2009, pp. 353–356
11. Y.-H. Liu, H.-H. Liu, T.-H. Lin, A super-regenerative ASK receiver with $\Delta\Sigma$ pulse-width digitizer and SAR-based fast frequency calibration for MICS applications, in *IEEE Symposium on VLSI Circuits*, 2009, pp. 38–39
12. F. Merli, Implantable antennas for biomedical applications, Ph.D. thesis, EPFL, 2011
13. T. Morrison, F. Zhang, S. Rai, J. Pandey, J. Holleman, B. Otis, The Bumblebee: a 0.3 gram, 560 μW, 0.1 cm^3 wireless biosignal interface with 10 m range, in *IEEE 47th DAC/ISSCC Student Design Contest,* June 2010
14. J. Pandey, B. Otis, A 90 μW MICS/ISM band transmitter with 22 % global efficiency, in *IEEE Symposium on Radio Frequency Integrated Circuits*, 2010, pp. 285–288
15. A.-S. Porret, T. Melly, D. Python, C.C. Enz, E.A. Vittoz, An ultra low-power UHF transceiver integrated in a standard digital CMOS process: architecture and receiver. IEEE J. Solid-State Circuits **36**(3), 452–466 (2001)
16. S. Rai, J. Holleman, J. Pandey, F. Zhang, B. Otis, A 500 μW neural tag with 2 μV_{rms} AFE and frequency-multiplying MICS/ISM FSK transmitter, in *IEEE International Conference on Solid-State Circuits*, Feb 2009
17. B. Razavi, J.M.J. Sung, A 6 GHz 60 mW BiCMOS phase-locked loop. IEEE J. Solid-State Circuits **29**(12), 1560–1565 (1994)
18. S. Roundy, E.S. Leland, J. Baker, E. Carleton, E. Reilly, E. Lai, B. Otis, J.M. Rabaey, P.K. Wright, V. Sundararajan, Improving power output for vibration based energy scavengers. IEEE Pervasive Comput. **4**(1), 28–36 (2005)
19. B. Sutton, P. Stadnik, J. Nelson, L. Stotts, Probability of interference between LP-LDC and LBT MICS implants in a medical care facility, in *IEEE Engineering in Medicine and Biology Society*, 2007, pp. 6721–6725
20. E.A. Vittoz, M.G.R. Degrauwe, S. Bitz, High-performance crystal oscillator circuits: theory and application. IEEE J. Solid-State Circuits **23**(3), 774–783 (1988)
21. R.J.M. Vullers, R.V. Schaijk, H.J. Visser, J. Penders, C.V. Hoof, Energy harvesting for autonomous wireless sensor networks. IEEE Solid State Circuits Mag. **2**, 29–38 (2010)
22. K. Wang, J. Koo, R. Ruby, B. Otis, A 1.8 mW PLL-free channelized 2.4 GHz ZigBee receiver utilizing fixed-LO temperature-compensated FBAR resonator, in *IEEE Conference on International Solid State Circuits* (ISSCC), Feb 2014
23. D.J. Yeager, J. Holleman, R. Prasad, J.R. Smith, B.P. Otis, NeuralWISP: a wirelessly powered neural interface with 1 m range. IEEE Trans. Biomed. Circuits Syst. **3**(6), 379–387 (2009)
24. D. Yeager, F. Zhang, A. Zarrasvand, B.P. Otis, A 9.2 μA gen 2 compatible UHF RFID sensing tag with -12 dBm sensitivity and 1.25 μV_{rms} input-referred noise floor, in *IEEE International Conference on Solid-State Circuits*, 2010, pp. 52–53
25. W. Zhuo, X. Li, S. Shekhar, S.H.K. Embabi, J.P. de Gyvez, D.J. Allstot, E. Sanchez-Sinencio, A capacitor cross-coupled common-gate low-noise amplifier. IEEE Trans. Circuits Syst. Express Briefs **52**(12), 875–879 (2005)
26. http://content.healthaffairs.org/content/27/2/w145.full.pdf
27. http://www.usgovernmentspending.com/federal_budget_fy13
28. http://www.who.int/mediacentre/factsheets/fs310/en/
29. http://wireless.fcc.gov/services/index.htm?job=service_home&idmedical_implant
30. http://scholar.lib.vt.edu/ejournals/JOTS/v35/v35n1/pdf/yildiz.pdf

31. http://www.microbattery.com/tech-duracell-hearing-aid-battery
32. http://www.dmcinfo.com/Portals/0/Blog%20Files/High%20pulse%20drain%20impact%20on%20CR2032%20coin%20cell%20battery%20capacity.pdf
33. http://www.excellatron.com/advantage.htm
34. http://edocket.access.gpo.gov/cfr_2008/octqtr/47cfr15.242.htm
35. http://focus.ti.com/lit/ds/swrs061f/swrs061f.pdf
36. http://w3.antd.nist.gov/ban/15-08-0519-01-0006.pdf

Architectures for Ultra-Low-Power Multi-Channel Resonator-Based Wireless Transceivers

Phillip M. Nadeau[$], Arun Paidimarri[$], Patrick P. Mercier, and Anantha P. Chandrakasan

Abstract This chapter explores the use of high-Q RF resonators as both channel filtering and frequency generation elements in ultra-low energy wireless transceivers. Design tradeoffs in using resonators are discussed and an example receiver and transmitter system are presented. In the receiver, direct filtering at RF improves the frequency selectivity of the design and enables a low-energy ring-oscillator based frequency plan. In the transmitter, FBAR-based oscillators can eliminate the need for a PLL, reduce the power consumption of the frequency generation, and improve the overall transmitter efficiency at low output powers.

Keywords FBAR • RF resonator • -10dBm • Resonant buffer • Pulse-shaping • Multi-channel • BAN • Body area networks • Low energy • ISM band

1 Introduction

Sensor networks have been improving with improved circuit, battery, energy harvesting, network protocol and algorithm design. With the proliferation of connected health and fitness devices, sensor networks around a human body are becoming increasingly common. These Body Area Networks (BAN) are characterized by their short distance (1–2 m) and moderate data rates (<1 Mbps). For example, applications like EEG are on the higher end of the data rates while breathing rate

[$]Author contributed equally.

P.M. Nadeau (✉) • A. Paidimarri • A.P. Chandrakasan
Massachusetts Institute of Technology, Cambridge, MA 02139, USA
e-mail: pnadeau@mit.edu; arun_p@mit.edu; anantha@mtl.mit.edu

P.P. Mercier
University of California San Diego, La Jolla, CA 92093, USA
e-mail: pmercier@ucsd.edu

P.P. Mercier, A.P. Chandrakasan (eds.), *Ultra-Low-Power Short-Range Radios*, Integrated Circuits and Systems, DOI 10.1007/978-3-319-14714-7_4

monitors might be at the low end. With current technologies, the wireless transceiver generally accounts for a large portion of the system power consumption [1].

In this work, we address some of the challenges with wireless communication for these short-distance, moderate data rate systems. A simple link-budget analysis will motivate some of the specifications for the radios. While there could be multiple competing technology solutions to address the power consumption challenge, some of which are discussed in other chapters of this book, our work focuses on the use of high-Q RF resonators. The resonators have inherently narrow bandwidth, which promises the design of stable low-phase noise oscillators and sharp channel select filters at RF. This could lead to lower power transceivers that do not require a PLL, or receivers with significantly improved robustness to interferers compared to similar architectures without a resonator.

We will discuss low-energy wireless circuit architectures that take advantage of the characteristics of RF resonators. We also will discuss options for extending the frequency range of such systems. As we will show, the main disadvantage of resonators is that their tuning range is small (less than 10 MHz), resulting in limited dynamic frequency range. In an increasingly congested band such as the 2.4 GHz ISM, with networks such as Bluetooth, Zigbee, and WiFi overlapping with each other, multi-channel operation is necessary for reliable communication. In addition, fading effects may result in frequencies with much worse performance than others at any given time. Hence the desire for techniques, such as those discussed in this chapter, to cover additional frequencies in resonator-based transceivers.

1.1 Network Topology and Link Margin Analysis

Short-distance mesh topologies have a large energy overhead due to the protocols used. On other hand, BANs typically operate in a star-topology configuration [2], as shown in Fig. 1. Star topologies, in addition, enable links with asymmetric energy budgets. The base-station (for example, a cellular phone) is relatively energy abundant compared to a sensor node that must have a long life-time or that must run from energy harvesting sources.

The path loss in a typical around-the-human-body link is about 60–70 dB, approaching 80 dB in certain cases [3]. Chapter "Channel Modeling for Wireless Body Area Networks" discusses channel modeling in BANs in more detail. The receive sensitivity of typical 2.4 GHz ISM-band radios, like Bluetooth, is at least −90 dBm [4], which implies that a nominal transmit power of −10 dBm on the sensor node is sufficient for reliable communication. This output power level is also the recommended transmit power in the IEEE802.15.6 BAN standard [2]. Similarly, typical transmitters around +10 dBm can easily fit the energy budgets of base-stations. Hence, a nominal receiver sensitivity of −70 dBm on the sensor node is sufficient.

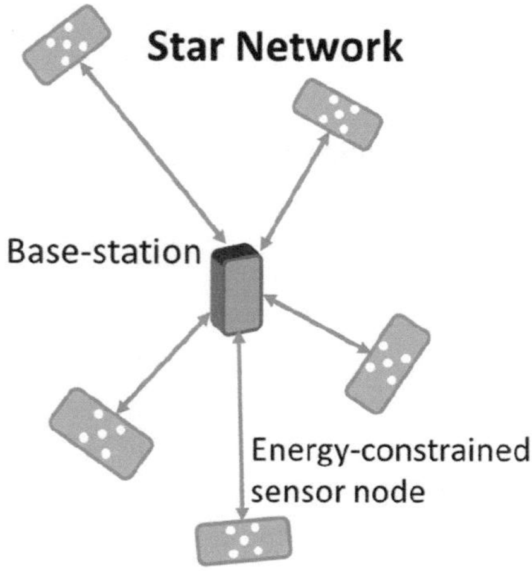

Fig. 1 Star topology, a common network architecture for body area networks

The challenge is to maintain good transmit efficiencies at −10 dBm and to minimize power consumption of receivers at −70 dBm sensitivity. The link margin, in atypical conditions, can be further enhanced through simple coding [5] or data rate reduction.

1.2 RF Resonators

Film Bulk Acoustic Wave Resonators (FBAR) or Surface Acoustic Wave (SAW) resonators are typically used in ladder configurations to generate sharp band-select filters for cellular communication. The circuit model of an individual resonator is shown in Fig. 2 along with its impedance plot. There are two characteristic frequencies in the resonator: (1) the series resonance, where the resonator presents a low impedance, and (2) the parallel resonance, where the resonator presents a high impedance. The Q of the resonator indicates the sharpness (or bandwidth) of the impedance plots.

These individual resonators have been used in oscillators [6–8] to provide stable center frequencies. They have also been used as channel-select filters in receivers to ease receiver circuit designs, resulting in low-power operation [6, 9–11].

We now further describe the use of resonators in receivers and transmitters, especially for the BAN scenario discussed in Sect. 1.1. We then present the architecture, design and experimental results for a multi-channel resonator-based receiver and transmitter.

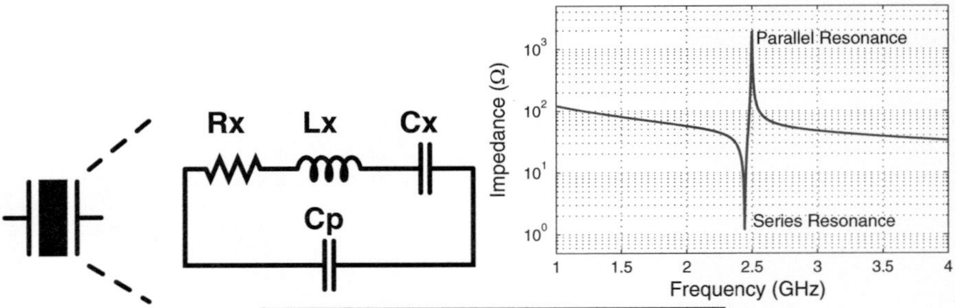

$R_x(\Omega)$	L_x(nH)	C_x(fF)	C_p(pF)	Q	R_p(kΩ)
1.05	107.2	38.99	1.292	1600	2.26

Fig. 2 Circuit model for RF resonators and the impedance profile of a representative FBAR

1.3 Receiver Tradeoffs and Related Work

In conventional receivers, two components typically dominate the power budget. The first is the LNA, which provides power-expensive RF gain, but overcomes the noise floor of the subsequent processing, thereby accessing sensitivity levels better than −90 dBm. The second is the LO generation hardware, which usually consists of an LC-based VCO locked to a crystal reference using a PLL. The crystal provides the long-term frequency stability necessary to receive the channel of interest, whereas the VCO is designed with sufficient phase noise stability to receive complex phase-based modulation schemes. A generic example of this kind of architecture is shown in Fig. 3a. Some example low energy radios of this variety are described in [12–14]. In general, these have worse energy efficiency than other techniques, but are more capable of peer-to-peer communication due to their excellent sensitivity, spectral efficiency, and multi-channel operation.

Various techniques have been proposed to reduce the energy consumption of receivers to the sub 10 nJ/b range, usually by compromising on a few key aspects of the system performance. Generally these are the sensitivity level, the supported modulation schemes, and the frequency selectivity. For example, if non-coherent amplitude-only detection (e.g. OOK or PPM) is employed, then super-regenerative architectures can enhance the energy efficiency significantly. Some example systems using this approach can be found in [15–17]. In addition, envelope-based detectors become very attractive from an energy perspective. Within this domain, recent work has explored relaxing the frequency synthesizer and front-end bandwidth requirements [18, 19] in order to lower the power consumption even further.

One category of ultra low energy OOK receivers which has emerged in the last 5–10 years are those that employ Bulk-Acoustic-Wave (BAW) resonators for directly filtering the desired RF channel. Figure 3b demonstrates some of the potential architectural advantages of using envelope detection with resonators. For one, if using the BAW as the source of frequency accuracy, the PLL can potentially be replaced with a simple VCO.

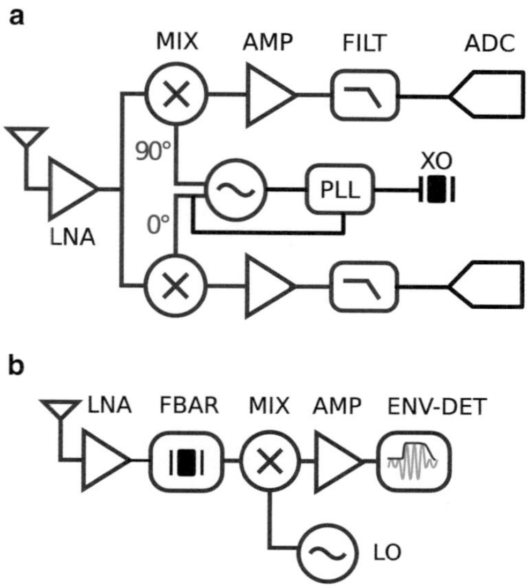

Fig. 3 Techniques for ultra-low energy RF receivers. (**a**) Conventional RX architecture. (**b**) Non-coherent amplitude-based detection

For example, the receivers in [6, 9–11] leveraged BAWs to provide much-improved channel filtering over architectures like [18, 19], complete with a very low energy design. However, a significant drawback to using resonators to filter individual channels is that the effective tuning range of the parallel resonance is quite small (less than 10 MHz), limiting channelization opportunities that are sorely needed in congested ISM bands.

The example receiver presented in this chapter (in Sect. 2) attempts to combine the best aspects of the previously proposed architectures for low-energy receivers [20]. For one, it keeps the LNA to improve the sensitivity level, and combines additional resonators to provide for more channel options. In addition, sticking with OOK or PPM modulation allows the IQ demodulation circuits to be replaced with simple envelope detection.

Although Sect. 2 is a description of a particular implementation of a resonator-based receiver, the discussion throughout focuses on general design criteria and their associated impacts on system-level performance.

1.4 Transmitter Tradeoffs and Related Work

Consider a generic transmitter architecture shown in Fig. 4. It requires a Local Oscillator (LO), modulation blocks, and a power amplifier. Traditional transmitters employ a full I/Q-based architecture as shown in Fig. 5a. These architectures, being very general, can support complex modulation schemes, but come at the cost of increased power consumption in the mixer and baseband amplifiers. The architectures also support multiple channels due to the PLL.

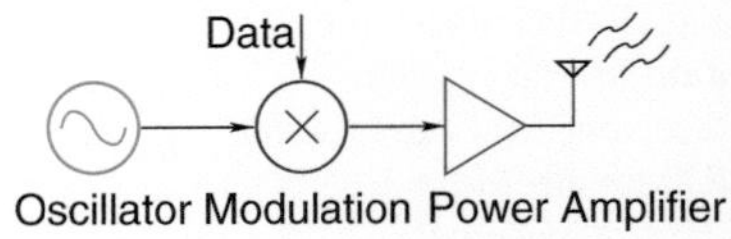

Fig. 4 The key blocks of a generic transmitter

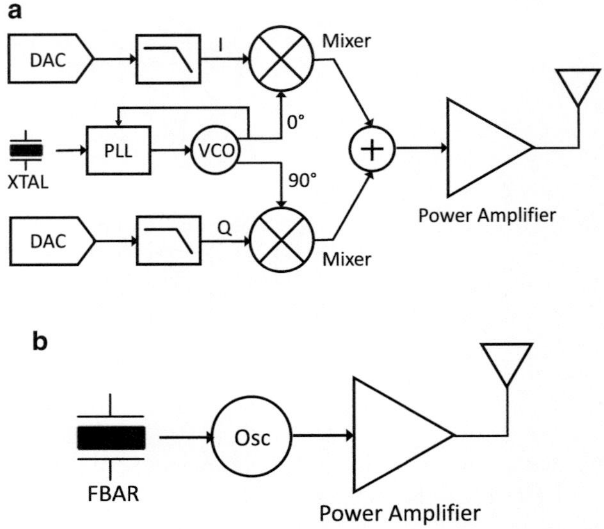

Fig. 5 Comparison of two transmitter architectures. (**a**) Traditional I/Q based transmitter. (**b**) Simplified FBAR-based transmitter

For a −10 dBm, or 100 μW output power, the PA is no longer the dominant power consumer, and the power of the remaining blocks can comprise a significant fraction of the total power consumption of the transmitter. If simple binary modulation schemes are used, the power of modulation can be greatly reduced, as is done in [7, 21]. Schemes such as OOK and MSK can all be implemented without significant impact on the energy consumption of the TX. This simplifies the problem to developing efficient LO generation schemes.

Multiple schemes have been employed to generate efficient LOs for low power transmitters. The PLL in [15], for example, is duty-cycled and turned on just before the packet transmission occurs. This reduces the peak power consumption of the transmitter. A PLL allows the LO to be set to any frequency, however, this comes with a slow startup time due to the finite bandwidth of the PLL. A slow frequency-correction loop is employed instead of a PLL in [21, 22]. The base-station, which has an accurate frequency reference, sends correction signals to the sensor node. The energy penalty on the sensor node is negligible. However, for very low duty-cycle applications this scheme is difficult to implement, since the drift in the sensor node can become too large.

Another approach is to use high-Q resonators that provide stable oscillation frequencies at RF, eliminating the need for a PLL, as shown in Fig. 5b. The high-impedance at parallel resonance results in ultra-low power LOs [23].

The parallel resonance frequency can be shifted lower by capacitive loading, but this also lowers the parallel resonance impedance, thus the power consumption and reducing the filtering provided by the resonator. Typically, frequency tuning of these oscillators is limited, and one resonator operates only in a single frequency channel [6]. Additional channels can be realized by adding resonators with different parallel resonance frequencies to define channels, and having an architecture that can efficiently select between them. Such two-channel implementations have been previously demonstrated in [6, 7].

In Sect. 3, we will describe a transmitter [24] that exploits the low power, frequency stability, and low noise aspects of the resonators, and that addresses the issue of single-channel operation by developing a scalable multi-resonator architecture. This architecture provides high transmit efficiencies for the low −10 dBm output power levels typical in BANs.

2 Multi-Channel Receiver Design

2.1 System Architecture

The receive chain for the example low-energy receiver presented in this chapter extends the low-power frequency plan previously presented in [10] and is shown in Fig. 6. The plan uses a power-efficient, but inaccurate ring oscillator to down-convert the RF signal into an IF range from 10 to 100 MHz, where the wide IF range tolerates the LO inaccuracy after a one-time calibration for process variation. The FBAR resonators provide high-Q channel filtering at RF, thereby eliminating the need to precisely tune the LO with a PLL. Provided the LO can be tuned to within 90 MHz, the desired signal can be placed within the bandwidth of the IF chain. Since filtering is provided by the FBAR, the IF bandpass filters seen in conventional super-heterodyne designs can be removed, saving power in the design.

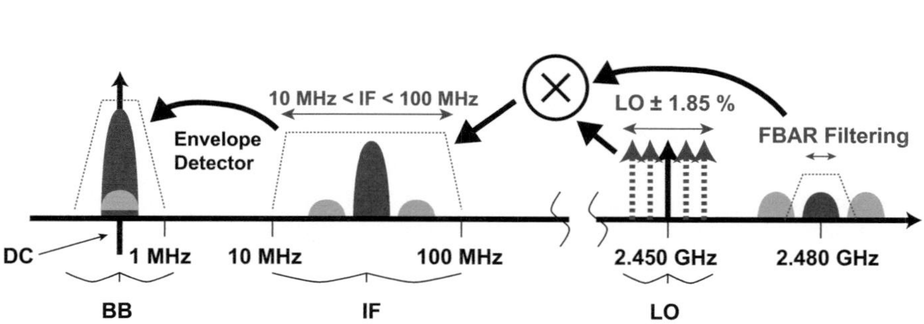

Fig. 6 Frequency plan for the FBAR-RX

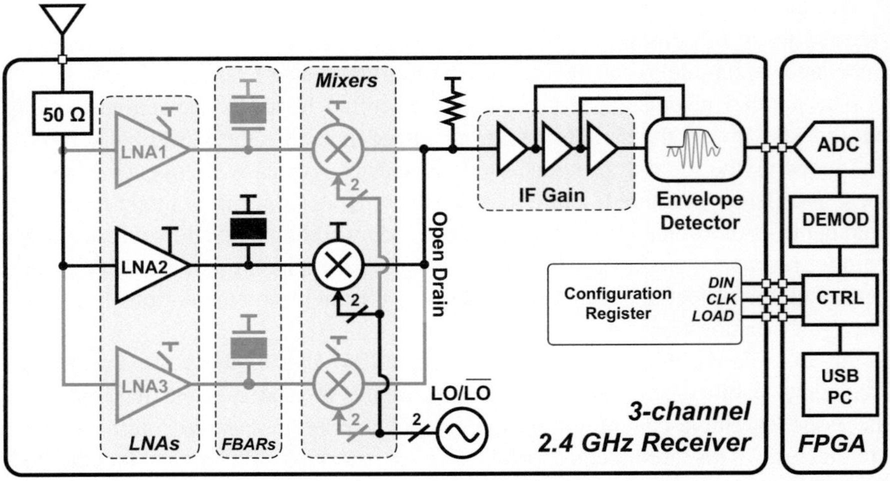

Fig. 7 Architecture of the FBAR-RX

The circuit architecture for the example receiver system is shown in Fig. 7. Separate LNA blocks for each channel provide isolation between the resonances of each FBAR, and an on-chip matching network tunes out the gate capacitance of the additional LNAs while simultaneously matching the input to 50 Ω. The LNAs have high gain at IF due to the high impedance of their bias networks at these frequencies, hence balanced mixers are used to prevent excessive noise from feeding through the mixer at IF. This improves the input-referred noise at the antenna and the overall sensitivity of the design. The signal pathways are recombined at the mixer stage in an open-drain fashion in order to share IF gain and envelope detection hardware. A low voltage of 0.7 V is used throughout the design in order to further improve energy efficiency.

The next sections will describe the design of the main circuit blocks in detail.

2.2 LNA

The LNA from the example implementation is shown in Fig. 8, and its constituent parts will be referenced throughout this section.

The LNA is one of the most critical blocks in a low-power receiver since: (1) as the first gain element in the signal chain, it has the most impact on the noise figure, and (2) achieving high-gain at RF is power-expensive, hence this block can dominate the power budget. The FBAR resonator provides certain advantages in both cases. For one, the high-Q passive filtering mitigates out-of-band noise and interferers and can be integrated into the resonant tank of conventional common-

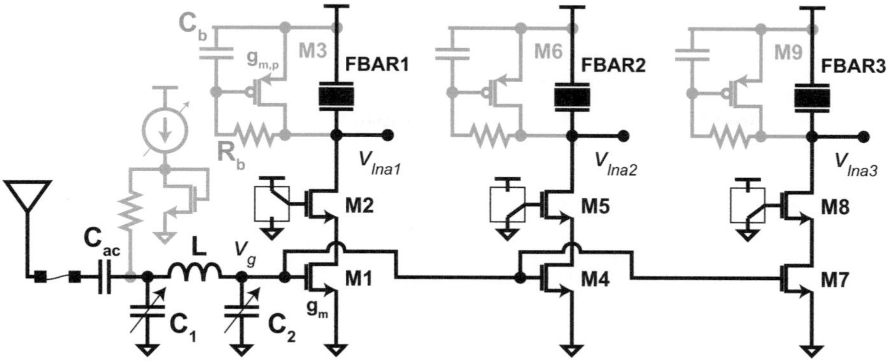

Fig. 8 LNA circuit diagram

source LNAs. Secondly, the high-impedance at parallel resonance produces high gain when combined with the g_m of the LNA transistor (such as M1).

The basis for the LNA design is a common-source (M1) with cascode (M2) approach. The following development discuses additional considerations for introducing an FBAR as the filtering element, beginning with the tank design.

2.2.1 Tank Design

In the standard common-source LNA design, a transconductance stage (such as M1 in Fig. 8) provides g_m conversion of an input signal voltage to a current, which is later converted to the output voltage by passing the current through a high-impedance load. In resonator-based receivers, the load is often an FBAR, whose impedance has a sharp filtering profile near 2.4 GHz, ideally covering the signal frequency of interest. A cascode (e.g. M2) is normally employed to increase output resistance and mitigate the Miller effect on the input side.

If a standard LC tank is used as the tuned load instead of a resonator, the equivalent parallel resistance is:

$$Q_p = \frac{R_p}{\omega L} \quad \Rightarrow \quad R_p = \omega L Q_p = (2\pi \times 2.4\,\text{GHz})(10)(5\,\text{nH}) = 754\,\Omega, \quad (1)$$

with reasonable parameters for L and Q. By way of comparison, for the FBAR resonators, the output impedance at parallel resonance is roughly $Z_p = 3\,\text{k}\Omega$. This translates into about 12 dB higher voltage gain with respect to the LC tank.

Next we discuss the filtering performance of the tank. The most important measure is the -3 dB bandwidth of the resonance. With the resonator as the narrowest filtering element, its bandwidth dominates the -3 dB bandwidth of the signal chain. Its bandwidth is determined by the natural Q of the resonator at its parallel resonance, and any additional impedances connected in series or parallel with it.

The second most important metric is the ratio of the tank's impedance magnitude at parallel resonance ($\|Z_p\|$) to its impedance magnitude off-resonance ($\|Z_o\|$). For the FBARs considered in this work, this ratio is

$$\frac{\|Z_p\|}{\|Z_o\|} \approx \frac{(3\mathrm{k}\Omega)}{(50\Omega)} = 60 \Rightarrow 36\ \mathrm{dB}. \tag{2}$$

In other words, a signal that is fully off-resonance should see about 36 dB attenuation with respect to a signal that is fully centered on the resonance. In reality, obtaining this entire attenuation range for an immediately adjacent signal is challenging due to the slow roll-off of the impedance when one moves far away from the −3 dB zone. The biasing network also plays a critical role.

Once the FBAR filtering model is understood, the next step is to design the biasing network. Unlike an LC tank, the resonator appears like a capacitor at DC, therefore, a network must be designed that will allow DC current to flow past the resonator and bias the common-source portion of the LNA. The main challenge is the resonator's extreme sensitivity to capacitive and resistive loading. This means that the parallel output impedance of the biasing network must be much higher than the FBAR impedance at RF, but at the same time, the impedance at DC must be low enough to avoid eating into the headroom of the LNA.

The basis for a biasing circuit for resonators is often the diode-connected PMOS "active inductor" [6, 25]. At DC, the capacitor C_b is open-circuited and the circuit self-biases through R_b at some fixed $V_{LNA,1}$ determined by the bias current and the transistor I-V characteristic. In small-signal RF, the capacitor is short-circuited and holds the gate potential of M3 fixed, hence the small signal output impedance is just the $\frac{1}{g_{ds,M3}}$ of transistor M3 at the bias point.

Though this is a good qualitative description that captures the basic characteristics of the biasing circuit at the two main frequencies of interest (DC and RF), a closer examination reveals a problem for the range of frequencies between these two points. The analytical expression for the small signal output impedance of the biasing network can be written as:

$$Z_{bias} = \frac{v_t}{i_t} = \frac{1}{g_{m,p}} \left(\frac{1 + sC_B R_B}{1 + s\frac{C_B}{g_{m,p}}} \right). \tag{3}$$

For a reasonable choice of parameters, Fig. 9 plots the biasing network impedance Z_{bias} combined in parallel with the FBAR impedance Z_{FBAR}. We can see from the plot that, aside from the desired resonance, the combined response has a wide and undesirable passband that unfortunately lies in the vicinity of the IF range. This generates gain at the IF frequency that leads to detrimental noise amplification since no further filtering of the signal is provided beyond the FBAR. Tuning the position of the pole and the zero in Eq. (3) is an exercise in trading off

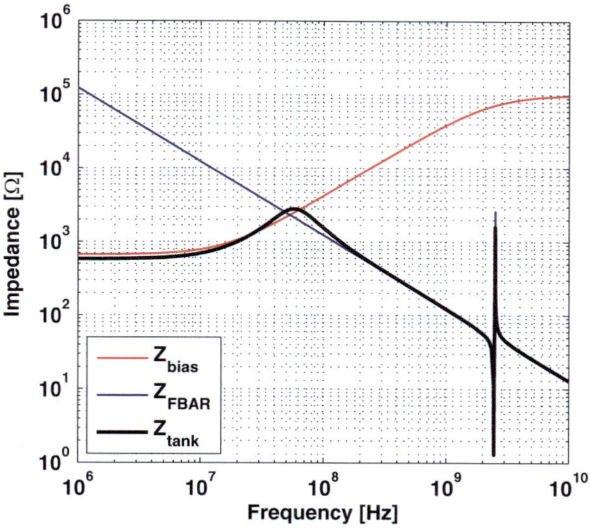

Fig. 9 Impedance of the LNA tank and its constituent parts

noise suppression with passband gain. To enable the best noise performance, the conclusion is that the subsequent mixer stage cannot be single-ended, as is often done in low power receivers to save power, but rather must be balanced in order to prevent the 10–100 MHz portion of the input spectrum from feeding through into the IF stages during the downconversion. This will be more fully demonstrated in Sect. 2.3.

2.2.2 Multiplexing

A scheme must be used to switch between the resonators dynamically in response to signals from a software or hardware-based controller.

One option is to place a switch in series with the resonator. In order to preserve filtering performance, this switch must have an ON-resistance of much less than the Z_o, and the capacitive load must be much less than C_p. A second option is to use a switch to conduct current. Here, degradation of the filtering can be avoided provided that the small-signal output impedance appearing in parallel with the resonator is much larger than the resonator's R_p (in this case, about 3 kΩ). It turns out that option two is much easier to implement in CMOS.

The simplest scheme that implements option two is shown in Fig. 10a. Although this configuration minimizes degradation of filtering performance of the FBAR, the capacitance at node **x** becomes a concern as the design is scaled to a higher number of channels. Since a matching network with a relatively large capacitance will be included, a better option is shown in Fig. 10b, where the capacitance of the extra common-source transistors is absorbed into the matching network.

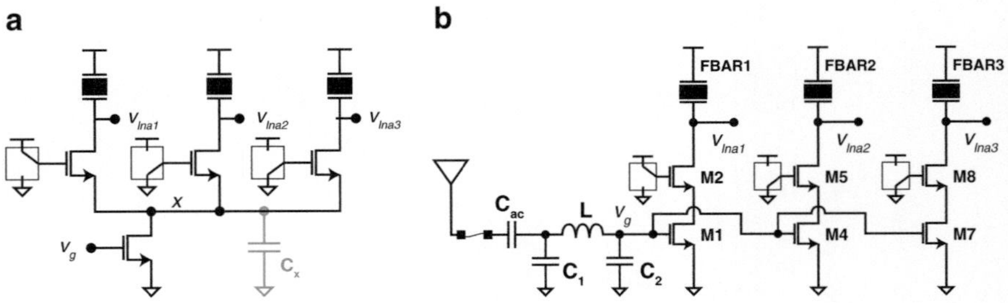

Fig. 10 LNA switching. (**a**) LNA switching with cascode only. (**b**) LNA switching, including common source transistor

2.2.3 Impedance Transformation

Impedance transformation is a key technique for extracting as much voltage-signal as possible from the weak incoming RF power-signal captured by the antenna. This desirable since subsequent amplifiers, such as the LNA, are common-source and rely more directly on voltage gain as opposed to power gain.

Extracting the maximum possible signal from the antenna is accomplished in two ways. The first is by capturing maximum power from the source antenna by matching its impedance to the input impedance of the network (in this case, 50 Ω). The second is by generating what some call "free" *passive gain* by matching to the high input resistance of the common source transistor(s).

It is worth noting that the gain is not really "free," but is merely the result of the impedance transformation process. In a lossless real impedance transformation, transforming the impedance upwards results in an increase in the voltage and a corresponding reduction in the current by the same factor, such that the total signal power is unchanged. So there is in fact no free power gain, but there is voltage gain, which is given as follows for an ideal lossless impedance transformation:

$$\frac{P_{out}}{P_{in}} = \frac{|V_{out,rms}|^2}{|V_{in,rms}|^2}\frac{R_{in}}{R_{out}} = 1 \quad \Rightarrow \quad \frac{|V_{out,rms}|}{|V_{in,rms}|} = \sqrt{\frac{R_{out}}{R_{in}}}. \tag{4}$$

Since the signal is narrowband, impedance transformation can be achieved with an on-chip matching network, such as the π-match topology. [26]. With a high input impedance on the LNA input transistor (about 4 kΩ), a voltage gain of up to 44 dB is achievable in theory, though the reality of lossy on-chip passives limits the gain to about 10 dB.

The complete LNA circuit with matching network, active gain stage, and tank is shown in Fig. 8.

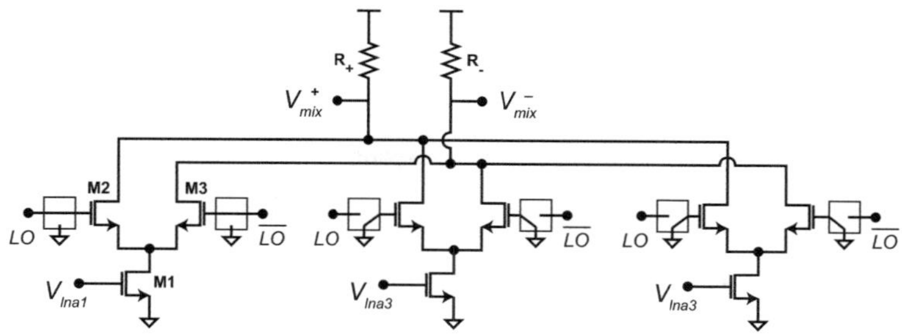

Fig. 11 Single-balanced mixer cells

2.3 Mixer

The mixer adopted for the receiver is the single-balanced design presented in Fig. 11. There are three mixer cells, one to capture the output from each of the three LNA branches.

In ultra-low energy receivers, high-frequency oscillator signals can dominate the power consumption, hence a single-phase clock is often desirable. Despite this, the unique frequency plan of this architecture necessitates the use of a balanced mixer with both LO and $\overline{\text{LO}}$ in order to improve the sensitivity.

To illustrate why, we can compare the down-conversion process in both the single-ended and differential cases:

$$\text{Single}-\text{ended} \quad \frac{v_{mix}(t)}{v_{lna}(t)} = g_m\left[\overbrace{0.5}^{\text{DC}} + \overbrace{\frac{2}{\pi}\sum_{n=1,3,5...}^{\infty}\frac{1}{n}\sin(n\omega_{LO}t)}^{\text{harmonics}}\right]R_L \tag{5}$$

$$\text{Differential} \quad \frac{v^{+}_{mix}(t) - v^{-}_{mix}(t)}{v_{lna}(t)} = g_m\left[\frac{4}{\pi}\sum_{n=1,3,5...}^{\infty}\frac{1}{n}\sin(n\omega_{LO}t)\right]R_L \tag{6}$$

In differential mode, the DC term is cancelled, thereby preventing any noise that may be present at the output of the LNA from translating to the very wide IF region (10–100 MHz) of the mixer output. This is illustrated in Fig. 12, which simulates the mixer gain from LNA to output. In the figure, we can also see the impact of un-matched duty-cycles, showing that the matching should be a few percent in order to achieve good rejection of the DC term.

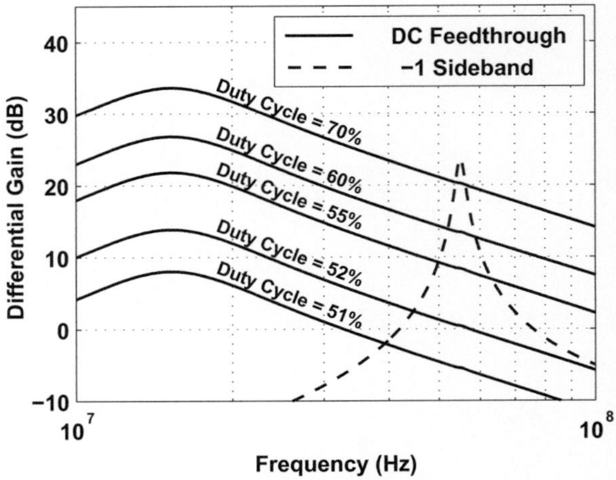

Fig. 12 Differential mixer gain for LO with non-50 % duty cycle

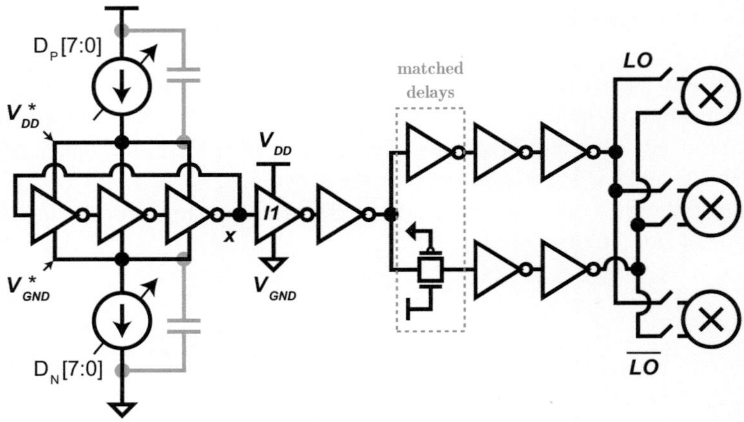

Fig. 13 Oscillator circuit

2.4 *Oscillator*

With the FBAR as the frequency stable element in the system, the downconversion oscillator can be simplified and the power consumption lowered compared to traditional architectures. In this system, a simple ring oscillator is sufficient to perform the down conversion.

The circuit designed for this system is shown in Fig. 13. It is a three-stage current-starved ring-oscillator with one branch for buffering the LO and an additional branch for generating and buffering $\overline{\text{LO}}$. In the $\overline{\text{LO}}$ branch, the inverted signal is generated by matching the delay of the pass-transistor to that of the inverter [27] in the LO branch.

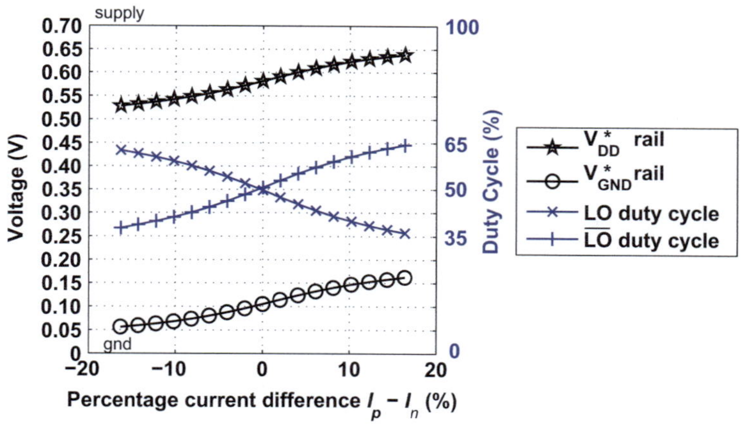

Fig. 14 Tuning the LO and $\overline{LO}$ duty cycles using mismatch in the current sources

The duty cycle is a function of the finite rise time of the oscillator signal at node x, and the midpoint of the virtual supply rail voltages with respect to inverter $I1$'s switching threshold. By adjusting the relative current difference between the top and bottom current sources, the virtual supply rails can be simultaneously shifted upwards or downwards in voltage without significantly affecting the frequency of the oscillation. This adjusts the time at which the oscillator crosses the inverter's switching threshold, and hence, tunes the duty cycle, as shown in Fig. 14.

2.5 IF Amplification

Efficient IF amplification can be provided by resistively-loaded differential band-pass amplifiers [28] biased in subthreshold for maximum $\frac{g_m}{I_D}$ efficiency. The circuit is shown in Fig. 15.

When biased in subthreshold, this style of amplifier is limited by the available headroom as demonstrated in the following relation:

$$A = g_m R_L = \left(\frac{I_D}{nV_T}\right) R_L = \frac{V_{swing}}{nV_T}. \tag{7}$$

The theoretical maximum gain, assuming V_{swing} = 200 mV, a typical n = 1.5, and $V_T = 26\,mV$, is 14 dB per stage, and simulated gain was 10.8 dB.

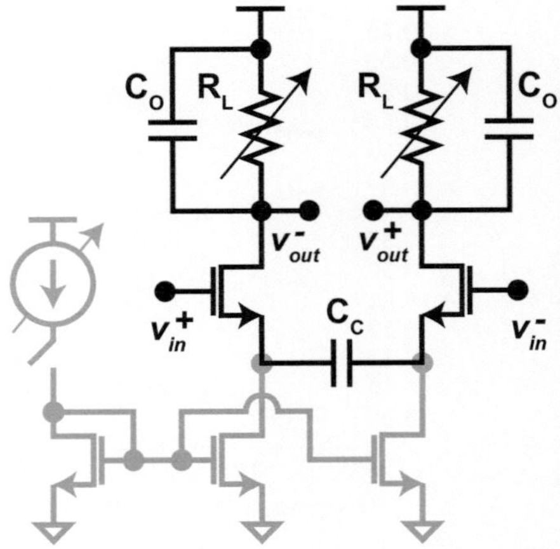

Fig. 15 Resistively-loaded, differential bandpass amplifier circuit for the IF

The filtering capabilities can be understood by examining the small-signal transfer function, which can be derived as

$$\frac{v_{od}}{v_{id}} = g_m R_L \left(\frac{s\frac{C_C}{g_m}}{s\frac{C_C}{g_m} + 1} \right) \left(\frac{1}{1 + sR_L C_O} \right). \tag{8}$$

The upper and lower corner frequencies are set by the first and second poles respectively, namely

$$\omega_{lower} = \frac{g_m}{C_C} \quad \text{and} \quad \omega_{upper} = \frac{1}{R_L C_O}, \tag{9}$$

and in this design, they are set at 10 MHz and 100 MHz respectively.

2.6 Envelope Detector

The final analog processing block of the design is the envelope detector. In this design, the exponential response of a subthreshold-biased MOSFET is used instead of a conventional diode-based rectifier. The circuit is shown in Fig. 16.

The design was adapted from [28]. In our implementation, only a single chain of IF amplifiers is used, with a controller-selectable tap-off point. This architecture allows multiple IF-gain stages to be multiplexed onto a single 20 pF output capacitor as a mechanism for coarse gain control on the order of 10 dB per stage, for a total of 40 dB of adjustable gain.

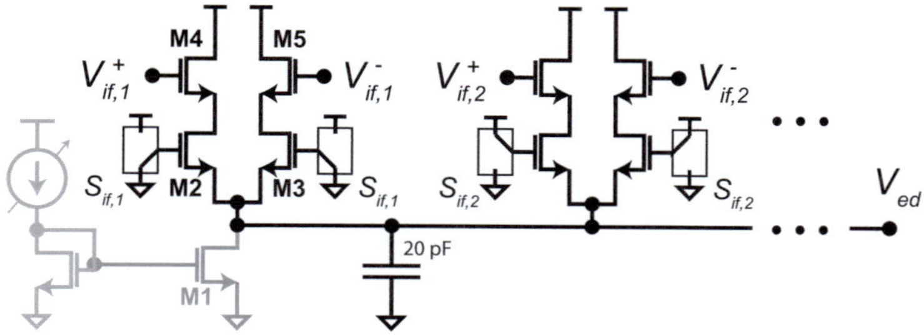

Fig. 16 Envelope detector circuit for selecting an IF stage

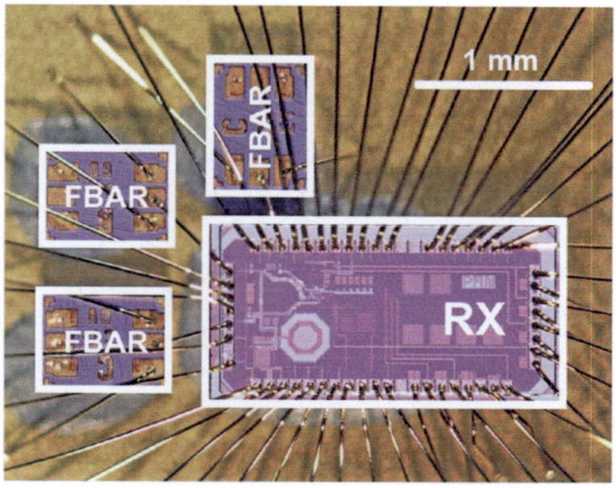

Fig. 17 Packaged receiver showing the three FBARs bonded to the 65 nm CMOS die

2.7 Receiver Performance

A prototype receiver system was fabricated in a 65 nm CMOS technology and wirebonded to three FBARs in a QFN package (Fig. 17). The following sections describe measurement and simulation results for this prototype with a focus on general key metrics for ultra-low-energy receivers.

2.7.1 Tuning the Signal Path

Firstly, to ensure power is indeed entering the 50 Ω RF port, we can examine the S_{11} by looking into the port with a network analyzer. For this measurement, the SMA connector, coplanar waveguide, and network analyzer were all designed to be 50 Ω without external matching components. The S_{11} measurement (Fig. 18a) demonstrates the impedance match at the receiver's RF input. The position of the notch can be made tuneable across the ISM band via the on-chip matching network, as was done in this design, in order to provide better coverage of the band. In this figure, the notch has been placed at the 2.413 GHz channel frequency of the selected FBAR.

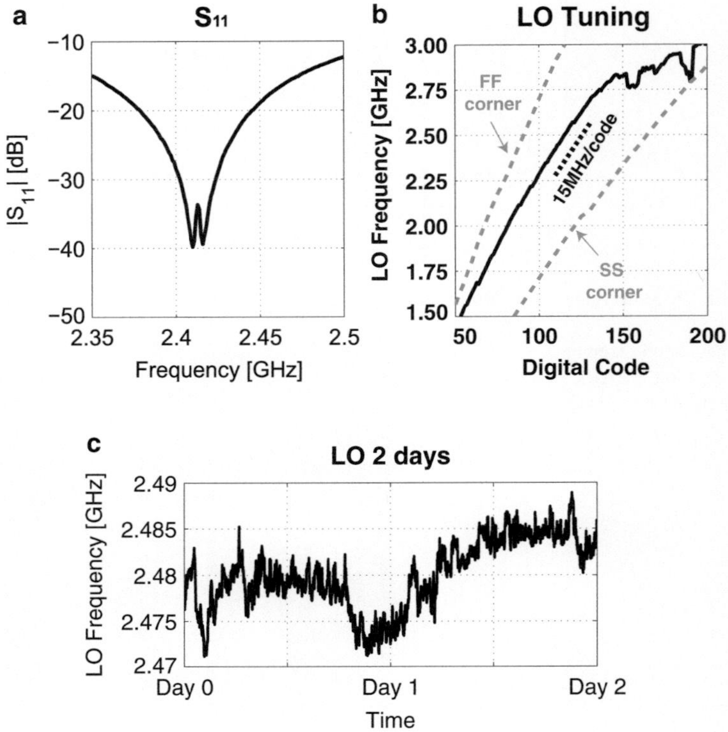

Fig. 18 Initial signal path tuning. (**a**) S_{11} measurement of input port. (**b**) LO tuning versus digital code. (**c**) LO drift over 2 days

Next, the LO's open-loop center frequency was characterized versus tuning word (Fig. 18b). In general, the total tuning range must be wide enough to cover PVT variations, but the frequency step size (15 MHz/code in this case) should remain less than the bandwidth of the IF chain in order to guarantee that all desired signals in the ISM band can be placed into the IF bandwidth.

The performance of the ring-oscillator over 2 days is characterized in Fig. 18c, showing about 20 MHz of total variation, mostly from temperature variations in the room in which the testing took place. In general, for robust performance over a wide temperature range, the system requires periodic frequency calibration, where the update rate is faster than the speed expected from the thermal transients.

2.7.2 Main RF Performance

Once the receiver was tuned, the full system RF performance was characterized by the plotting bit-error-rate (BER) curves for 1 Mbps OOK modulation (Fig. 19).

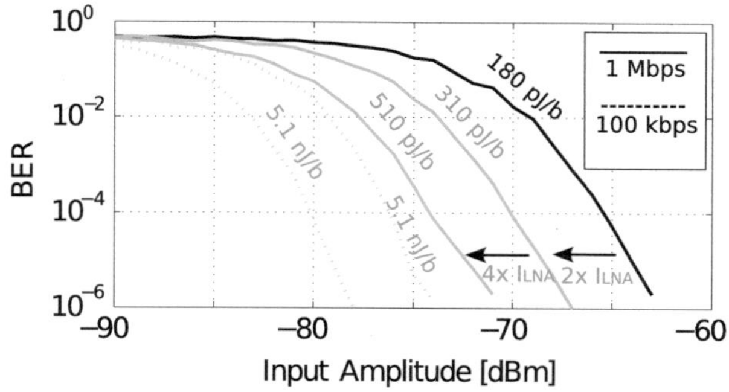

Fig. 19 The measured BER at under various data-rate and LNA gain settings

With a 0.7 V supply, the system achieved −67 dBm (BER = 10^{-3}) sensitivity at an overall energy consumption of 180 pJ/b.

If the nominal sensitivity is not satisfactory for the given RF conditions and additional energy is available, the total energy consumption can be traded for a better sensitivity by adjusting the LNA gain via its bias current (and using a 1 V LNA supply due to headroom concerns in this particular design). Also, at slower data rates (e.g. 100 kb/s), the ADC samples can be low-pass filtered before demodulation leading to further sensitivity improvement, shown up to −82 dBm in Fig. 19.

Next, the receiver's bandwidth and adjacent channel rejection were characterized. For the first part, the receiver's RF port was excited with a CW tone that was swept across each of the three FBAR-defined channels in turn. The plot in Fig. 20a shows the envelope detector's response. Normalized to the DC level with zero input power, the measured −3dB bandwidth using this method is 6 MHz. This bandwidth results from a combination of the FBAR's loaded −3 dB bandwidth of about 2.4 MHz and the non-linear detection characteristic of the envelope detector.

For the second part, the adjacent channel rejection was explored using a similar technique to that described in the IEEE standard for Medical Body Area Networks (802.15.6) [2]. In this test, the power of the desired signal was set to 3 dB above the nominal sensitivity level of the receiver (desired signal = −64 dBm), while the power of a similarly modulated interferer (1 Mbps OOK) was increased in an adjacent channel at various offsets until the BER of the desired signal degraded back to 10^{-3}. The Adjacent Channel Rejection (ACR) was taken as the ratio of this interfering signal strength divided by the desired signal strength and is plotted in Fig. 20b.

2.7.3 Block-Level Comparison

A block-by-block performance summary can lead to additional insight into the operating point of this receiver architecture and the tradeoffs that can be made to improve the performance.

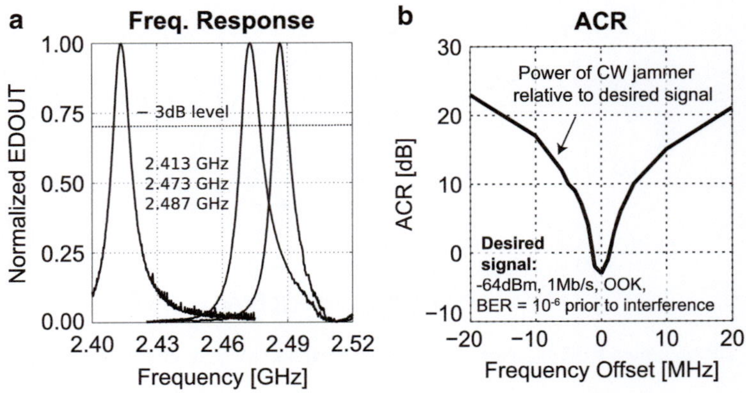

Fig. 20 Receiver bandwidth measurements. (**a**) Total response of the receiver chain to a CW sweep. (**b**) Adjacent channel rejection

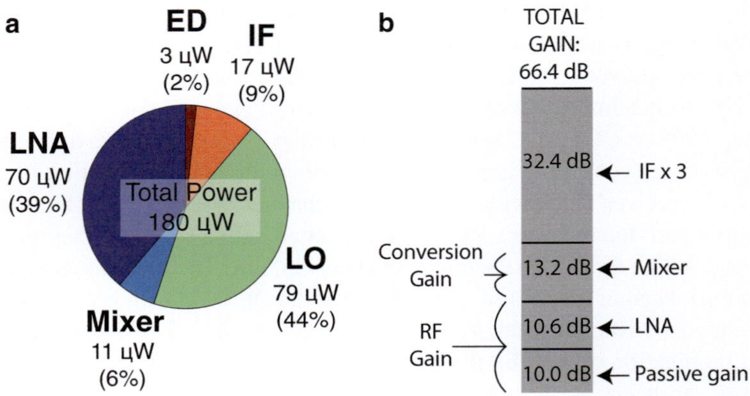

Fig. 21 Block-by-block specifications. (**a**) Measured power consumption by block. (**b**) Simulated gain of the signal chain

The measured power consumption of the sub-blocks is shown in Fig. 21a, where the LNA and the LO are seen to be the dominant power consumers in the system. The gain of the system is distributed according to Fig. 21b. It can be seen from these two figures that gain at RF is very-power expensive, and downconversion should be performed as early as possible in order to have an energy-efficient design.

The consequence of down-converting too early is a hit in noise performance, which manifests as a lower overall sensitivity level. Table 1 shows the simulated input-referred noise parameters for the blocks in the signal chain calculated according to the following equation:

Table 1 Simulation of noise performance of the receive chain

Block	Gain A_{block} (dB)	Noise PSD N_{block} (V^2_{rms}/Hz)	Noise bandwidth $B_{n,block}$ (MHz)	Input referred noise $P_{n,in,block}$ (fW)
Antenna	0	2.1×10^{-19}	5	20.7
LNA	20.6	9.6×10^{-17}	5	83.6
Mixer	13.2	7.0×10^{-15}	90	5250.0
IF stage 1	10.8	4.0×10^{-15}	90	250.0
IF stage 2	10.8	4.0×10^{-15}	90	20.8
IF stage 3	10.8	4.0×10^{-15}	90	1.7
Total noise				5626.8
(in dBm)				–82.5 dBm

$$P_{n,in} = \underbrace{kTB_{n,lna}}_{P_{n,in,ant}} + \underbrace{\frac{N_{lna}B_{n,lna}}{A_{lna}(50\Omega)}}_{P_{n,in,lna}} + \underbrace{\frac{N_{mix}B_{n,mix}}{A_{mix}A_{lna}(50\Omega)}}_{P_{n,in,mix}} + \underbrace{\frac{N_{if1}B_{n,if1}}{A_{if1}A_{mix}A_{lna}(50\Omega)}}_{P_{n,in,if1}} + \cdots. \quad (10)$$

It can be seen that the mixer, as the first wideband stage, dominates the noise figure calculation for this architecture. This suggests the flexibility to increase the sensitivity of the design by adding additional gain ahead of the mixer, for example by increasing the LNA gain. The demonstration of this tradeoff can be seen Fig. 19, where doubling the LNA current (and hence it's g_m and gain) improved the sensitivity by 5 dB compared with a maximum theoretical improvement of 6 dB.

2.8 Summary of Receiver Design

Figure 22 compares the results of this work with previously published 2–3 GHz receivers of various types, all targeted for low power applications. In summary, the design compares favorably in energy per bit versus sensitivity at 1 Mb/s, however, a compromise has been made in using an FBAR resonator (which cannot be appreciably tuned in frequency) to help achieve the lower power consumption. This lack of tuning capability can be partially mitigated by providing an architecture to use additional resonators to define additional channels, and the concept has been demonstrated in a system-in-package form factor. In the future, having integrated resonators in silicon beside the circuits will greatly expand the flexibility in designing receiver architectures that take advantage of these very high-Q devices.

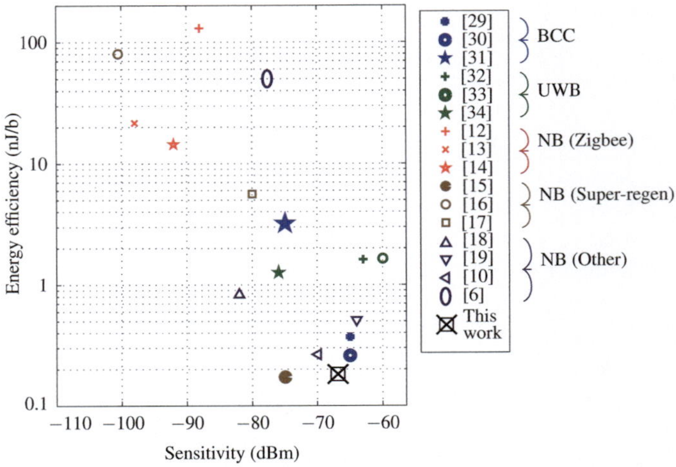

Fig. 22 Previously published low-energy receivers of various architecture styles

3 Multi-Channel Transmitter Design

3.1 Transmitter Architecture

The multi-channel FBAR-based transmitter architecture is shown in Fig. 23. An FBAR-based oscillator provides a stable center frequency that defines the communication channel. Multi-channel capability is achieved by multiplexing separate FBAR oscillators. A key to efficient multiplexing is a low input capacitance resonant buffer that then drives a −10 dBm power amplifier. The architecture supports BPSK, OOK and MSK modulation at data rates up to 1 Mbps. Pulse-shaping for each of the modulation schemes improves spectral efficiency.

The challenge of maintaining high transmit efficiency while delivering a low output power is addressed by the use of FBAR oscillators along with a high efficiency PA. Low voltage operation with rail-to-rail swing on all RF nodes minimizes short-circuit currents and provides the maximum possible drive. The PA, in addition, has a digitally tuneable impedance transformation network that maintains full-swing operation while tuning the output power. The tuneability of the output power can also be used for amplitude pulse-shaping (Gaussian for OOK and Square Root Raised Cosine for BPSK). This implementation of the pulse shaping is in contrast to a linear mixer and linear PA approach. The digital logic implementing the pulse shaping is implemented in an FPGA.

Each of the individual circuit blocks in the architecture are explained in detail in the following sections, followed by the measurement results.

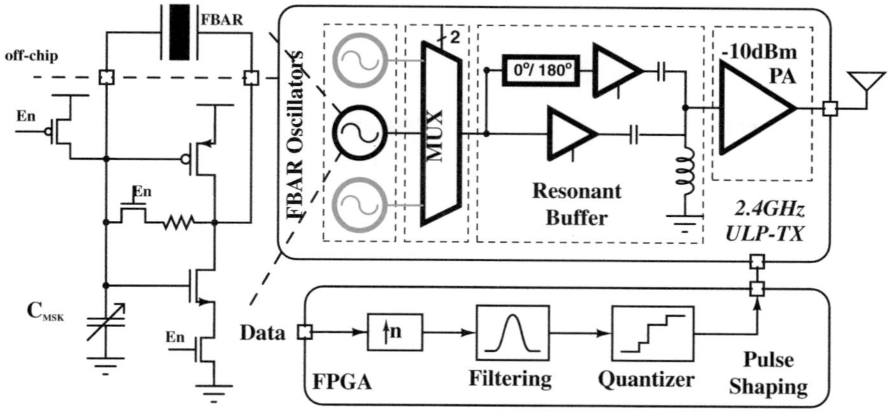

Fig. 23 TX architecture with the FBAR oscillator of one channel

3.2 *LO Generation and Frequency Modulation*

Resonator-based oscillator designs have been studied extensively in the literature [8, 23, 35–38]. The high-Q and frequency stability of the resonator enables stable LO generation. It also results in low phase noise and allows for PLL-less systems, leading to faster startup times.

Some oscillator topologies operate the resonators at their series resonance [8, 37], while others [23, 35, 38] use the parallel resonance. There is a range of performance for these oscillators in terms of phase noise and power consumption, though all provide excellent long-term stability. In this work, because of the simple modulation schemes and low output power, phase noise requirements are relaxed while power consumption is the most important specification.

Low-area biasing circuits are important because these circuits are repeated for each channel used. An inverter-based Pierce oscillator topology is used in our transmitter [35] (Fig. 23) because it has simple, low-area biasing circuits and provides rail-to-rail output.

The current of the circuit at its bias point should be such that the startup condition, $g_{\rm m} > g_{\rm m,crit}$ is met, where $g_{\rm m,crit}$ is the critical transconductance below which oscillations cannot be sustained [36]. Low-voltage operation (0.7 V) helps reduce undesired short-circuit power in the oscillator and subthreshold operation improves transconductance efficiency.

MSK Modulation Minimum Shift Keying (MSK) is implemented with frequency modulation instead of an I/Q architecture because of the simplicity and fine frequency tuning capability of high-Q FBAR oscillators. In order to achieve MSK modulation at 1 Mbps, center frequency tuning of ±250 kHz must be provided. A 150 fF digitally controlled capacitor bank $C_{\rm MSK}$, shown in Fig. 23, implements the required tuning. The frequency tuning for an FBAR oscillator is given by:

$$\frac{\Delta f_o}{C_{\text{MSK}}} = -\frac{f_o - f_s}{4C_p} \tag{11}$$

where f_s is the series resonant frequency and C_p is the effective parallel capacitance of the FBAR plus the loading from the circuits. In contrast, an LC oscillator has a frequency tuning given by:

$$\frac{\Delta f_o}{C_{\text{MSK}}} = -\frac{f_o}{2C} \tag{12}$$

For typical LC oscillators in this context ($C \approx 500\,\text{fF}$), and for the desired 250 kHz tuning, the C_{MSK} should be about 100 aF, whereas for the FBAR ($C_p \approx 1.2\,\text{pF}$, $f_o - f_s \approx 50\,\text{MHz}$), the required C_{MSK} is much larger at about 25 fF. This shows the relative ease of fine-frequency tuning of FBAR oscillators in contrast to LC tanks, though it comes with the cost of single channel operation.

3.3 *Oscillator Multiplexing for Channel Selection*

This section describes the circuits used and trade-offs involved in multiplexing FBAR oscillators to the power amplifier. The large area consumed by the PA precludes full replication of the entire circuit for channelization purposes; as a result, a single PA must be shared amongst all channels. The primary concern is the loading (capacitive and resistive) of the pierce oscillator and the requirement of a strong rail-to-rail signal at the input of the power amplifier, whose input capacitance is about 200 fF. There are two primary means to multiplex the FBAR oscillators to the PA, as shown in Fig. 24. The following two sections discuss the relative merits of each approach.

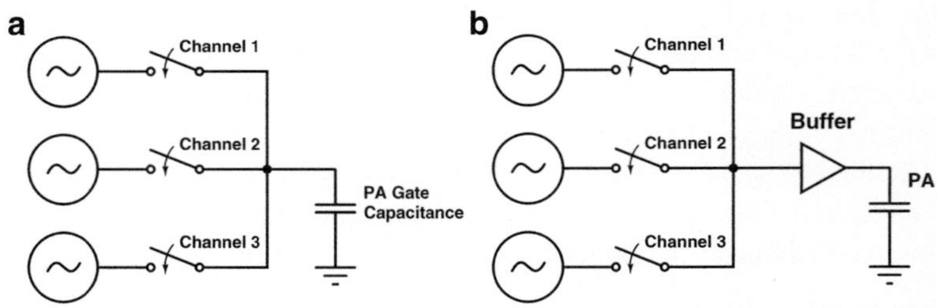

Fig. 24 Two multiplexing schemes: (**a**) Direct multiplexing. (**b**) Multiplexing with buffer

3.3.1 Direct Multiplexing to the PA

Shown in Fig. 24a, the direct multiplexing scheme ideally has near zero power overhead, since the PA capacitance is in resonance with the oscillator. The effective Q of this capacitance presents an equivalent resistive load to the oscillator, which de-Qs it, increasing the power consumption. The effective resistive loading presented by the multiplexor (R_p) needs to be compared to the 2 kΩ impedance of the FBAR at parallel resonance.

The following analysis calculates the effective resistance seen by the oscillator from the multiplexer loading. Let C_L be the capacitance of the load that is being multiplexed. Let C_P be the parasitic capacitance of each transmission gate as seen on the load side. Thus, if the number of channels multiplexed is n, the total capacitance presented to the oscillator would be $C_L + nC_P$. Let R be the series resistance of an 'on' transmission gate and note that the product $k = RC_P$ is approximately a constant for a given process. The Q of this load to the oscillator is $\frac{1}{\omega R(C_L+nC_P)}$. The effective loading resistance presented to the oscillator is thus (assuming high-Q):

$$R_P \approx R \cdot Q^2 = \frac{1}{\omega^2 R(C_L + nC_P)^2} \tag{13}$$

$$= \frac{C_P}{\omega^2 k(C_L + nC_P)^2} \tag{14}$$

For a given C_L and channel number n, the maximum R_P is obtained when $C_P = C_L/n$. Thus, the maximum resistance presented to the oscillator is given by:

$$R_{P,max} = \frac{C_L/n}{\omega^2 k(2C_L)^2} = \frac{1}{n} \cdot \frac{1}{4\omega^2 k C_L} \tag{15}$$

Thus, given a specification for a minimum R_P that avoids oscillator performance degradation, the maximum number of channels possible is given by:

$$n_{max} = \left\lfloor \frac{1}{4\omega^2 k} \cdot \frac{1}{R_P C_L} \right\rfloor \tag{16}$$

While multiplexing to the $C_L = 200$ fF PA, if an effective loading of $R_P > 5$ kΩ is desired, only one channel is supported in the 65 nm technology used in this work. For three channel operation, the $R_{P,max} = 2.9$ kΩ only. Thus, self-loading of the switches limits scaling to a large number of channels.

In addition, the input capacitance to the PA is not a constant as it can vary via the Miller multiplication of the gate-drain capacitance, C_{gd}, of the transistors in the PA (see Sect. 3.4). This changing load can cause instability in the center frequency of the transmitter.

3.3.2 Multiplexing with a Buffer

Alternatively, a buffer (Fig. 24b) could be used to decouple the joint problem of multiplexing in the presence of large PA input capacitance. The buffer drives the large capacitance, while presenting a much reduced load to the multiplexing circuit, making it much easier to size the transmission gates and scale to a large number of channels.

However, if a simple inverter buffer chain is used, the switching power consumption, $C_{PA}V_{DD}^2 f_c$ is 250 μW. This is prohibitive from an efficiency standpoint, considering that the target output power of the PA is 100 μW. In addition, the power consumption is sensitive to layout parasitics.

3.3.3 Resonant Buffer

This buffer power consumption problem can be alleviated through the use of a resonant buffer topology, shown in Fig. 25. The load capacitance is resonated with an on-chip inductor, similar to techniques used in resonant clock distribution networks [39]. The inverter only needs to provide for the losses in the tank. Since it is driven by the rail-to-rail swinging output of the oscillator, no special biasing is required for the circuit. The power consumption of the resonant buffer is:

$$P_{\mathrm{Buf}} = V_{\mathrm{DD}} I_{\mathrm{short-circuit}} + \frac{V_{\mathrm{DD}}^2}{8R_{\mathrm{p,\,ind}}} \tag{17}$$

$R_{\mathrm{p,\,ind}} \approx \omega L Q_{\mathrm{ind}}$. The value is about 1 $k\Omega$ in the technology used in this design.

Since the inverter needs to drive only the effective load resistance, and not the total capacitance, the size of the inverter can be made very small. A sizing ratio of <1:10 compared to the PA was sufficient with the input capacitance of the buffer

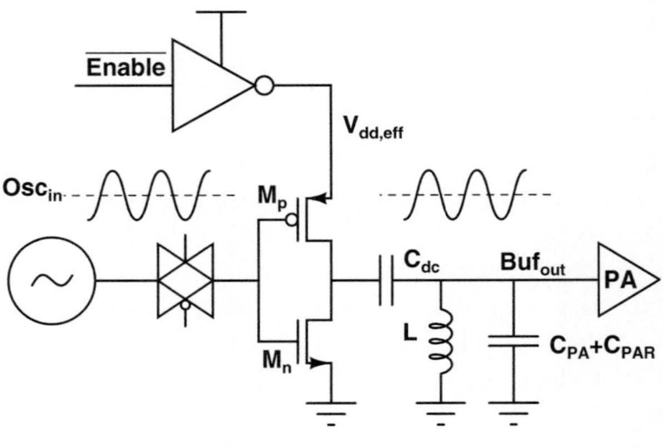

Fig. 25 Resonant buffer used in multiplexing the oscillators to the PA

being just 20 fF. With this, the theoretical maximum number of channels increases to $n_{max} = 17$ for $R_P > 5\,\text{k}\Omega$ (from Eq. (16)).

The buffer is tristated with the $\overline{enable}$ signal for OOK operation. The buffer circuit consumes $\approx 100\,\mu\text{W}$ from the 0.7 V supply, a 2.5× improvement over the inverter buffer chain.

3.3.4 Alternate Buffer Paths for BPSK Capability

The TX block diagram in Fig. 23 shows an alternate buffer path with a 0°/180° option for BPSK modulation. Matched, but phase-inverted delays, implemented with a pass-gate/inverter-pair are used [27, 40], while the full circuit implementation is described in detail in [24]. The overhead of the alternate buffer paths compared to the single-stage resonant buffer is about 30 μW.

3.4 Integrated Pulse-Shaping Power Amplifier

The power amplifier is targeted for a nominal output power of −10 dBm while operating from a 0.7 V power supply. The resonant buffer stage provides a strong rail-to-rail signal as an input to the PA, with the ability to drive a large capacitance. All modulation schemes (OOK, BPSK and MSK) are provided from the previous stages. However, the ability to perform amplitude modulation to improve spectral efficiency of BPSK and OOK is required. In this section, the design of the PA is discussed. It integrates pulse shaping capability while maintaining efficiency at low output powers of −10 dBm. Since high efficiency is the primary goal, we first look at PAs operated in saturation and then consider ways to incorporate pulse shaping.

On a single-ended 50 Ω antenna, −10 dBm amounts to a swing of $\pm 100\,\text{mV}_{\text{p-p}}$. An impedance transformation network is therefore necessary for a high efficiency power amplifier operating at 0.7 V. We now consider two possible topologies.

3.4.1 NMOS-Only vs. Push-Pull Topology

Figure 26a shows a typical NMOS-only power amplifier topology. This is a linear PA whose output is proportional to its input. However, when driven strongly with the rail-to-rail output of the resonant buffer at the peak efficiency operating point, the drain node has a swing of $2V_{DD}$. If R_L is the impedance provided by the matching network, the output power is given by

$$P_{\text{out, max}} = \frac{V_{DD}^2}{2R_L} \tag{18}$$

For the peak efficiency point of the PA to be at −10 dBm, $R_L = 2.45\,\text{k}\Omega$. The required 50× impedance transformation is not practical for efficient on-chip implementations [26].

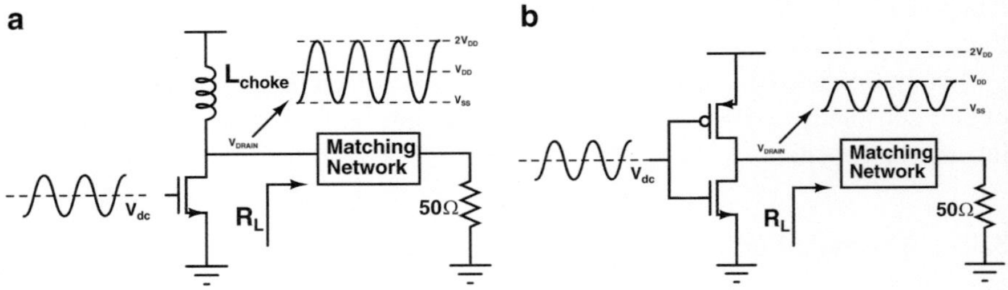

Fig. 26 Two canonical PA topologies considered. (**a**) NMOS-only topology. (**b**) Inverter-based push-pull topology

On the other hand, for the push-pull PA shown in Fig. 26b the drain node swings from 0 to only V_{DD} when operating at its peak efficiency, thus effectively reducing the output power delivered for the same load impedance. The output power in this case is:

$$P_{\text{out, max}} = \frac{V_{DD}^2}{8R_L} \tag{19}$$

Hence, for -10 dBm output, $R_L = 612\,\Omega$. This impedance transformation ratio of 12$\times$ is amenable to efficient on-chip implementations.

3.4.2 Tunable Impedance Transformation Networks

Once the supply voltage and load impedance are fixed, the power amplifier has a constant output power given by Eq. (19). In order to change the output power, while maintaining a high efficiency, the knobs available are the supply voltage and the effective impedance seen by the PA. If the former method is used for pulse shaping, the supply modulator would need to operate with a bandwidth greater than 1 MHz to accommodate data rates of 1 Mbps. These supply modulators, while ideally 100 % efficient [41], need external components and have significant control overhead, which is prohibitive for low output power applications.

The alternate method of adjusting the effective impedance provided to the PA through a tuneable impedance transformation network is explored in this work. The network must achieve resonance at multiple settings, while transforming the impedance to different values. The settings of the network should be made digitally tuneable in order to enable streamlined pulse shaping. At each of the settings, the PA operates at its peak efficiency, with the drain swinging rail-to-rail. Amplitude modulation is achieved even with a constant envelope input, thus simplifying the design of the previous stages.

Fixed-inductor designs are preferred because of the large area and low-Qs of on-chip inductors. This eliminates the a simple L-match since a fixed-valued inductor can only provide a fixed impedance transformation. Similarly, matching networks

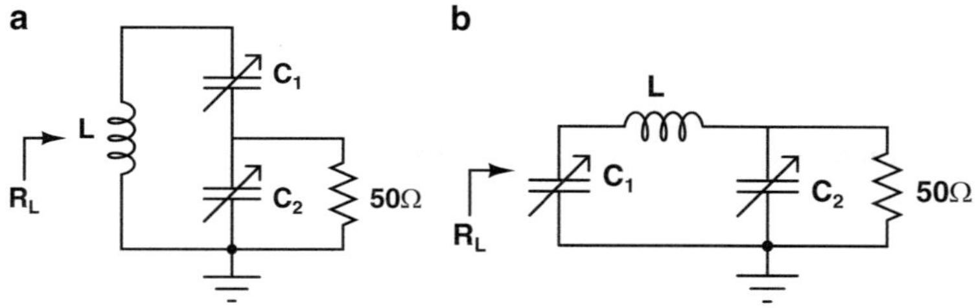

Fig. 27 Tunable impedance transformation networks. (**a**) Tapped-capacitor match. (**b**) π-Match

with two inductors are avoided because of the increased area. Thus, a π-matching network and a tapped-capacitor (Fig. 27) matching network are considered.

The design equations for the tapped capacitor match are given below, derived through narrowband parallel-to-series transformations [26]:

$$Q_2 = \omega \cdot C_2 \cdot 50 \tag{20}$$

$$C_{2,\mathrm{s}} = C_2 \cdot \frac{(Q_2^2 + 1)}{Q_2^2}, \quad R_s = \frac{50}{Q_2^2 + 1} \tag{21}$$

$$C_{\mathrm{s,eff}} = \frac{C_1 C_{2,\mathrm{s}}}{C_1 + C_{2,\mathrm{s}}}, \quad Q_\mathrm{L} = \frac{1}{\omega \cdot C_{\mathrm{s,eff}} \cdot R_\mathrm{s}} \tag{22}$$

$$C_\mathrm{eff} = C_{\mathrm{s,eff}} \cdot \frac{Q_\mathrm{L}^2}{Q_\mathrm{L}^2 + 1}, \quad R_\mathrm{L} = R_\mathrm{s} \cdot (Q_\mathrm{L}^2 + 1) \tag{23}$$

$$\omega^2 = \frac{1}{LC_\mathrm{eff}} \tag{24}$$

Q_2 is the loaded quality factor of the capacitor C_2, R_s is the series impedance seen by the effective series capacitance, $C_{\mathrm{s,eff}}$ formed by C_1 and C_2. Q_L is the loaded quality factor of this effective series capacitance, and also the loaded quality factor of the inductance. Thus, for a desired value of R_L, the required values of C_1 and C_2 can be calculated. A similar analysis can be done for the pi-match.

Since the analysis above assumes ideal passives, the "loaded quality factor" indicates the ratio of energy stored in the element to the energy radiated by the antenna per cycle [26]. But, in reality, the passives have a finite intrinsic quality factor ($Q_\mathrm{intrinsic} = \frac{\omega L}{R_{\mathrm{ind,\,parasitic}}} \text{or} \frac{1}{\omega C R_{\mathrm{cap,\,parasitic}}}$), which implies that a fraction of the stored energy is also dissipated by these losses every cycle. The power loss in a passive is:

$$P_\mathrm{loss} = P_\mathrm{out} \cdot \frac{Q_\mathrm{loaded}}{Q_\mathrm{intrinsic}} \tag{25}$$

a

$L(nH)$	$C_1(fF)$	$C_2(fF)$	$R_L(\Omega)$	Q_L
	958	1012	300	3
	890	1460	450	5.4
6.44	851	1807	600	7.1
	805	2366	900	9.7
	779	2820	1200	11.9

b

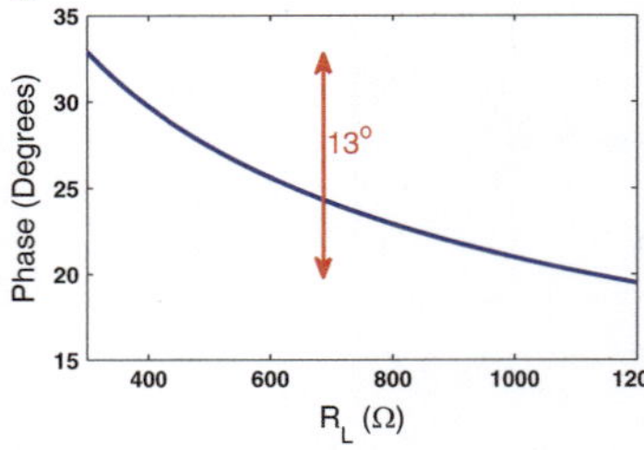

Fig. 28 Tunable tapped capacitor matching at 2.5 GHz. (**a**) Design values. (**b**) Phase variation

This indicates that for a given intrinsic Q, the efficiency of the matching network reduces as the loaded Q seen by the passive element increases. Thus, matching networks that minimize loaded Q are more preferable.

The design of the tapped-capacitor matching network using a 6.44 nH on-chip inductor is shown in Fig. 28a for transformation from 300 Ω to 1.2 kΩ. The maximum loaded-Q for this case is $Q_{\text{loaded}} = 12$. A similar analysis with the pi-match shows the loaded-Q to be as high as $Q_{\text{loaded}} = 17$ [24], which would lead to higher matching network losses.

Figure 28b plots the phase of the RF signal at the antenna relative to the phase of the RF input into the PA. The variation of this phase difference would indicate the amount of degradation of BPSK modulation with pulse shaping applied. The tapped-capacitor match has a 13° variation, but the equivalent pi-match has a much higher variation of 30° [24].

The pi-match has the desirable property that the tuneable capacitors are both ground referenced, making their design easier. However, given the low output swing ±100 mV for −10 dBm, tuning of the floating-capacitor in the tapped-capacitor match is feasible. Hence, given the other advantages of better efficiency and lower phase degradation, the tapped-capacitor match is the network of choice.

3.4.3 Design of Capacitor Banks

The capacitor banks C_1 and C_2 require a wide tuning of about 2–3×, while also maintaining a high Q. They are designed with some fixed capacitance and a binary-weighted capacitor bank built with MIM-capacitors. Large switches improve Q while the associated drain capacitance reduces tuning range. In order to improve this trade-off, a boosted 1 V supply voltage is used to drive the switches.

These capacitor banks, being digitally switched, can be changed at rates greater than 10 MHz, sufficient for up to 10× oversampled pulse shaping of 1 Mbps OOK and BPSK modulations.

3.4.4 Pulse Shaping Logic

In this work, the logic for driving the capacitor banks to apply pulse-shaping is implemented off-chip on an FPGA, as indicated in Fig. 23 and Sect. 3.1. However, it is important to estimate the power overhead of this operation. In [5], the Gaussian filter consumes only 15 μW in 65 nm CMOS. Similarly, in [15], the digital baseband including packet generation and raised-cosine pulse-shaping consumes only 62 μW in 90 nm CMOS.

3.4.5 Final PA Design

The final design of the proposed push-pull PA is illustrated in Fig. 29. Transistors M1 and M2 are biased at $\approx V_{TN}$ and $\approx (V_{DD} + V_{TP})$ respectively with on-chip DACs to trade-off short-circuit current and on-resistance. This is done to make the most use of the 0.7 V swing from the resonant buffer.

A total of 7.5 dB output power tuning range is achieved in the final design through the tuneable impedance transformation network by implementing a slightly wider capacitance tuning range as compared to the design values in Fig. 28a. Further tuning of output power, if required, can be achieved by statically varying the supply voltage of the PA alone through high-efficiency DC-DC converters. Hence, in the most flexible system, a DC-DC converter sets the average power of the transmitter while the pulse shaping is provided by the matching network.

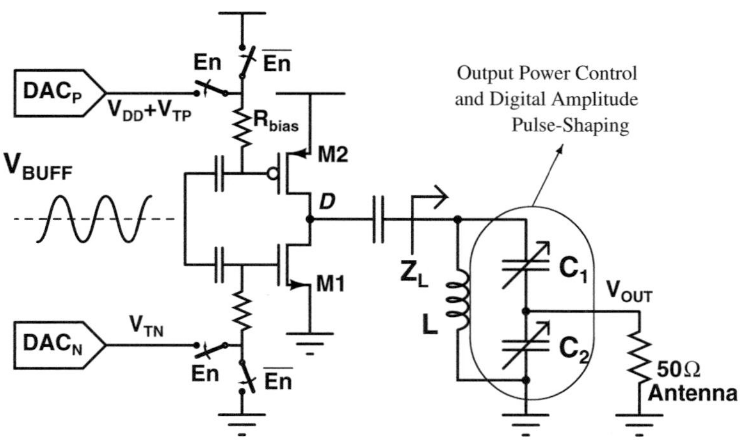

Fig. 29 Integrated PA with pulse-shaping using tuneable impedance transformation

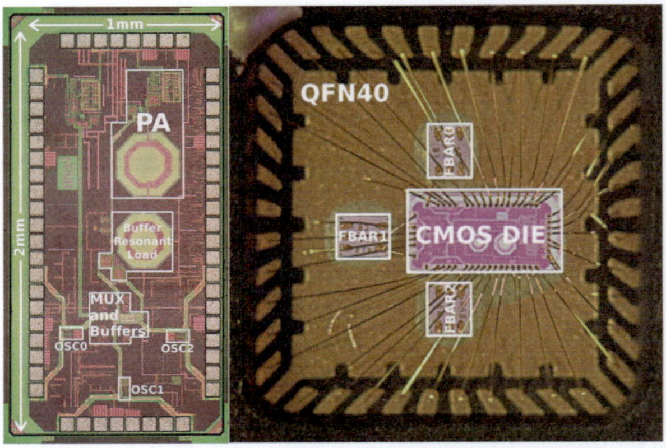

Fig. 30 Die Photo of TX and the packaged chip with 3 FBARs

3.5 Measurement Results

The transmitter is fabricated in a 65 nm CMOS process and is co-packaged with three FBARs as shown in Fig. 30. The TX core area is 0.324 mm^2. All the RF circuits are nominally powered from a 0.7 V supply, while all the digital switches in the multiplexers and capacitor banks are powered with a 1 V supply. An FPGA configures the serial interface of the chip as well as provides data for modulation and pulse shaping.

3.5.1 Oscillators and Resonant Buffer

Each FBAR oscillator consumes 150 μW. The center frequencies of the three channels as defined by the FBARs in one of the measured chips are at 2.421, 2.480 and 2.491 GHz. The oscillator can be tuned over a 600 kHz range, sufficient for the 1 Mbps MSK modulation. The phase noise of the oscillator as measured at the output of the PA is -132 dBc/Hz at a 1 MHz offset.

After the oscillator enable signal is turned on, the RF output at the antenna stabilizes in under 4 μs as shown in Fig. 31. The frequency accuracy at 4 μs is better than $\pm$20 ppm. The startup time corresponds to only 4 bit periods, ensuring efficient operation of the transmitter for even very short-length packets.

The resonant buffer consumes 100 μW and the alternate path with BPSK capability consumes an additional 30 μW.

3.5.2 Power Amplifier

The power amplifier is characterized at a nominal voltage of 0.7 V as well as at scaled voltages of 0.5 and 1 V. The rest of the circuits are always operated at 0.7 V. At each PA voltage, the DAC voltages are adjusted to optimize the short-circuit current

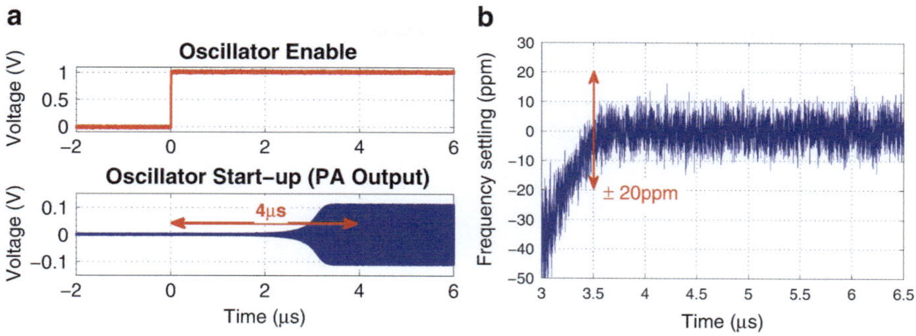

Fig. 31 Startup transients for FBAR oscillator. (**a**) Amplitude settling. (**b**) Frequency settling

Fig. 32 Efficiency of the PA for various $V_{\text{DD, PA}}$ and impedance transformation settings

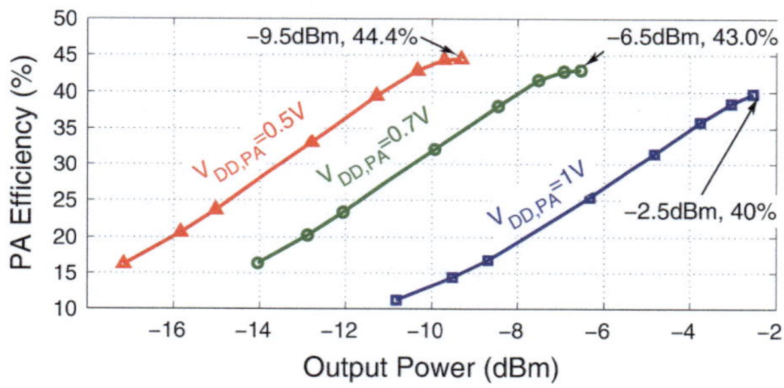

and on-resistance trade-off to maximize efficiency. The output power is swept only by varying the impedance transformation network.

Figure 32 shows the measured efficiency of the PA alone. At 0.7 V, the output power tuning is centered at around −10 dBm and has a peak efficiency of 43 % at −7 dBm. At 0.5 V, the peak efficiency increases to 44.4 % at −9.5 dBm output power and at 1 V, the peak efficiency goes to 40 % for −2.5 dBm.[1] It can be seen that the peak efficiencies remain approximately constant at each supply voltage. The losses in the tank and switches scale with the square of the supply voltage. Similarly, the output power also scales the same way, giving similar efficiency. Overall, we see a 14.5 dB tuning range from −17 to −2.5 dBm.

The global transmitter efficiency is the ratio of radiated power to power consumption of the entire transmitter. This is a metric that takes into account the power overhead of the various circuits in the transmitter, including LO generation and modulation. At 0.7 V PA supply, the peak TX efficiency is 28.6 %. At 1 V PA supply, it increases to 33 %, while at 0.5 V, it drops to 23 %. These are illustrated in Fig. 33. Unlike the PA efficiency, the peak TX efficiency does not remain constant at each

[1]With the supply voltage increased to 1 V, the peak output power increased by 4 dB instead of the expected 3 dB. This is attributed to a change in PA bias voltages which are adjusted to re-optimize short-circuit current and on-resistance.

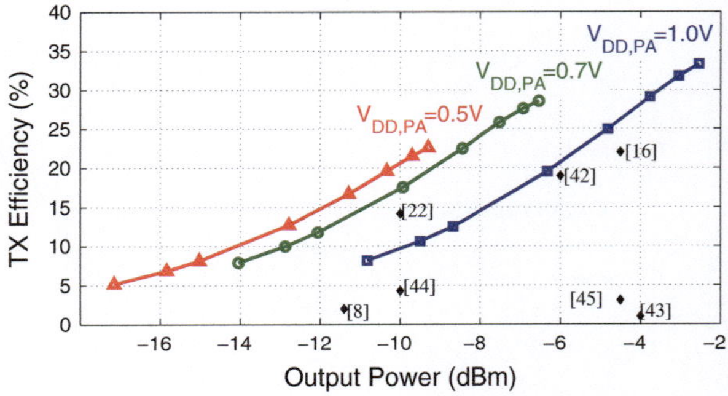

Fig. 33 Overall efficiency of the transmitter plotted for various $V_{\text{DD, PA}}$ and impedance transformation settings. Results are also compared to previous frequency-stable TXs

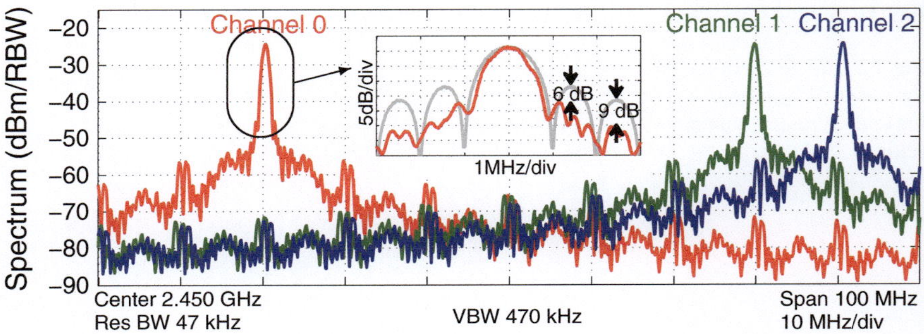

Fig. 34 Spectra of Gaussian pulse-shaped and phase-scrambled 1 Mbps OOK for the three channels measured from the chip

PA supply voltage because of the constant fixed power consumed by other blocks. The figure also compares the transmitter to other frequency-stable transmitters with sub-mW outputs operating at GHz frequencies.

3.5.3 Modulation and Pulse Shaping

OOK with Gaussian Pulse Shaping The TX is capable of OOK modulation at 1 Mbps through turning on and off the resonant buffer and PA. Gaussian pulse shaping with α=0.3 is applied with 10× oversampling. In order to avoid the spurs that arise from the feed-through component of OOK, the phase of the output is scrambled by a pseudo random sequence using the BPSK paths [40]. Figure 34 shows the superimposed spectra for 1 Mbps OOK of the three channels measured on the chip. This shows the multi-channel capability of the architecture. Spurs from the 10× oversampling are below −30 dBc. The figure also shows the first sidelobe and second sidelobe reduced by 6 dB and 9 dB respectively.

Overall, for a −12.5 dBm average output power, the entire transmitter consumes 440 pJ/bit.

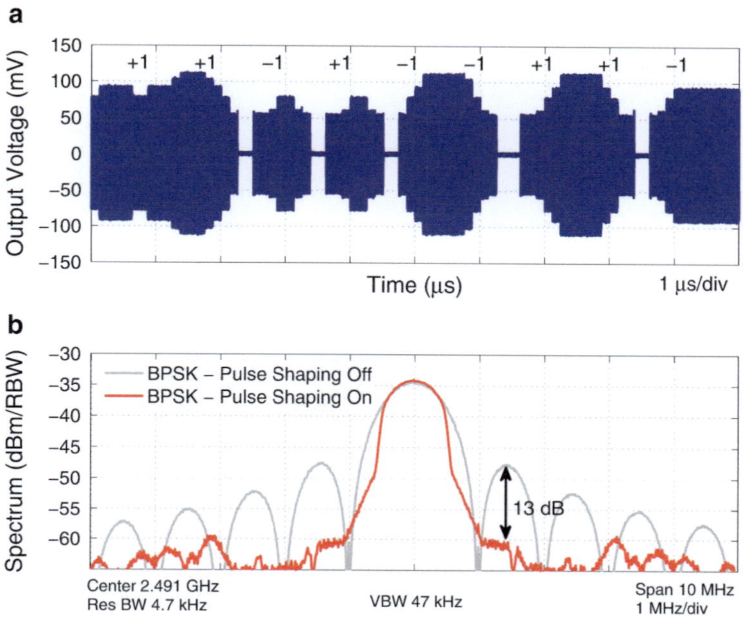

Fig. 35 1 Mb/s BPSK with SRRC filtering. (**a**) Time-domain waveform. (**b**) Spectrum

BPSK with SRRC Pulse Shaping Square Root Raised Cosine (SRRC) pulse shaping with $\beta = 0.3$ and oversampling of $8\times$ is used for BPSK modulation. The first side-lobe is reduced by 13 dB, effectively reducing the −20 dBc bandwidth of the signal from 6 MHz down to only 1.5 MHz. Figure 35 shows the time-domain waveform and frequency spectra. The transmitter consumes 530 pJ/bit at 1 Mbps while transmitting an average of −11 dBm output power in this mode.

MSK with Gaussian Pulse Shaping $10\times$ oversampled Gaussian pulse shaped MSK is applied to the tuning capacitors on the FBAR oscillators. The first sidelobe is reduced by 7 dB and the second sidelobe by >20 dB with the filtering. In this mode, the TX consumes 550 pJ/bit while delivering −10 dBm output power.

3.6 *Summary of Transmitter Design*

A transmitter architecture optimized for the short-distance link budgets of Body Area Networks has been presented. Specifically, the power amplifier has been optimized for operation at −10 dBm, and the tunable impedance transformation network results in efficient integrated pulse shaping, achieving high spectral efficiency. Further, high transmitter efficiency is achieved at these low output power levels by the use of a high-Q FBAR-based LO generation scheme. Multi-channel operation

Table 2 Transmitter performance summary

Technology	65 nm CMOS
Supply	0.7 V (RF), 1 V(Switch)
Num. channels	3
Startup time	4 μs
Data rate	1 Mb/s
Phase noise	−132 dBc/Hz (1 MHz off.)
PA peak Eff.	44.4 %
TX peak Eff.	33 %
P_{OUT}	−17 to −2.5 dBm
Energy per bit and Average P_{OUT}	
OOK (Gauss.)	440 pJ/bit at −12.5 dBm
BPSK(SRRC)	530 pJ/bit at −11 dBm
GMSK	550 pJ/bit at −10 dBm

has been achieved using inherently single-channel resonators through efficient oscillator multiplexing. In addition, low voltage operation at 0.7 V and maximum use of the swing available at all RF nodes improves energy efficiency. Overall, the transmitter has been measured to consume 440 pJ/b for 1 Mbps Gaussian pulse-shaped OOK at an average output power of −12.5 dBm. The performance of the chip is summarized in Table 2.

4 Conclusions

In this chapter, we have discussed some of the trade-offs in the design of low power short-distance RF transceivers using resonators. In particular, we examined some of the architecture choices that RF resonators provide. In transmitters, the advantages are primarily the low phase noise and good frequency stability, while still consuming low enough power to enable a high system-efficiency at low output power levels. In the example implementation presented in this chapter, we have shown that it is possible to have a 20% system efficiency at -10 dBm output power.

In receivers, resonators have been shown to improve the filtering capabilities of the RF front-end without sacrificing too much in the way of additional power. Furthermore, the presence of resonators allows the architecture to be simplified to the point where a ring-oscillator LO can be used, further lowering the power consumption. In the implemented receiver, we have demonstrated 180 pJ/b energy efficiency for a sensitivity level of −67 dBm and a front-end 3 dB bandwidth of 6 MHz.

Though we have demonstrated three channel capability in this work, fully integrated on-chip resonators such as the resonant body transistor [46] may, in the future, lead to additional frequency coverage.

Acknowledgements This work was supported by the Interconnect Focus Center, one of six research centers funded under the FCRP, an SRC entity. Chip fabrication was provided by the TSMC University Shuttle Program and FBARs were provided by Avago Technologies. Additional funding was provided by the Natural Sciences and Engineering Research Council of Canada fellowship.

References

1. N. Verma, A. Shoeb, J. Bohorquez, J. Dawson, J. Guttag, A. Chandrakasan, A micro-power EEG acquisition SoC with integrated feature extraction processor for a chronic seizure detection system. IEEE J. Solid-State Circuits **45**(4), 804–816 (2010)
2. IEEE standard for local and metropolitan area networks - part 15.6: wireless body area networks. IEEE Std 802.15.6-2012 (February 2012), pp. 1–271
3. E. Reusens, W. Joseph, B. Latré, B. Braem, G. Vermeeren, E. Tanghe, L. Martens, I. Moerman, C. Blondia, Characterization of on-body communication channel and energy efficient topology design for wireless body area networks. IEEE Trans. Inf. Technol. Biomed. **13**(6), 933–945 (2009)
4. C. Cojocaru, T. Pamir, F. Balteanu, A. Namdar, D. Payer, I. Gheorghe, T. Lipan, K. Sheikh, J. Pingot, H. Paananen, et al., A 43mw bluetooth transceiver with-91dbm sensitivity, in *IEEE International Solid-State Circuits Conference (ISSCC) Digest of Technical Papers* (2003), pp. 90–480
5. G. Angelopoulos, A. Paidimarri, A. Chandrakasan, M. Medard, Experimental study of the interplay of channel and network coding in low power sensor applications, in *Communications (ICC), Conference on 2013 IEEE International* (June, 2013), pp. 5126–5130
6. B.P. Otis, Y.H. Chee, R. Lu, N.M. Pletcher, J.M. Rabaey, An ultra-low power MEMS-based two-channel transceiver for wireless sensor networks, in *Symposium VLSI Circuits Digest of Technical Papers* (2004), pp. 20–23
7. Y. Chee, A. Niknejad, J. Rabaey, A 46% efficient 0.8 dbm transmitter for wireless sensor networks, in *Symposium VLSI Circuits Digest Technical Papers* (2006), pp. 43–44
8. D. Daly, A. Chandrakasan, An energy-efficient OOK transceiver for wireless sensor networks. IEEE J. Solid-State Circuits **42**(5), 1003–1011 (2007)
9. N. Pletcher, S. Gambini, J. Rabaey, A 65 μW, 1.9 GHz RF to digital baseband wakeup receiver for wireless sensor nodes, in *IEEE Custom Integrated Circuits Conference, 2007. CICC '07* (2007), pp. 539–542
10. N.M. Pletcher, S. Gambini, J.M. Rabaey, A 2GHz 52 μW wake-up receiver with -72dBm sensitivity using uncertain-IF architecture, in *IEEE ISSCC Digest Technical Papers* (2008), pp. 524–633
11. M. Contaldo, B. Banerjee, D. Ruffieux, J. Chabloz, E.L. Roux, C.C. Enz, A 2.4-GHz BAW-based transceiver for wireless body area networks. IEEE Trans. Biomed. Circuits Syst. **4**(6), 391–399 (2010)
12. A. Balankutty, S.-A. Yu, Y. Feng, P.R. Kinget, A 0.6-V zero-IF/low-IF receiver with integrated fractional-N synthesizer for 2.4-GHz ISM-Band applications. IEEE J. Solid-State Circuits **45**(3), 538–553 (2010)
13. M. Camus, B. Butaye, L. Garcia, M. Sie, B. Pellat, T. Parra, A 5.4 mW/0.07 mm^2 2.4 GHz front-end receiver in 90 nm CMOS for IEEE 802.15.4 WPAN standard. IEEE J. Solid-State Circuits **43**(6), 1372–1383 (2008)
14. A. Liscidini, M. Tedeschi, R. Castello, A 2.4 GHz 3.6mW 0.35mm^2 quadrature front-end RX for ZigBee and WPAN applications, in *IEEE ISSCC Digest of Technical Papers* (2008), pp. 370–620

15. M. Vidojkovic, X. Huang, P. Harpe, S. Rampu, C. Zhou, L. Huang, K. Imamura, B. Busze, F. Bouwens, M. Konijnenburg, J. Santana, A. Breeschoten, J. Huisken, G. Dolmans, H. de Groot, A 2.4GHz ULP OOK single-chip transceiver for healthcare applications, in *IEEE ISSCC Digest of Technical Papers* (2011), pp. 458–460
16. B. Otis, Y.H. Chee, J. Rabaey, A 400 uW-RX, 1.6mW-TX super-regenerative transceiver for wireless sensor networks, in *IEEE ISSCC Digest of Technical Papers*, vol. 1 (2005), pp. 396–606
17. J.-Y. Chen, M.P. Flynn, J.P. Hayes, A fully integrated auto-calibrated super-regenerative receiver in 0.13-μm CMOS. IEEE J. Solid-State Circuits **42**(9), 1976–1985 (2007)
18. S. Drago, D.M.W. Leenaerts, F. Sebastiano, L.J. Breems, K.A.A. Makinwa, B. Nauta, A 2.4GHz 830pJ/bit duty-cycled wake-up receiver with $-$82dBm sensitivity for crystal-less wireless sensor nodes, in *IEEE ISSCC Digest of Technical Papers* (2010), pp. 224–225
19. X. Huang, S. Rampu, X. Wang, G. Dolmans, H. de Groot, A 2.4GHz/915MHz 51μW wake-up receiver with offset and noise suppression, in *IEEE ISSCC Digest of Technical Papers* (2010), pp. 222–223
20. P. Nadeau, A. Paidimarri, P. Mercier, A. Chandrakasan, Multi-channel 180pJ/b 2.4GHz FBAR-based receiver, in *Proceedings IEEE RFIC Symposium* (2012), pp. 381–384
21. J. Bohorquez, A. Chandrakasan, J. Dawson, A 350μW CMOS MSK transmitter and 400μW OOK super-regenerative receiver for medical implant communications. IEEE J. Solid-State Circuits **44**(4), 1248–1259 (2009)
22. J. Bae, L. Yan, H. Yoo, A low energy injection-locked fsk transceiver with frequency-to-amplitude conversion for body sensor applications. IEEE J. Solid-State Circuits **46**(4), 928–937 (2011)
23. B.P. Otis, J.M. Rabaey, A 300μw 1.9-GHz CMOS oscillator utilizing micromachined resonators. IEEE J. Solid-State Circuits **38**(7), 1271–1274 (2003)
24. A. Paidimarri, P. Nadeau, P. Mercier, A. Chandrakasan, A 2.4 GHz multi-channel FBAR-based transmitter with an integrated pulse-shaping power amplifier. IEEE J. Solid-State Circuits **48**, 1042–1054 (2013)
25. M. Flatscher, M. Dielacher, T. Herndl, T. Lentsch, R. Matischek, J. Prainsack, W. Pribyl, H. Theuss, W. Weber, A bulk acoustic wave (BAW) based transceiver for an in-tire-pressure monitoring sensor node. IEEE J. Solid-State Circuits **45**(1), 167–177 (2010)
26. T.H. Lee, *The Design of CMOS Radio-Frequency Integrated Circuits*, 2nd edn. (Cambridge University Press, New York, 2004)
27. B. Razavi, K.F. Lee, R.-H. Yan, A 13.4-GHz CMOS frequency divider, in *IEEE ISSCC Digest of Technical Papers* (1994), pp. 176–177
28. D. Daly, Energy efficient RF transceiver for wireless sensor networks, S.M. Thesis, EECS Department, M.I.T., Cambridge, MA, (May 2005)
29. N. Cho, J. Lee, L. Yan, J. Bae, S. Kim, H.-J. Yoo, A 60kb/s-to-10Mb/s 0.37nJ/b adaptive-frequency-hopping transceiver for body-area network, in *IEEE ISSCC Digest of Technical Papers* (2008), pp. 132–602
30. A. Fazzi, S. Ouzounov, J. van den Homberg, A 2.75mW wideband correlation-based transceiver for body-coupled communication, in *IEEE ISSCC Digest Technical Papers* (2009), pp. 204–205,205a
31. L. Yan, J. Bae, S. Lee, B. Kim, T. Roh, K. Song, H.-J. Yoo, A 3.9mW 25-electrode reconfigured thoracic impedance/ECG SoC with body-channel transponder, in *IEEE ISSCC Digest of Technical Papers* (2010), pp. 490–491
32. N.V. Helleputte, M. Verhelst, W. Dehaene, G. Gielen, A reconfigurable, 130 nm CMOS 108 pJ/pulse, fully integrated IR-UWB receiver for communication and precise ranging. IEEE J. Solid-State Circuits **45**(1), 69–83 (2010)
33. M. Crepaldi, C. Li, K. Dronson, J. Fernandes, P. Kinget, An ultra-low-power interference-robust IR-UWB transceiver chipset using self-synchronizing OOK modulation, in *IEEE ISSCC Digest of Technical Papers* (2010), pp. 226–227

34. D.C. Daly, P.P. Mercier, M. Bhardwaj, A.L. Stone, Z.N. Aldworth, T.L. Daniel, J. Voldman, J.G. Hildebrand, A.P. Chandrakasan, A pulsed UWB receiver SoC for insect motion control. IEEE J. Solid-State Circuits **45**(1), 153–166 (2010)
35. Y. Chee, A. Niknejad, J. Rabaey, A sub-100μw 1.9-ghz CMOS oscillator using FBAR resonator, in *IEEE RFIC Symposium Digest Papers* (June, 2005), pp. 123–126
36. E. Vittoz, M. Degrauwe, S. Bitz, High-performance crystal oscillator circuits: theory and application. IEEE J. Solid-State Circuits **23**(3), 774–783 (1988)
37. J. Hu, L. Callaghan, R. Ruby, B. Otis, A 50ppm 600MHz frequency reference utilizing the series resonance of an FBAR, in *2010 IEEE Radio Frequency Integrated Circuits Symposium (RFIC)* (May, 2010), pp. 325–328
38. B. Otis, J. Rabaey, *Ultra-Low Power Wireless Technologies for Sensor Networks* (Springer, New York, 2007)
39. S. Chan, P. Restle, K. Shepard, N. James, R. Franch, A 4.6GHz resonant global clock distribution network, in *IEEE International Solid-State Circuits Conference (ISSCC) Digest of Technical Papers*, vol.1 (2004), pp. 342–343
40. P. Mercier, D. Daly, A. Chandrakasan, An energy-efficient all-digital UWB transmitter employing dual capacitively-coupled pulse-shaping drivers. IEEE J. Solid-State Circuits **44**(6), 1679–1688 (2009)
41. F. Raab, P. Asbeck, S. Cripps, P. Kenington, Z. Popovic, N. Pothecary, J. Sevic, N. Sokal, Power amplifiers and transmitters for RF and microwave. IEEE Trans. Microw. Theory Tech. **50**(3), 814–826 (2002)
42. A. Molnar, B. Lu, S. Lanzisera, B. Cook, K. Pister, An ultra-low power 900 MHz rf transceiver for wireless sensor networks, in *Proceedings IEEE Custom Integrated Circuits Conference (CICC)* (2004), pp. 401–404
43. D. Ruffieux, J. Chabloz, M. Contaldo, C. Muller, F. Pengg, P. Tortori, A. Vouilloz, P. Volet, C. Enz, A narrowband multi-channel 2.4 GHz MEMS-based transceiver. IEEE J. Solid-State Circuits **4**(1), 228–239 (2009)
44. A. Wong, D. McDonagh, G. Kathiresan, O. Omeni, O. El-Jamaly, T. Chan, P. Paddan, A. Burdett, A 1V, micropower system-on-chip for vital-sign monitoring in wireless body sensor networks, in *IEEE International Solid-State Circuits Conference (ISSCC) Digital Technical Papers* (2008), pp. 138–602
45. P. Bradley, An ultra low power, high performance medical implant communication system (mics) transceiver for implantable devices, in *Proceedings of IEEE Biomedical Circuits and Systems Conference* (2006), pp. 158–161
46. R. Marathe, B. Bahr, W. Wang, Z. Mahmood, L. Daniel, D. Weinstein, Resonant body transistors in IBM's 32 nm SOI CMOS technology. J. Microelectromech. Syst. **23**, 636–650 (June, 2014)

Ultra-Low Power Wake-Up Radios

Nathan E. Roberts and David D. Wentzloff

Abstract There is a growing class of event-driven devices that require instant-on wireless connectivity, but only use the radio to communicate intermittently throughout their lifetime. Home automation devices and most wellness monitors fall into this class, only using or needing their radios when prompted by an event. In these applications, the radios dominate the amount of energy consumed from the batteries. More specifically, the energy spent synchronizing the radios, or maintaining a connected state, dominates, as opposed to the energy spent communicating data.

Keywords Ultra-low power • Wake-up radio • Short range communication • Energy detection • Event-driven interrupt • Synchronization

1 Definition of a Wake-Up Radio (WRX)

1.1 Introduction

There is a growing class of event-driven devices that require instant-on wireless connectivity, but only use the radio to communicate intermittently throughout their lifetime. Home automation devices and most wellness monitors fall into this class, only using or needing their radios when prompted by an event. In these applications, the radios dominate the amount of energy consumed from the batteries. More specifically, the energy spent synchronizing the radios, or maintaining a connected state, dominates, as opposed to the energy spent communicating data.

Because of the intermittent nature of event-driven communication, ultra-low power short range radios spend much of their time in a sleep state, where the radios are not capable of sending or receiving packets, in an effort to conserve energy

N.E. Roberts (✉)
University of Michigan, Ann Arbor, MI, 48109, Now at PsiKick Inc., Charlottseville, VA, 22902
e-mail: nathan@psikick.com

D.D. Wentzloff
University of Michigan, Ann Arbor, MI, 48109

P.P. Mercier, A.P. Chandrakasan (eds.), *Ultra-Low-Power Short-Range Radios*,
Integrated Circuits and Systems, DOI 10.1007/978-3-319-14714-7_5

and reduce their average power consumption. This creates a challenge when two radios need to establish communication with each other. Both radios need to wake up and communicate at the same time, which requires the radios to be synchronized with one another. The most efficient way to synchronize is to have local clocks that are synchronized between all radios that need to communicate, enabling them to all operate with the same reference timing. However, local timing references drift and inevitably lose synchronization, requiring periodic re-sync using the radios. Furthermore, high-accuracy timing references can in some cases consume more power than the radios themselves. Asynchronous communication does not require a synchronized clock, but instead relies on radios to communicate intermittently to establish communication. While simpler in hardware, the drawback to asynchronous communication methods is that energy is wasted every time a radio wakes up to try and establish communication and it fails. This also doesn't solve the power problem of event-driven communication, where the device must periodically turn on its radio to check for wireless messages triggered by a random event. Increasing the frequency of checks with the radio reduces the latency for the device to respond to an event, however increases the average power consumption, leading to a trade-off.

Asynchronous methods for synchronization are typically lower power than their synchronous counterparts, and are therefore commonly found in short range low power radio protocols like Bluetooth Low Energy (BLE) [1]. To quantify the energy requirements of BLE, a state of the art BLE transceiver from Dialog Semiconductor is used as a reference [2]. In the BLE standard, a receiver establishes communication with a transmitter by turning on quickly and listening on the channel(s) for the transmitter. If a transmitter is detected then a communication link can be established; if not, then the receiver turns back off. This state is called the 'scan' state. Short scan times and long intervals between the scans results in a low average power, but also limits the probability of establishing communication quickly. According to the BLE specification, the shortest amount of time a receiver can be in the scan state is 2.5 ms and the longest interval between scans is 10.24 s. This would result in a duty cycle of 0.024 %. Such a low duty cycle rate would place the energy burden on the transmitter to transmit at a greater duty cycle rate, and therefore consume more energy, in order to increase the probability of establishing communication. The Dialog Semiconductor SmartBond™ DA14580 BLE transceiver has an advertised power consumption of 14.7 mW in receive mode, which means it consumes 36.75 μJ of energy for each scan operation and can reduce its average power down to 3.5 μW, not including the power of any circuits that remain active between radio intervals, such as a timer, voltage regulators, and memory retention power.

A wake-up radio (WRX) can be used as an alternative to these types of synchronization methods. The WRX acts as a secondary receiver within an asynchronous protocol. While the other radios are conserving power in an ultra-low power sleep state, the WRX stays on to continuously monitor the channel for activity and enables the main communication receiver when it detects another radio trying to communicate. Because the WRX remains on continuously it must be very low power, which is the main specification that drives WRX design. This event-driven synchronization methodology is energy efficient because the high power receiver

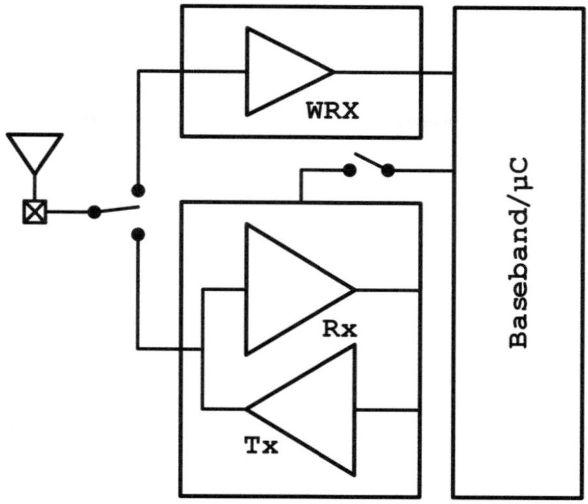

Fig. 1 WRX integrated into a transceiver

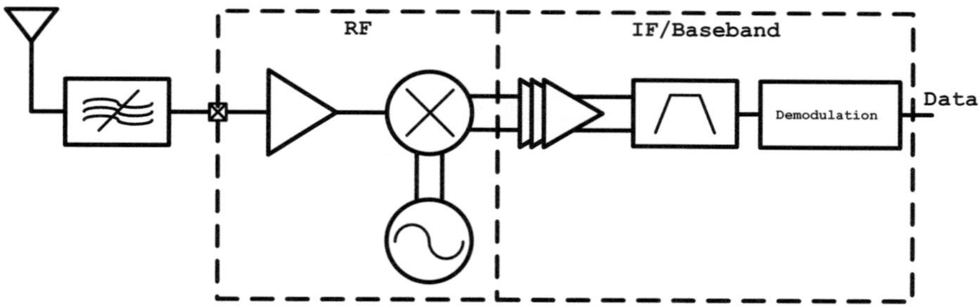

Fig. 2 Traditional receiver architecture

remains off as long as possible and only wakes when communication is necessary. Figure 1 shows a simplified example of a transceiver that employs a WRX for synchronization.

In order for a WRX to reach power levels that are orders of magnitude lower than a traditional receiver, the traditional receiver architecture needs to be re-examined. Figure 2 shows a generalized block diagram on a traditional receiver that would be found in a commercial low power radio. The radio is divided into two main sections based on the frequency of operation required of the circuit blocks: RF and IF/baseband. Due to the energy required for amplification at RF frequencies, the left section dominates in terms of power. The LNA needs to amplify signals at RF while meeting noise requirements, which comes at the expense of power. A PLL is used to create a stable local oscillator near the RF frequency (depending on the RF radio architecture), and the mixer must meet noise and linearity requirements while down converting the RF signal. It makes sense then that a WRX looks to eliminate some, if not all, of these RF components to achieve power reduction.

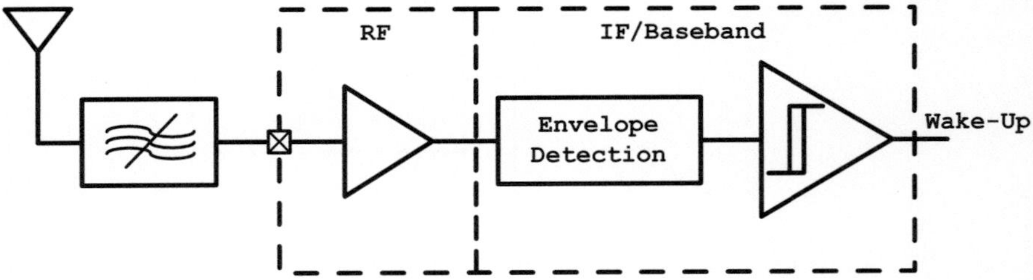

Fig. 3 Generic WRX architecture

The following sections provide a brief history of the WRX by describing the typical architecture for a WRX and showing the first and then most recent publication from the International Solid State Circuits Conference (ISSCC). The rest of the chapter will describe the shortcomings present in published WRXs and will provide solutions to these issues. The chapter will conclude with a detailed description of a demonstrated WRX that solves the stated issues and is compared with other state of the art work.

1.2 Generic WRX Architecture

A generic WRX uses a simple energy detection scheme to keep power as low as possible. This also restricts the WRX to simple modulations like on-off keying (OOK) or frequency-shift keying (FSK). Demodulating received signal amplitude is simpler than demodulating signal power in two different frequency bands, so OOK is the most common modulation seen in WRX designs. As seen in Fig. 3, the generic WRX consists of channel filtering and, optionally, gain at RF, followed by mixing and then rectification, or direct rectification without the additional mixing element (as shown in Fig. 3). The more energy contained in the frequency band of interest, the greater the voltage change at the output of the rectifier which can then be sensed by a demodulating comparator, sometimes referred to as a 1-bit ADC. When enough energy is present to trip the comparator, the WRX is considered to have issued a wake-up signal. Comparing this to the traditional receiver architecture in Fig. 2, most, if not all, of the high power RF circuits are eliminated. Many WRX designs keep the LNA to improve sensitivity, but often the envelope detector that performs signal detection and down conversion to baseband eliminates the need for a PLL and mixer.

1.3 First ISSCC Publication [3]

The first WRX published at ISSCC was in 2008 by Nathan Pletcher et al. at UC Berkeley [3]. The WRX achieved a power consumption of 52 μW, which is much

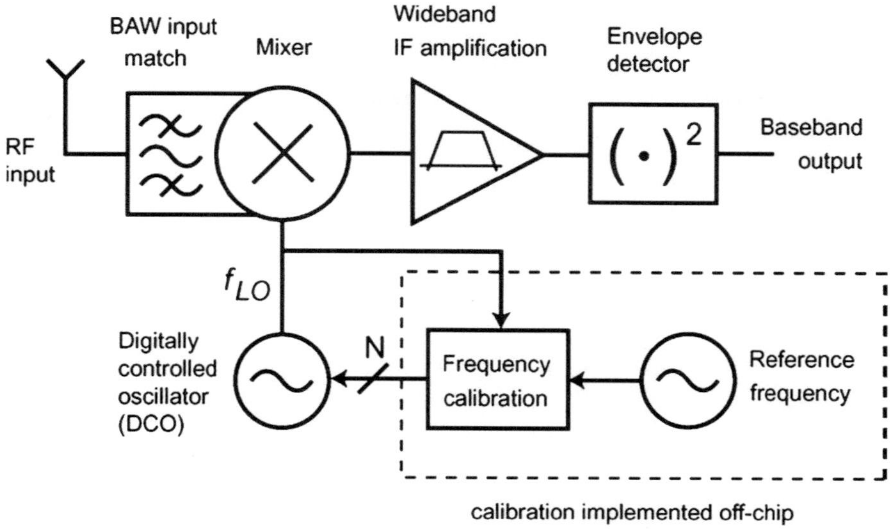

Fig. 4 Block diagram [3]

lower than Dialog Semi's 14.7 mW BLE receiver. In order to reduce the average power of Dialog's BLE receiver to match the WRX, the Dialog receiver would have a duty cycle rate of 0.35 % meaning it can perform a scan operation at most once every 714 ms. This original WRX looks more like a traditional receiver than the generic WRX in that it has a mixing circuit and an oscillator. The main strategic decisions that reduced the power was the use of an "uncertain-IF" architecture as well as a MEMS-based resonator at the RF front-end to provide high-Q filtering and passive voltage gain. The WRX operates at 2 GHz and was implemented in a 90 nm CMOS technology and has a sensitivity of −72 dBm and a data rate of 100 kbps. Further power reduction was realized using a 0.5 V supply.

The "uncertain-IF" architecture is aimed at exploiting the potential energy savings found in a common ring oscillator compared to an LC oscillator while removing the receiver's dependency on phase noise and frequency accuracy that make LC oscillators superior to their ring counterparts.

The block diagram in Fig. 4 shows how the architecture was designed to reduce the dependency on high quality oscillators. The low power oscillator mixes the RF signal to IF, but the bandwidth of the IF amplifiers is wide enough to cover the variation caused by the ring oscillator. Envelope detection down converts the IF frequency to baseband. The matching network for the WRX uses a BAW resonator with a capacitive transformer to provide both filtering and a 50 Ω match. In addition, using the BAW provides a 12 dB voltage gain without the need for a power-hungry LNA. A single-ended mixer is chosen as the RF front-end in an effort to maximize conversion gain as well as be compatible with a single-ended ring oscillator topology. The IF amplifiers filter out any LO feed through before it reaches the envelope detectors. The image is filtered at RF and the LO only needs to be

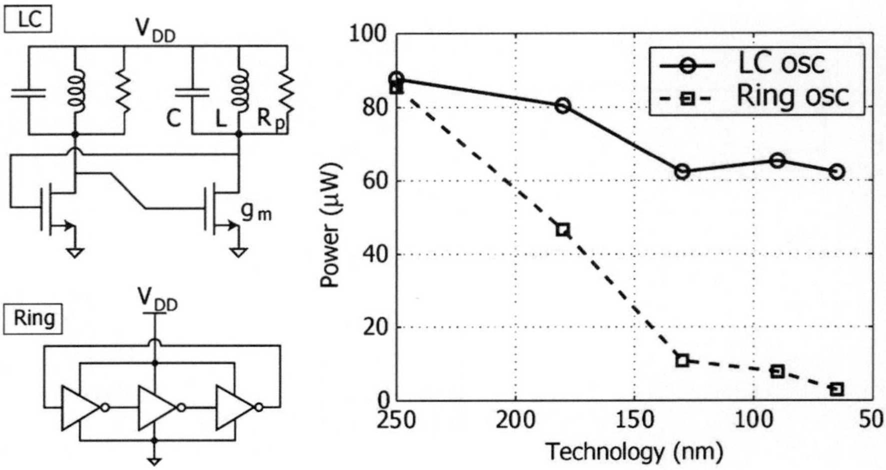

Fig. 5 Ring vs. LC oscillator [3]

accurate within 100 MHz around the RF frequency. This is possible because the envelope detector directly converts the IF signal to DC [3].

A single-ended 3-stage ring oscillator is chosen in the design to keep power consumption as low as possible compared to an LC oscillator. The fundamental need for sufficient g_m in an LC oscillator with limited quality passives necessitates a certain amount of current consumption. On the other hand, technology scaling allows the ring oscillator to reduce its CV^2 power quicker than its LC counterpart. In deep-submicron processes, the authors show that the power difference between a ring and LC is substantial, as illustrated in Fig. 5.

The DCO is implemented using the lowest power, full-swing ring oscillator possible: a 3-stage ring using inverters from a standard cell library. Two resistive DACs modify the virtual supply rail, as seen in Fig. 6. The two DACs are used to keep the common mode of the ring within the dynamic range of the LO buffers so they can be restored to full VDD before going to the mixer. Low-threshold devices ensure adequate frequency from the ring despite the 0.5 V VDD supply.

To summarize the important decisions that reduced the power to 52 μW: the WRX was analyzed from the architectural level where the authors created an architecture that allowed for a low power oscillator that would produce an uncertain-LO frequency instead of a higher power oscillator with very fine accuracy, like a PLL. In addition, the BAW resonator on the front of the WRX provides a passive 12 dB of voltage gain, removing the need for an LNA and allowing the signal to go straight into a mixer. Full measurement results will be shown at the end of the chapter in Table 2.

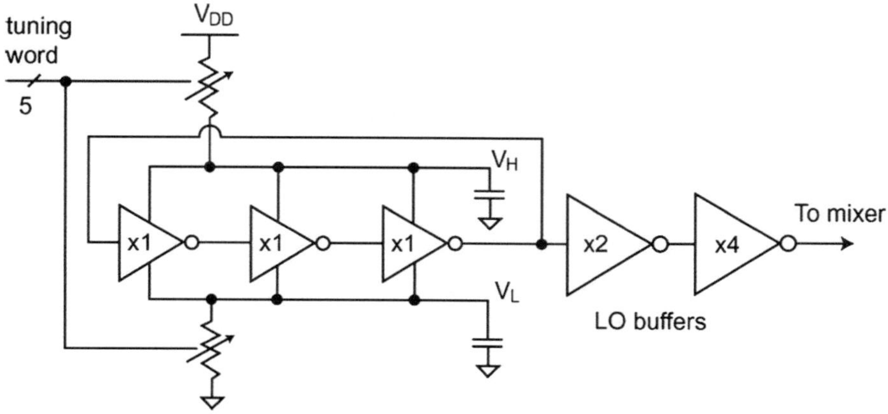

Fig. 6 DCO [3]

1.4 Most Recent ISSCC Publication [4]

The most recent WRX publication at ISSCC was in 2012 by Jeongki Choi et al. from KAIST [4] and it addresses a major shortcoming present in nearly all WRX designs: in-band and out-of-band interference. The WRX was designed for the 5.8 GHz ETCS band in China, for short range communication.

In order to block interferers, filtering must be used, but the authors argue that typical approaches like using off-chip RF SAW filters are not sufficient in the GHz range because of their wide bandwidth and their insertion loss degrades sensitivity [4]. The authors argue against using a narrowband bandpass filter (BPF) because they are often implemented as R-C filters to reduce power and therefore have poor selectivity. For this WRX the authors propose a delay-based BPF that they claim has a narrow and sharp frequency response sufficient to reject interferers.

Figure 7 shows a block diagram of the WRX, which matches up well with the generic WRX architecture of Fig. 3. An external balun is used to increase gain prior to the RF envelope detector. The envelope detector is a pseudo-differential amplitude detector so it outputs a single-ended signal that is then split in two signals: one driven straight into the positive terminal of a hysteretic comparator (1-bit ADC) and the second signal passed through an RC filter before driving into the negative terminal of the same hysteretic comparator. The output of the comparator is the wake-up signal.

To cancel out-of-band interference, a delay-based BPF was used and is shown in Fig. 8 and is implemented after the demodulating comparator in digital logic. It consists of a low pass filter (LPF) and a high pass filter (HPF). The filter is low power because it is designed using standard digital cells and passive components so, other than leakage, the filter consumes no power when a signal is not present.

To summarize, the authors of this WRX used digital processing to implement filtering for this WRX providing strong out-of-band interference rejection which is a

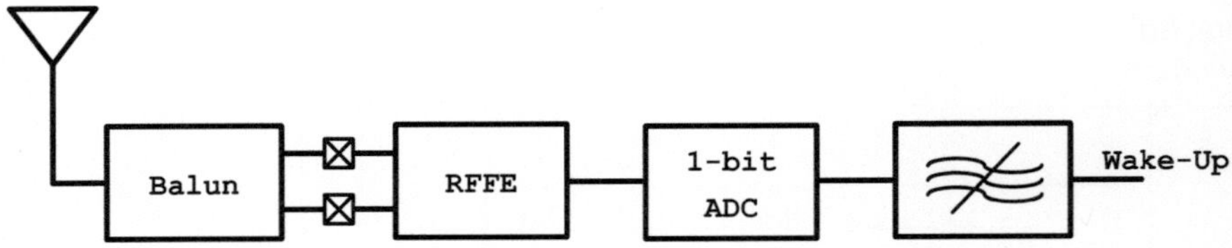

Fig. 7 WRX block diagram

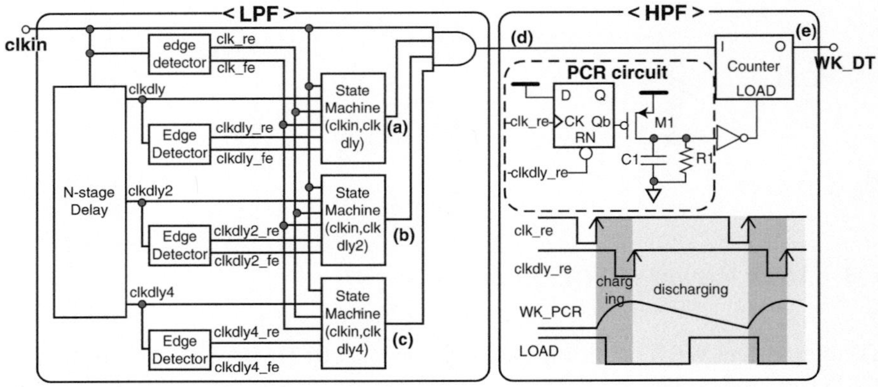

Fig. 8 Delay-based BPF [4]

major drawback in the generic WRX architecture most commonly seen. In addition, in-band interference is also addressed since the wake-up signal is not asserted until 15–17 cycles of a 14 kHz square wave is detected [4]. Power consumption of the WRX is roughly 45 μW with a 14 kHz data rate. A full breakdown on the technical specifications will be shown in Table 2 at the end of the chapter.

1.5 Summary

The generic WRX architecture and the two ISSCC WRXs show common similarities and all are aimed at reducing power as low as possible: OOK-modulated signals envelope detected and demodulated using a comparator. This has proven to be an optimal design when low power is prioritized. Other low power architectures like FSK are more complicated because they require filtering around two different frequencies and then require two envelope detectors following the filters. Phase-shift keying (PSK) is more complicated yet and resembles the architecture seem in Fig. 2.

While [3] demonstrated very low power, [4] kept the power very similar while adding additional interference rejection in-band, through the detection of a 14 kHz square wave, and out-of-band, by filtering around the frequency of interest. While the evolution of WRXs has been important, further features are needed to address the shortcomings described in the next section.

2 WRXs in Short Range Radio Applications

2.1 *Shortcomings of the Traditional WRX Architecture*

The generic WRX block diagram, as well as the WRX radios presented previously, would struggle to operate effectively as part of a larger system in a multi-user environment due to several limitations. This section will describe some of their shortcomings and the following section will explore ways to mitigate these issues.

2.2 *False Wake-Ups and Interference*

Most published WRXs use the simple OOK modulated energy detection architecture discussed previously to keep power low. Any undesired ambient signal with enough energy at the proper frequency can trigger a false wake-up of these radios, and false wake-ups result in significant amounts of wasted energy on the node. In order to prevent false wake-ups, a WRX must have enough local processing to differentiate a wake-up event from ambient interference, both in-band and out-of-band, without use of the node's main processor. An example of this is [4] which addressed out-of-band interference rejection through digital filtering and in-band interference through the detection of a 14-kHz square wave. If an ambient signal is strong enough to trip the demodulating comparator in Fig. 3, then the generic WRX will assert a wake-up. If an ambient signal is stronger than the desired signal then even a pattern, like a 14 kHz square wave, will not be detected because the demodulating comparator has saturated.

2.3 *Unique Wake-Ups*

If all the nodes in a sensor network rely on the same WRX strategy using energy detection architectures, then when the transmitter tries to wake a node, it will wake all the nodes at the same time producing significant wasted energy due to false wake-ups. Using [4] as an example; if five WRXs are deployed and all receive the same 14 kHz square wave, then all will wake up, whether that is the intention of the transmitter or not. It's possible four of those nodes woke up unintentionally and each turned on their higher power main receivers without purpose.

2.4 Power

A WRX with ~50 μW of power is considered ultra-low power. However, when integrated on a sensor node or microcontroller SoC, the WRX power still dominates over the sleep power of the node. With sleep power in sensor nodes in the single digit μW range, due to leakage in SRAM and digital logic, even state-of-the-art WRXs dominate the overall power budget when listening for packets. In order to be practical, the WRX's active power should be below that of the sensor node's sleep power.

3 New Approaches for WRXs

In order to function in a multi-user environment, a WRX needs to prevent false wake-ups by rejecting both in-band and out-of-band interferers, as well as constant jammers. The ability to wake-up a single node in the presence of many is necessary as well as further power reduction. This section will take a look at these issues and offer solutions for each one.

3.1 Power

3.1.1 Ultra-Low Power Radio Academic Survey

A survey of published ultra-low power radios was conducted to characterize the relationship between sensitivity and power. Figure 9 shows this survey spanning major conferences and journals from 2006 to 2013, comparing the radio's power versus sensitivity. It should be noted that these are all custom ultra-low power radios that use some form of energy detection architectures and therefore commercial radios like Bluetooth and Zigbee are not shown here. Also, note that this survey includes radios of different architectures, different operation frequencies, and different data rates; none of which is separated in the plot. With sensitivity, in dBm, on the x-axis and power, in μW, on the y-axis, two distinct trends can be seen. First, when looking at sensitivity greater than −60 dBm (−60 dBm and to the right on the x-axis) it can be seen that changing sensitivity is not correlated to the power of the receiver. This is mainly because at these sensitivity numbers the radio is not noise-limited and therefore does not need to spend extra power amplifying the signal. However, there is a floor around 50 μW which suggests there is a minimum power required of the radio regardless of sensitivity. Increasing sensitivity from −60 dBm (−60 dBm and to the left on the x-axis) there is a linear trend with a slope of roughly −1/2, suggesting a correlation between sensitivity and power. Slope-fitting from the plot, it can be seen empirically that a 20 dB change in sensitivity results in a 10× change in power.

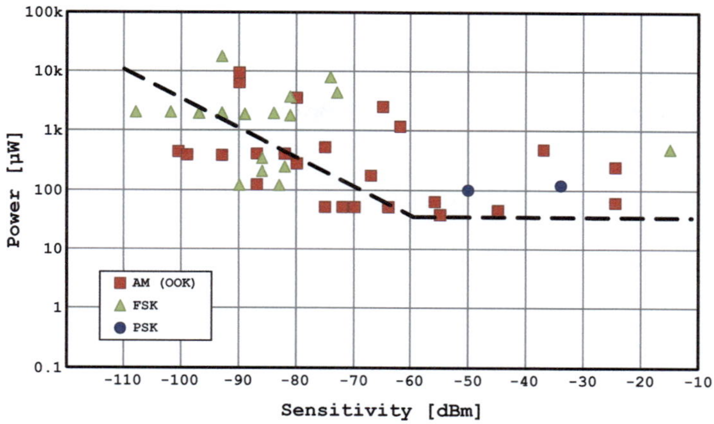

Fig. 9 Low power radio survey

For short range radios, this knowledge can be used to make key design tradeoffs. For example, most radio designs push the sensitivity low to provide as much communication range as possible, but that may often lead to power-costly overdesign, especially with the emergence of sensor networks and short communication ranges. Sensitivity needs to be evaluated from the context of sensor networks.

3.1.2 Channel Path Loss (for Body-Worn Sensors)

Short range communication devices differ from more traditional long range communication devices in two significant ways. First, as seen in Fig. 10, small movements between two nodes a long distance away do not seem significant over short time periods. For short range applications, many that include body-worn sensor nodes, the distance and position of the nodes relative to each other changes significantly in short time periods. Second, the amount of path loss between nodes at long range is much more significant than the path loss between nodes that are close together. Because of these changes, receivers that are suitable for longer range communication might not be optimal, or may be over-designed, for short range applications. To see this in more detail the wireless channel a sensor network featuring at least one body-worn node is analyzed.

The channel for wireless sensor nodes involves the environment, and electromagnetic propagation, around the human body which is complicated due to the body's effect on antenna performance [5]. One reason the channel for a body sensor network is unique compared to other channel models is the channel's dynamic characteristics due to the inherently limited motion of a person's body and the proximity of the sensors to the body itself. The distance and angle of antennas mounted to sensors worn on the body will constantly change in relation to one another as a person performs daily activities. Intuitively, the channel should oscillate

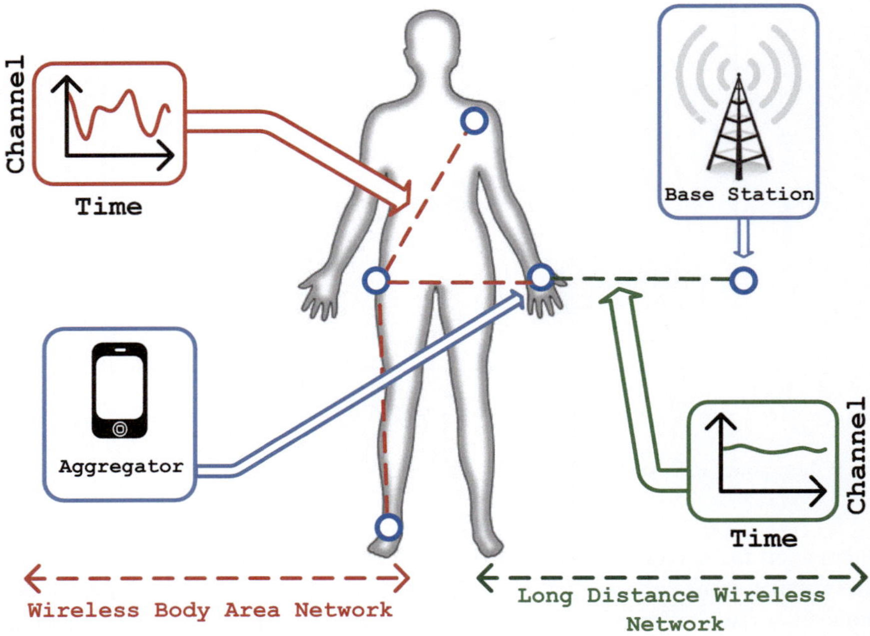

Fig. 10 Short vs. long range communication

between strong and weak conditions as the person walks or runs. Furthermore, these oscillations should be bounded by the physical limitations of the body's movement or lack of movement (i.e. the human body is never perfectly still) [6].

For most applications of wireless sensor nodes, streaming data is not necessary and therefore the radios are not required to be in constant contact with an aggregator or another node in the network. This knowledge can be used to design radios that don't need to communicate in the entire channel. To better understand the dynamics of the channel, custom hardware using discrete components was designed and records RSSI between two sensor nodes at a rate of 1.3 kS/s, which is then converted to path loss [6]. Figure 11 shows an example of such data. The lighter colored periodic waveform in the background was taken with a user wearing a sensor on their chest and their wrist while running. One can see the periodic movement that would be expected from such a repetitive exercise. Also note the peak-to-peak variation and average path loss in the figure. If it is assumed that the transmitted power and antenna gain are 0 dBm and 0 dB, respectively, then the sensitivity of the receiver is equivalent to the measured path loss. Using Fig. 11 as an example, sensitivity of the radio can be reduced to −40 dBm and still be able to communicate almost half the time.

The background waveform shown in Fig. 11 is a controlled example, where channel periodicity can is guaranteed. A more realistic channel can be seen in the darker colored foreground waveform in Fig. 11 which shows path loss measured

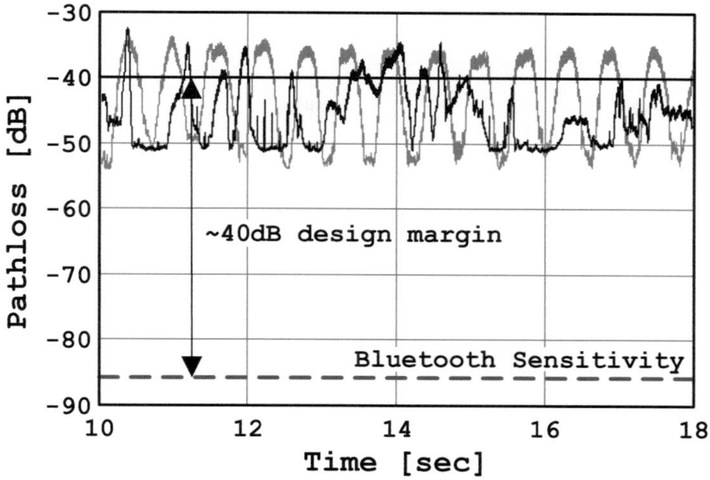

Fig. 11 Periodic channel path loss

between a sensor on someone's wrist and hip while they played tennis. Also on the plot is the average sensitivity of a Texas Instruments Bluetooth receiver [2]. As can be seen, for this specific application the Bluetooth receiver is significantly over-designed. Reducing the sensitivity of the receiver by 30 dB would allow for reduced design constraints in the receiver while still allowing for communication in the entire channel. Allowing for intermittent communication, design constraints can be relaxed even more. According to the channel model, if the sensitivity of the receiver is set to −40 dBm, same as for the running example, then there will be many instances in time where communication is possible, and when possible, the receiver can communicate while realizing the power benefits from the sensitivity reduction [6].

3.1.3 Sensitivity Reduction

A radio's sensitivity dictates the smallest signal that it can correctly detect and demodulate and it directly impacts the overall power consumption of the radio as well as the radio's range. A lower sensitivity number (more negative) means the radio can detect smaller signals and therefore communicate farther. Improving sensitivity requires amplification and good noise performance, most coming at RF or IF frequencies, which always means extra power. If the sensitivity is too low then communication becomes difficult and/or unreliable. Because of this, sensitivity traditionally is a specification that has stringent requirements and often leads to over-design to ensure communication even in adverse conditions. However, with the advent of sensor networks, it is important to re-evaluate even some of the most traditional specs. With communication ranges shrinking (like on-body

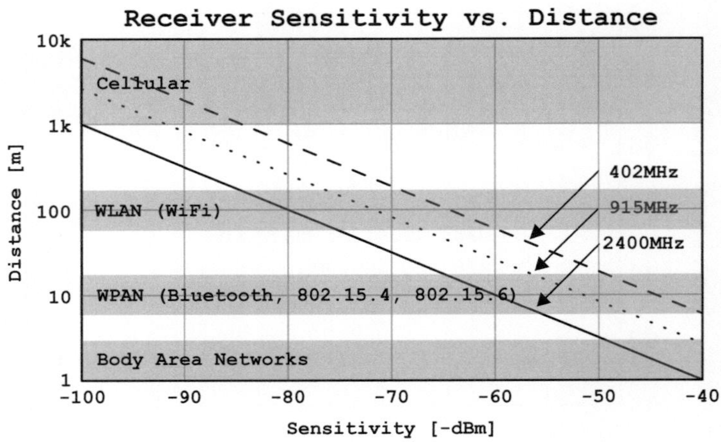

Fig. 12 Calculated communication range

communication), and without the need to communication constantly, sensitivity can become a useful knob with which we can tune power.

Figure 12 shows the communication distance vs. path loss using Friis free space path loss equation. Three communication frequencies are used: the 2.4 GHz and 915 MHz ISM bands and the 402 MHz MICS band. If one assumes the transmitter power and antenna gain is 0 dBm and 0 dBi respectively, then the calculated path loss equals communication distance. As Fig. 12 shows, for technologies like cellular and WLAN (WiFi) which require relatively long communication range, the need for sensitivity is significant, but as communication range is reduced towards applications like WPAN and sensor nodes, the burden on sensitivity is reduced.

Returning to the discussion of WRXs, sensitivity directly impacts power. By removing the dependency on sensitivity, and therefore the need for gain at RF/IF frequencies, the power can be reduced below the $\sim$50 μW limit seen in Fig. 9. A design utilizing this methodology will be presented later in the chapter.

3.2 Removing False Wake-Ups: Interference and Unique Wake-Ups

The previous section addressed the power issue with WRXs, and the next section will address the issue of false wake-ups. Codes can be used to reject both in-band and out-of-band interferers as well as enable the ability for individual WRX wake-up in the presence of multiple WRXs.

3.2.1 CDMA Codes

The WRX in [4] relied on a 14 kHz square wave to assert a wake-up, but this method can be generalized to address both interference and individual wake-ups. Instead of a square wave, the use of CDMA codes with large cross-correlations are ideal for receivers utilizing OOK modulation, like WRXs do. Large cross-correlations within specific CDMA codes (for example 15-bit Kasami codes) allow multiple codes to be used while reducing the risk of false wake-ups.

3.3 Jamming

The use of CDMA codes with OOK modulation still leaves a WRX useless if a signal with enough power jams the desired signal. In this instance, the ability to control the demodulating comparators hysteresis levels will allow the WRX to sit above the jamming signal. As long as the desired signal has more energy than the jamming signal (an added benefit of sensitivity reduction is that ambient interferers have to be much stronger) then the WRX should still be able to receive the wake-up code.

3.4 Summary

This section described ways to address all the issues presented in the previous section about using WRXs in a multi-user environment. Sensitivity reduction can be exploited to remove the reliance on gain at RF and therefore reduce power levels beyond ~50 μW. CDMA codes can be used to reject both in-band and out-of-band interferers as well as allow for unique wake-ups. Finally threshold controlling the demodulating comparator can help the WRX overcome potential jammers. The next section will detail a state-of-the-art WRX that utilizes these techniques.

4 A 116 nW 31-bit CDMA WRX [7]

A WRX that addresses the power, false wake-up, interference, and jamming issues is presented below. The WRX uses sensitivity reduction to lower its power consumption to just 116 nW. False wake-ups are mitigated by a subthreshold baseband processor capable of demodulating a selectable 31-bit OOK modulated code before toggling the wake-up signal. Interference is corrected through the use of an automatic threshold controller (ATC) which modulates the receiver's sensitivity based on the detected level of interference. The RF front-end operates over a broad frequency range, tunable by an off-chip band-select filter and matching network, and demonstrated in the 403 MHz MICS band, as well as the 915 MHz and 2.4 GHz ISM bands. The WRX radio has a raw OOK chip-rate of 12.5 kbps. The radio is 0.35 mm^2

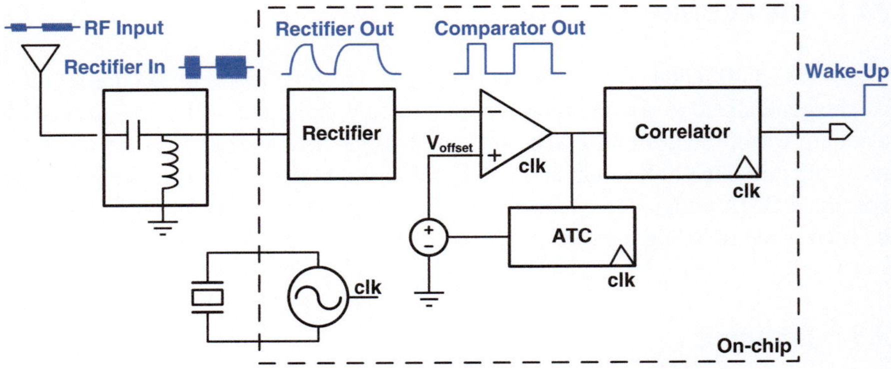

Fig. 13 Block diagram of 116 nW CDMA WRX

and operates using a 1.2 V supply for the crystal reference and RF demodulation, and a 0.5 V supply for baseband processing in the subthreshold region.

4.1 System Architecture

Figure 13 shows a block diagram of the CDMA WRX. An OOK modulated RF signal passes through a passive input matching network that filters and boosts the signal before going on-chip. A 30-stage rectifier down-converts the RF signal to baseband, which is then sensed by the dynamic comparator clocked at 4× the chip-rate. The offset voltage of the comparator is controlled by the ATC (Automatic Threshold Controller) which compensates for interferers. A bank of 4 parallel, bit-shifted 31-bit correlators continuously compare the received chip sequence with a programmable wake-up code, and toggles the wake-up signal only when a correlation result exceeds a programmable correlator threshold. The reference clock for the receiver is generated using an off-chip 50 kHz crystal with an integrated oscillator. The oscillator and comparator operate from a 1.2 V supply while all the digital logic operates in subthreshold at 0.5 V. Because of the broadband design in the front end rectifier, the WRX can operate from the 400 MHz to 2.4 GHz bands, and can be tuned using a band-select filter and matching network off-chip. Sleep power was carefully designed using thick-oxide header devices on all the circuits. It should be noted that the first three blocks make up the generic WRX architecture of Fig. 3. The other circuit blocks could plug into any other WRX that follows that same architecture.

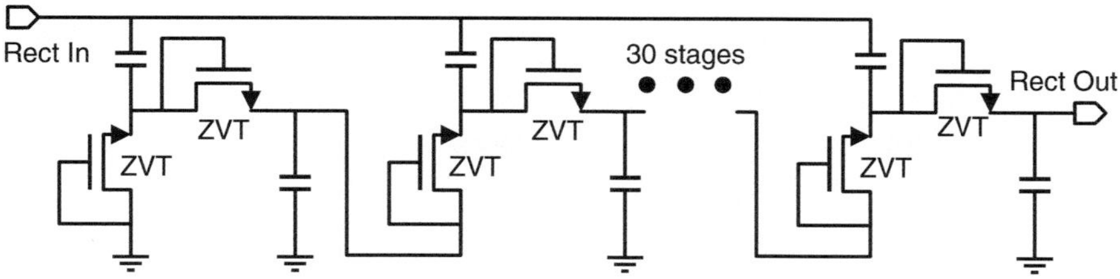

Fig. 14 RF rectifying front end

4.2 Circuit Descriptions

In this section, the major circuit blocks of the WRX are explained in detail. All circuits use a thick-oxide PMOS header to improve sleep power.

4.2.1 Off-Chip Matching Network

For the WRX, a 2 element off-chip matching network was used and provided a passive 5 dB voltage boost. The input impedance of the chip was measured on a network analyzer to be 23-j35 Ω at 400 MHz so a 12 pF series capacitor and a 15.7 nH shunt inductor were used. Devices like BAW or FBAR resonators can also be used to tune to the desired frequency of operation.

4.2.2 RF Rectifier

Because the sensitivity has been reduced, an LNA is not necessary to amplify the received signal. Instead a zero-power RF rectifier replaces the LNA, saving significant power and allowing communication in the nanowatt range. As seen in Fig. 14, the rectifier's structure is the same as the Dickson Multiplier [8], but the output voltage calculation is different because all transistors operate in subthreshold due to the small RF input [9].

This subthreshold rectifier uses zero-threshold transistors and 30 stages to achieve sufficient RF gain with a fast charging time. The input impedance of the chip is 23-j35 Ω at 400 MHz, 12-j13 Ω at 900 MHz and 88-j5.8 Ω at 2.4 GHz. The Q factor of the input impedance is low, due to a voltage limiter that prevents the rectified voltage from exceeding the breakdown voltage of the FETs, so a broadband matching network could be implemented.

4.2.3 Comparator with ATC

The clocked comparator, shown in Fig. 15, applies regenerative feedback clocked by the 50 kHz oscillator. Two separate current biases are each controlled by 4-

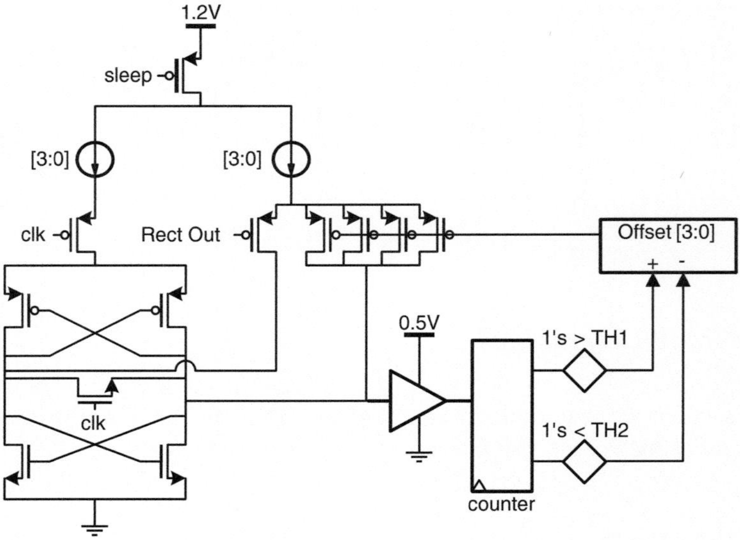

Fig. 15 Dynamic comparator and ATC

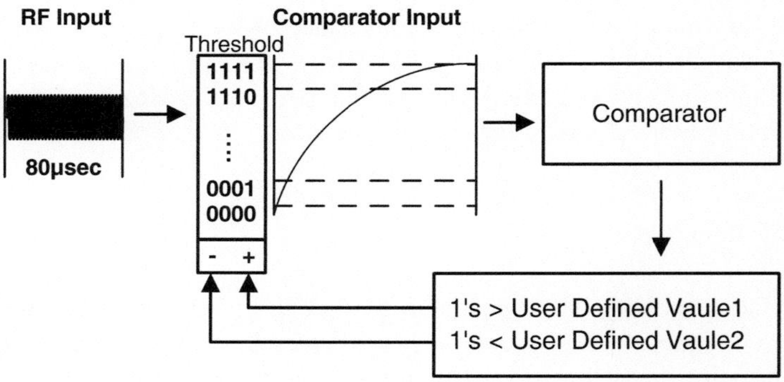

Fig. 16 ATC signaling

bit binary-weighted current DACs. In addition, the comparator threshold can be programmed to a 4-bit binary-weighted value to tune the sensitivity of the receiver. The threshold of the comparator is controlled in a feedback loop by the ATC which dynamically changes the comparator's offset voltage to overcome interference signals.

A diagram showing operation of the ATC can be seen in Fig. 16. As RF input signal comes in, the RF rectifier outputs the signal for the comparator. The comparator compares this signal with its threshold. The ATC monitors the samples coming from the comparator output for one 31-bit code period. If the number of 1's is greater than a user defined Value1 (indicating the comparator threshold is continuously exceeded by an interferer), then the ATC will increase the comparator

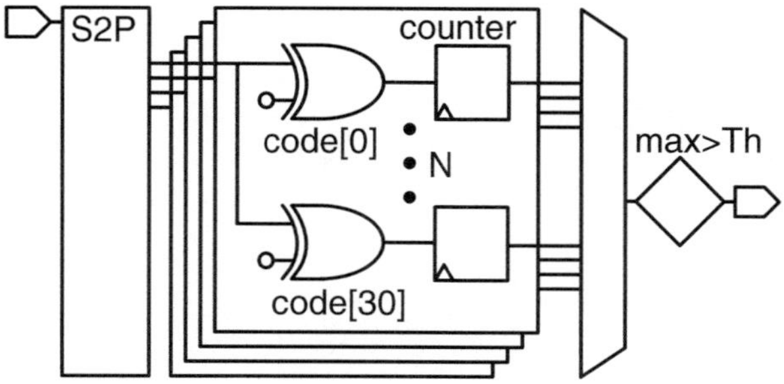

Fig. 17 Block diagram of correlators

threshold to bring the sensitivity of the receiver above that of the interfering signal. When the number of 0's at the output of the comparator reaches a separate, user defined Value2 (indicating the interferer is gone and the comparator threshold is never exceeded), the ATC then reduces the threshold to increase the sensitivity of the receiver. Hysteresis is added between these values to eliminate limit-cycles. With this mechanism, the comparator can reject interference signals, and even if the interference signal is modulated by BPSK, or OOK, the comparator threshold is set to above the maximum level of interference so the comparator will produce the correct output.

4.2.4 Correlators

A bank of four correlators continuously correlates the 4×-oversampled comparator bit-stream with a programmable, 31-bit CDMA code. This synchronizes to the transmitted code and only issues a wake-up output when the desired code is received. The digital baseband processor was synthesized in subthreshold in order to save power. Eight different Gold codes can be selected using control bits in a scan chain. This will allow for a single transmitter to uniquely wake up 8 possible WRXs. Gold codes are a set of binary sequences whose cross-correlation among the set is bounded into three values [10]. Gold codes are commonly used when implementing CDMA and they are easily implemented with 2 LFSRs (Linear Feedback Shift Register) and an XOR gate. In this work, 31-bit Gold codes with 3 configuration bits are implemented.

The correlator compares the last two samples in each bit slice. Therefore, each 31-bit code results in a total of 62 comparisons. A programmable correlator threshold allows the user to define a value between 1 and 61 that must be exceeded in order to declare a code received indicating a valid wake-up event. A lower threshold value would mean fewer bits have to match the code, tolerating a higher BER and resulting in better sensitivity, but leads to more false wake-ups. A higher threshold would prevent false wake-ups, but also reduce the sensitivity of the receiver.

Figure 18 shows a basic block diagram of the correlator. To synchronize the receiver to the transmitted code, four correlators operate at the same time and each correlator receives shifted samples of each bit slice since the receiver is 4×-

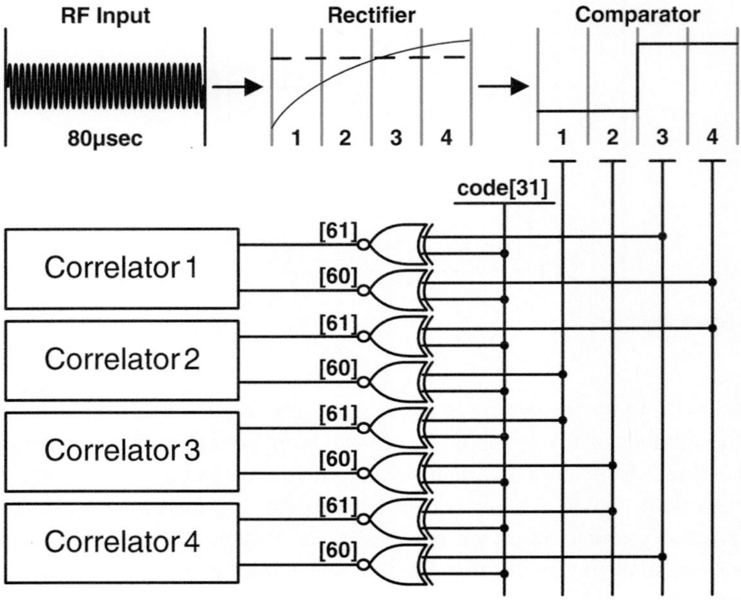

Fig. 18 Correlator signaling

oversampled. In each correlator, all possible shifts of the 31 bit Gold code are simultaneously correlated with the incoming bit stream, so that after a single 31 bit sequence, the receiver is guaranteed to synchronize to the wake-up signal. Each parallel correlator will have a different number of correct comparisons based on the code shifts and phase difference between the WRX and the transmitter. If any of the four correlators results are greater than the correlator threshold, the wake-up signal will be asserted. An example showing how the correlator handles a single chip slice is shown in Fig. 18. This example shows what happens if the WRX receives a '1' as the last chip in the sequence. The OOK signal is gradually rectified during the chip window.

4.2.5 Crystal Oscillator

A 50 kHz crystal oscillator serves as the reference clock of the radio [11]. As seen in Fig. 19, an off-chip crystal is used, and the oscillator's primary amplifier is an inverter with resistive feedback. Figure 19 also shows the intrinsic values of the 50 kHz crystal. Using these values the critical g_m can be calculated using Eq. (1)

$$\text{critical } g_m = \frac{W}{QC} \cdot \frac{(C_1C_2 + C_2C_0 + C_0C_1)^2}{C_1C_2} \tag{1}$$

which is the transconductance of the amplifier that must be produced to achieve sustained oscillations. C and C0 represent the motional and shunt capacitance of the crystal, ω and Q are the resonant frequency and Q factor of the crystal, and C1 and C2 are the load capacitance in the circuit. If the primary amplifier is biased in the near threshold region where the g_m/i_D ratio is around 10, then the current consumption to reach this critical gm value is around 20 nA.

Initially, the transconductance of the primary amplifier is much greater than the critical g_m of the crystal, which is needed to quickly increase the oscillation amplitude. However, as the oscillation amplitude increases, the DC level of the oscillation also drops and this common-mode signal is used in feedback to starve the primary amplifier until it settles with sustained oscillations. Measured results show the total power consumption is 30 nW when sustaining oscillations using a 1.2 V supply. The oscillation is then buffered to provide the reference clock for the WRX. Measured results show the oscillator has an RMS jitter of 6 ns.

4.2.6 Sleep Power

Sleep power in the WRX was carefully considered during the design process to support a duty cycled wakeup strategy. To improve sleep mode energy, thick oxide power gating devices were used throughout the design. Wake-up time is dominated by the slow startup time of the crystal oscillator, which takes about 1.1 s to oscillate. Putting the WRX into a full sleep mode, where every circuit is power-gated, results in a sleep power of 18 pW. Clock gating the crystal oscillator, and putting all the circuits except the crystal oscillator into sleep mode, results in a sleep power of 30 nW.

4.3 Measurement Results

The WRX was fabricated in IBM's 130 nm CMOS process and has an active area of 0.35 mm^2 without pads. It uses two separate voltage supplies; a 1.2 V supply for the

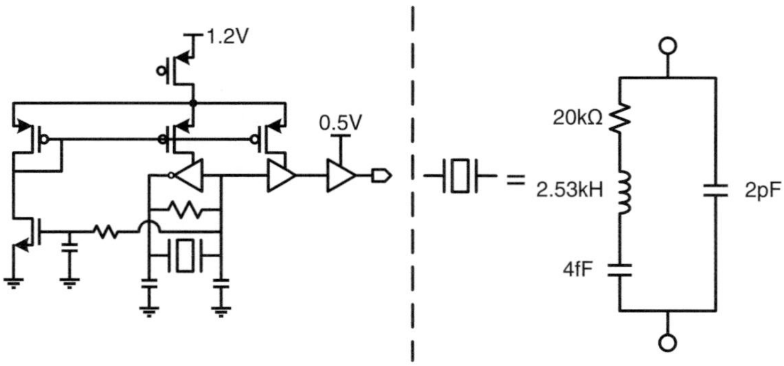

Fig. 19 50 kHz crystal oscillator clock reference

crystal oscillator and demodulating comparator and a 0.5 V supply for all baseband processing. A die micrograph can be seen in Fig. 20.

Transient operation of the WRX receiving a 31-bit code is shown in Fig. 21. Because of the bit-shifted parallel correlators, the WRX is able to automatically synchronize to the incoming bit stream upon receiving the first wake-up code. The top two traces show the RF input signal and the RF rectifier converting the signal to baseband. The third trace shows the output of the comparator being clocked at 4× the data-rate by the oscillator and the final trace is the wake-up signal being toggled by the correlator. The WRX is capable of CDMA by selecting different codes used by the correlator block.

If an interfering signal is strong enough to exceed the comparator threshold, then the ATC increases the comparator's threshold until it is above the interfering signal. A transient of this operation can be seen in Fig. 22. The top signal is the received RF signal, which is jammed by a 2.4 GHz tone at 8 ms. With the interferer present, the receiver cannot initially demodulate the code. After 15 ms, the ATC has raised the threshold of the comparator above that of the interferer, and the WRX regains synchronization.

The WRX has an active power of 116 nW with a sleep power of 18 pW. It has a raw OOK chip-rate of 12.5 kbps and a symbol rate of 50 kbps and sensitivities of −45.5 dBm, −43.4 dBm, and −43.2 dBm at 403 MHz, 915 MHz, and 2.4 GHz, respectively. The top of Fig. 23 shows the chip error rate (BER) curves for the 403 MHz, 915 MHz, and 2.4 GHz bands. The bottom of Fig. 23 shows the BER as the correlator threshold is varied. The measurements were taken using a −40 dBm

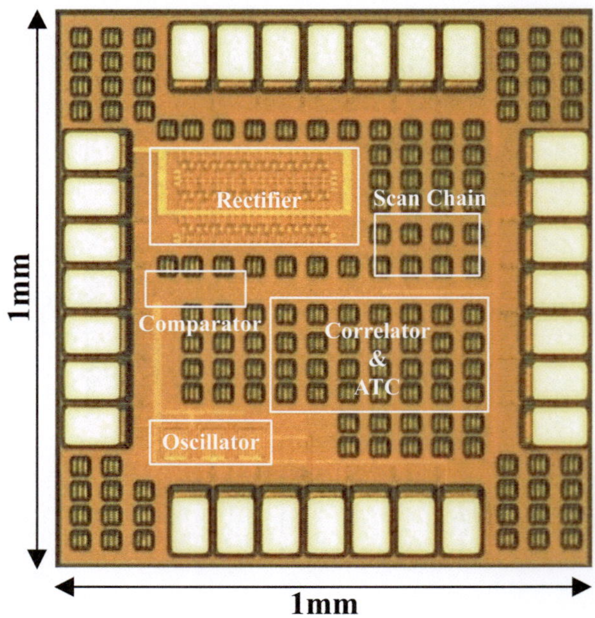

Fig. 20 Die photo of CDMA WRX

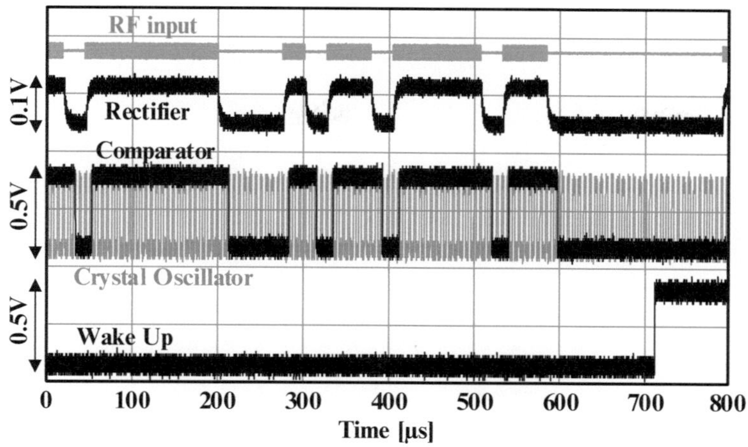

Fig. 21 Transient operation

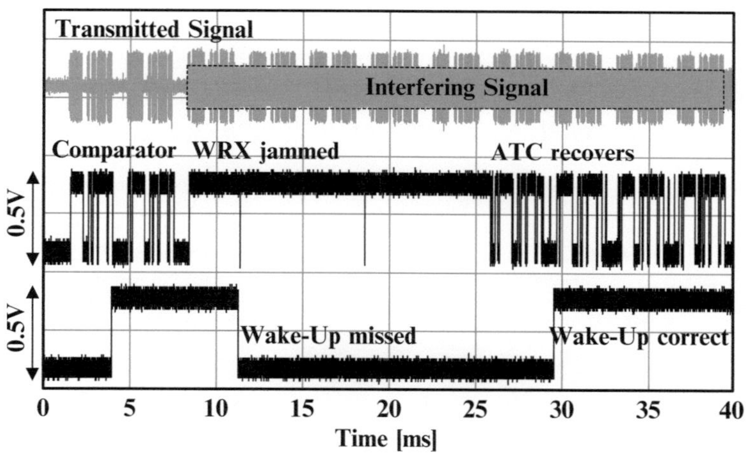

Fig. 22 Interference

received signal in the 2.4 GHz band. The figure also shows the impact this threshold has on false wake-ups. From these two data sets, the correlator threshold can be set to maximize sensitivity while minimizing the possibility of a false wake-up.

To demonstrate the ability of the WRX to be selective in its wakeup code, therefore allowing different codes to wake-up different WRXs, Fig. 24 shows an experimental setup where the input signal was connected to two different WRXs, each with a different wakeup code programmed. The top of Fig. 24 shows that code 1 and code 2 were transmitted back to back and the resulting output from the WRXs correctly toggles when the proper code is sent, but ignores the incorrect code.

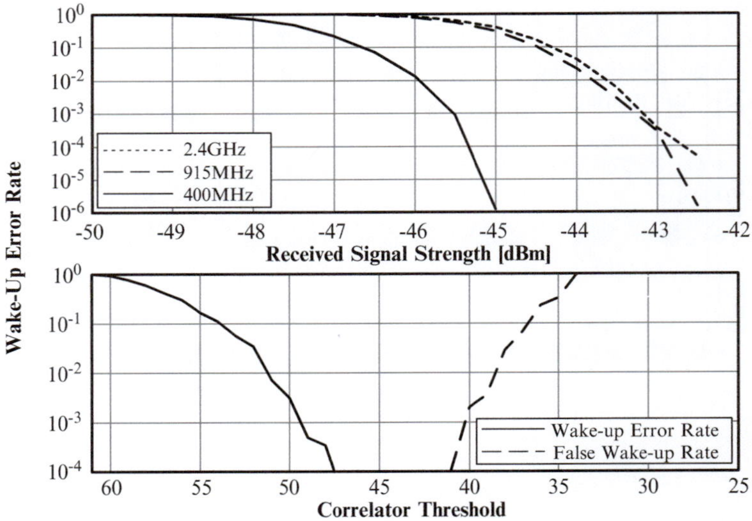

Fig. 23 BER

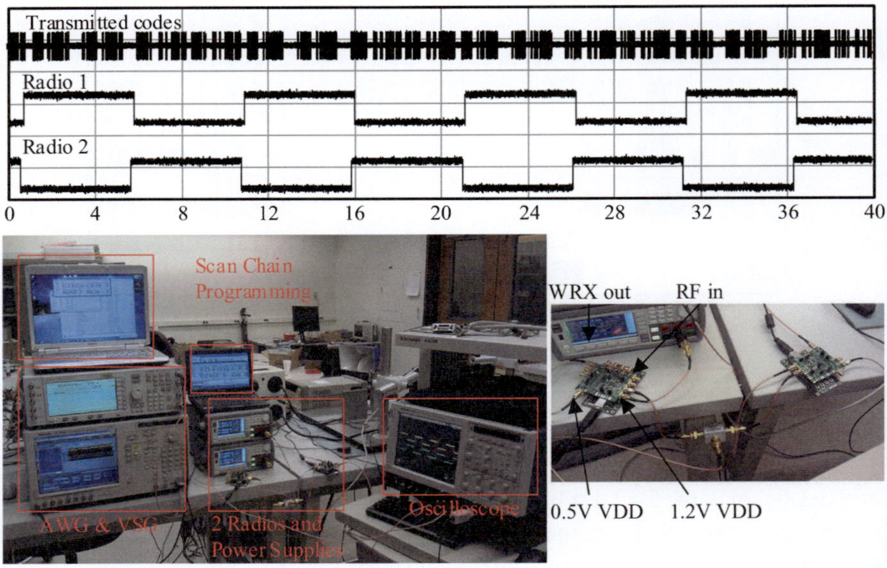

Fig. 24 WRXs receiving different codes

Table 1 shows a power breakdown on the WRX on the left. Digital logic consisting of the four parallel correlators and a clock divider all synthesized in subthreshold consumes the bulk of the 116 nW power. Second is the crystal oscillator consuming 38.4 nW. The demodulating comparator consumes 8.4 nW and the RF rectifier front end consumes no power, for a total of 116.3 nW.

Table 1 CDMA WRX specs

Power breakdown (nW)		Receiver specifications	
RF rectifier	0	Energy/bit	9.28 pJ
Comparator	8.4	Energy/wake-up	287.7 pJ
Digital logic	69.5	Max signal level	−15 dBm
Crystal oscillator	38.4	Max interferer level	−20 dBm
Total	116.3	Code length	31
Sleep (pW)	18	# Pre-defined codes	8

Table 2 Summary of WRXs

	[3]	[4]	[8]		
Power (μW)	52	45	0.116		
Sleep (pW)	N/A	N/A	20		
Frequency (MHz)	2,000	5,800	403	915	2,400
Data rate (kbps)	100	14	12.5/31		
Sensitivity (dBm)	−72	−45	−45	−43	−41
SIR (dB)	N/A	N/A	3.3	1.7	1.7

The right side of Table 1 shows the receiver's specs. It requires 9.28 pJ/bit and with a 31-bit wake-up code, the total energy to wake-up the node would be 287.7 pJ. The maximum signal level tolerable is −15 dBm and the maximum interferer level is −20 dBm.

5 Summary

A WRX can act as a synchronizing receiver within an asynchronous protocol and helps to preserve the main radio's power consumption by allowing it to remain in a low power sleep state for as long as possible. To do this, the WRX stays active and continuously monitors the channel for activity and wakes the main radio when it detects another radio trying to communicate. Since it is on all the time, the WRX must be very low power; lower than the sleep power of the sensor node.

Referring back to the introduction section of this chapter the Dialog Semiconductor BLE radio required 36.75 μJ of energy to enter the scan state for the minimum 2.5 ms time window. To achieve an average power from the BLE radio that equals the active power of the WRX, the BLE radio would only be able to scan once every 706 ms for [3] and 816 ms for [4]. By utilizing sensitivity reduction, the BLE radio can only scan once every 5.3 min to match the active power in [7]! Table 2 shows the performance metrics of the three WRXs mentioned in this chapter.

The CDMA WRX presented addressed false wake-ups caused by interference and multiple users as well as jamming and power consumption that hinder the performance of past WRXs. False wake-ups are addressed through the use of 8 selectable 31-bit CDMA codes. Jamming is resolved through the ATC which controls the threshold level of the demodulating comparator and adjusts it as necessary. With reduced sensitivity specifications, the use of a zero-power RF energy harvester was used as the RF front end of the receiver and subthreshold processing was implemented to the keep entire radio in the nanowatt power region. With power that is less than a typical sensor node's sleep power, the WRX is not the energy dominant circuit when the node is asleep which makes it a very suitable synchronization technique for sensor nodes. The technique of using CDMA codes and ATC is not confined to the sensitivity reduced WRX, but fits any WRX that matches the generic architecture, which allows a designer to optimize the WRX for power, range, and data rate while maintaining ability to reject interference and handle multiple users.

References

1. Bluetooth SIG, Specification of the Bluetooth System. Bluetooth 4.0 standard, vol. 0, 30 June 2010
2. Dialog Semiconductor, DA14580 Low Power Bluetooth Smart SoC, DA14580 datasheet, February 2014
3. N.M. Pletcher et al., A 52 μW wake-up receiver with -72 dBm sensitivity using an uncertain-IF architecture. JSSC, January 2009
4. J. Choi et al., An interference-aware 5.8 GHz wake-up radio for ETCS. ISSCC, February 2012
5. P.S. Hall et al., Antennas and propagation for on-body communication systems. IEEE Antenn. Propag. Mag. (June 2007)
6. N.E. Roberts et al., Exploiting channel periodicity in body sensor networks. IEEE Emerg. Selected Top. Circuits Syst. **2**(1), 4–13 (2012)
7. S. Oh et al., A 116 nW multi-band wake-up receiver with 31-bit correlator and interference rejection. CICC, September 2013
8. J.F. Dickson, On-chip high-voltage generation in MNOS integrated circuits using an improved voltage multiplier technique. IEEE J. Solid-State Circuits **11**(3), 374–378 (1976)
9. S. Oh, D.D. Wentzloff, A -32 dBm sensitivity RF power harvester in 130 nm CMOS. RFIC 2012, June 2012
10. J.K. Holmes, Spread Spectrum Systems for GNSS and Wireless Communications. Artech House, ISBN 978-1-59693-093-4, 2007
11. H. Oporta, An ultra-low power frequency reference for timekeeping applications, Master's Thesis, Oregon State University, December, 2008

Commercially Viable Ultra-Low Power Wireless

Gangadhar Burra, Srinath Hosur, Subhashish Mukherjee, Ashish Lachhwani, and Sankar Debnath

Abstract This chapter looks at various practical aspects of architecting and designing low power wireless radios and systems-on-chip for applications such as consumer wearables, industrial automation etc. The chapter starts with a discussion on the need for industry accepted protocols for low power wireless and aspects in these protocols that lend themselves to low power implementations. With these protocols in place, we then look at practical design techniques of the RF/analog components, followed by a look at the Physical layer and the MAC and conclude the section by looking at the overall SoC design techniques for proper energy management. The chapter concludes by looking at the upcoming IEEE 802.11ah standard and discuss how this is adapted in an advantageous manner for low power wireless applications.

Keywords IoT • Internet of things • Low power wireless • IEEE 802.15.6 • IEEE 802.11ah • BLE (Bluetooth Low Energy) • BAN (Body Area Networks) • Polar modulation • SoC • Energy management

G. Burra (✉)
Qualcomm Inc., San Jose, CA
e-mail: gburra@qca.qualcomm.com

S. Hosur
Texas Instruments Inc., Dallas, TX
e-mail: hosur@ti.com

S. Mukherjee • S. Debnath
Texas Instruments India Ltd., Bangalore, India
e-mail: s-mukherjee@ti.com; s-debnath@ti.com

A. Lachhwani
Qualcomm India Pvt. Ltd., Bangalore, India
e-mail: ashishi@qti.qualcomm.com

P.P. Mercier, A.P. Chandrakasan (eds.), *Ultra-Low-Power Short-Range Radios*, Integrated Circuits and Systems, DOI 10.1007/978-3-319-14714-7_6

1 Introduction

Internet of Things (IoT) or Internet of Everything (IoE) is a new growth area for wireless connectivity and sensor sub-systems. IoT involves bringing wireless connectivity to consumer and industrial applications. IoE applications range across many broad categories. Correspondingly, the wireless eco-system associated with these applications also needs to have a standardized and inter-operable set of protocols and interfaces. Broadly, the categorization can be across the following segments.

- Industrial automation
- Consumer/Wearable
- Home automation
- Gaming and entertainment

The commercial motivation for these segments is reasonably clear. A 2013 study by Gartner indicates that the installed base of "things", excluding PCs, tablets and smartphones will grow to 26 billion units by 2020 [1], with a total economic value-add from IoT across industries reaching $1.9 trillion as a combination of both semiconductor devices and monetizable applications and services.

Shown below in Figs. 1 and 2, courtesy of IHS [2], are two sets of data points that show IoT market segmentation in Industrial automation, while Fig. 3 shows IoT market projection in consumer applications.

Table ES.1
World Market for Industrial Internet Connected Devices - Installed Base
Connected Devices (M)

	2011	2012	2013	2016	2019	2022	2025	CAGR 11-25
Building Automation	225	263	328	767	1,220	1,825	2,615	19.1%
% of Total	15.0%	11.9%	8.2%	5.6%	4.4%	3.8%	3.5%	
Commercial Transportation	140	184	233	636	1,056	1,543	2,113	21.4%
% of Total	9.3%	8.3%	5.9%	4.6%	3.8%	3.2%	2.8%	
EFT-POS, Smart Cards	36	79	132	561	1,915	4,770	9,760	49.3%
% of Total	2.4%	3.6%	3.3%	4.1%	7.0%	10.0%	13.1%	
Industrial Automation	708	1,156	2,581	10,082	20,382	35,222	53,947	36.3%
% of Total	47.2%	52.1%	64.8%	73.4%	74.0%	73.6%	72.1%	
Lighting	53	71	101	318	673	1,065	1,509	27.1%
% of Total	3.5%	3.2%	2.5%	2.3%	2.4%	2.2%	2.0%	
Power & Energy	10	13	17	30	42	56	70	14.9%
% of Total	0.7%	0.6%	0.4%	0.2%	0.2%	0.1%	0.1%	
Security	86	111	138	343	554	696	850	17.8%
% of Total	5.7%	5.0%	3.5%	2.5%	2.0%	1.5%	1.1%	
Test & Measurement	1	1	2	3	4	5	8	14.8%
% of Total	0.1%	0.1%	0.0%	0.0%	0.0%	0.0%	0.0%	
Other Industrial	240	337	453	1,002	1,709	2,701	3,911	22.0%
% of Total	16.0%	15.2%	11.4%	7.3%	6.2%	5.6%	5.2%	
Total Internet Connected Devices	**1,499**	**2,216**	**3,985**	**13,741**	**27,555**	**47,881**	**74,782**	**32.2%**
Growth Rate		47.8%	79.8%	244.9%	100.5%	73.8%	56.2%	

Source: IHS **Apr-14**

Fig. 1 Market projection for Industrial automation IoT

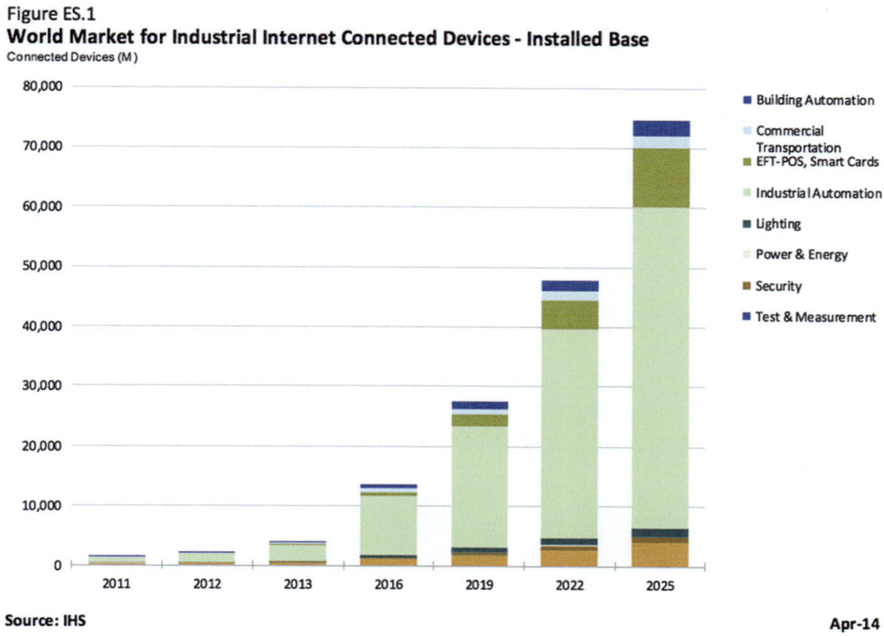

Fig. 2 Another view of IoT for Industrial automation

	CY13	CY14	CY15	CY16	CY17E	CY18E	CAGR
IMS - Healthcare and Medical	*2.7*	*4.7*	*8.6*	*15.3*	*27.1*	*48.2*	*87%*
Fitness and Wellness	*22.7*	*30.7*	*41.3*	*55.6*	*76.7*	*111.1*	
Activity Monitors	*9.1*	*14.6*	*22.0*	*31.4*	*44.7*	*63.8*	*57%*
Fitness & Heart Rate Monitors	*13.5*	*15.8*	*18.1*	*20.5*	*23.2*	*26.2*	*15%*
Smart Glasses	*0.0*	*0.0*	*0.0*	*0.0*	*0.0*	*0.0*	*68%*
Smart Clothing	*0.0*	*0.0*	*0.1*	*0.2*	*0.3*	*0.5*	*97%*
Sleep Sensors	*0.1*	*0.3*	*1.0*	*3.5*	*8.5*	*20.6*	*143%*
Emotional Measurement	*0.0*	*0.0*	*0.0*	*0.0*	*0.0*	*0.0*	*115%*
Infotainment	*0.3*	*2.5*	*10.4*	*21.1*	*43.0*	*91.3*	
Smart Watches	*0.2*	*2.3*	*9.9*	*19.0*	*36.6*	*70.5*	*409%*
Smart Glasses	*0.0*	*0.1*	*0.3*	*0.7*	*1.6*	*4.1*	*146%*
Head-up Displays	*0.0*	*0.0*	*0.2*	*1.3*	*4.6*	*16.5*	*263%*
Wearable Imaging Devices	*0.0*	*0.1*	*0.1*	*0.1*	*0.2*	*0.2*	*29%*

Fig. 3 Market projection for IoT in consumer applications

From a technology standpoint, irrespective of the applications, an IoT system can be generalized as shown in Fig. 4. It consists of the following major components.

- Microcontroller Unit (MCU) sub-system—in a typical IoT SoC, the MCU would be the primary processing element. Typically, the design care-about for the MCU sub-system address a performance/power trade-off in both active mode—where the selection of the processing element is done based on the complexity of the processing application and in "sleep" or "retention" mode—where the design of

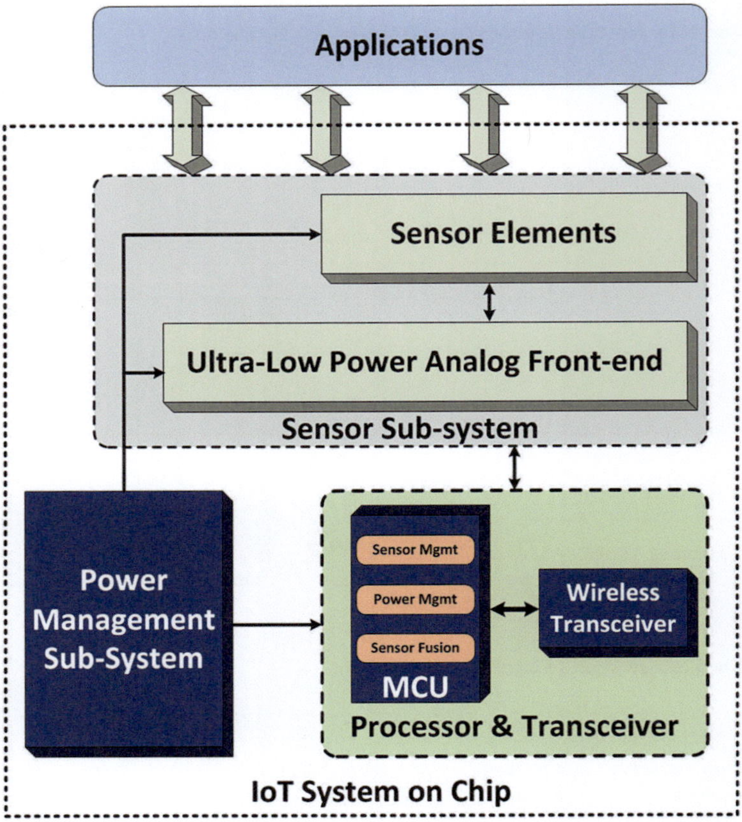

Fig. 4 Conceptualized and generalized view of an IoT system

the processor and memory sub-systems are focused the lowest possible current consumption so as to elongate the battery life to the maximum. The design of an MCU sub-system is fairly involved, but is not the focus of this chapter.

- Low power analog front-ends (AFE's)—the AFE's are the analog interface elements to the various sensors, depending on the application. Again, the AFE is not the focus of this chapter, but typical considerations for a commercial design of the AFE involve power consumption, mode flexibility (for e.g.: the ability to have a common design that can accept a current/voltage/resistive etc. sensor interfaces) as well as the overall dynamic range. Typically, sensor AFEs incorporate a 12 or a 16 bit SAR ADC as the quantizing element before data is moved to the processing sub-system.
- Power management sub-system—battery management of a sensor IoT SoC is the key to optimal life cycles for a given application. Power management techniques can involve a wide gamut of circuits from energy harvesting blocks to high efficiency buck and boost converters and extremely low quiescent current low

Table 1 Comparison of various standards based wireless technologies

Parameter	BLE	Bluetooth	IEEE 802.15.4	Wi-Fi (both 11n and 11ah)
Network	PAN	PAN	LAN	LAN
Connection	Star	Star	Mesh/Star	Star
Range	$<$10 m	$<$10 m	$\sim$500 m	$\sim$50–500 m
Bandwidth	1 MHz	1 MHz	2 MHz	4–20 MHz
Power	<25 mW	<60 mW	<50 mW	$<$150 mW (RX)
Market/application	Wearables	Legacy fitness/PAN applications	Home automation	Network gateway/cloud connected sensors

drop out regulators. Please refer to chapter "Energy Harvesting Opportunities for Low-Power Radios" for more information on this topic.

- Wireless sub-system—designing standards based and ultra low power wireless transceivers is key to making IoT systems commercially viable. While proprietary wireless radios are indeed possible, creating a broader eco-system necessitates a standards based approach. This chapter chooses two such standards as an example (Bluetooth low energy and IEEE 802.15.6) [31]. It also uses the example standards to indicate techniques needed for a robust radio design—in the physical layer, MAC, radio and digital architecture design. This chapter also briefly identifies some differences between a short range ultra-low power standard such as BLE and another upcoming ultra-low power, but a wider range standard viz. IEEE 802.11ah.

The goal of this chapter is to provide a view on the design considerations for standards based wireless IoT systems. The challenges associated with the design of these systems are different compared to design criteria for wireless sub-systems in either laptops or smartphones. Sleep and active power, low latency wake-up, low memory foot-print etc. are some of the parameters that are key to an optimal design.

The next few sections address each of the transceiver sub-systems. Before we delve into these, it is useful to look at a brief table of the various standards based wireless technologies that can be used in a sensor IoT. Table 1 shows this comparison. This table shows the need and opportunity to further drive low power design techniques in each of the widely accepted wireless standards.

2 Physical Layer and MAC: Requirements & Architecture

Before we dwell into depth on the architectural requirements, it is useful to see where the current solutions readily available in the marketplace stand. Figure 5 shows the breakdown in power for a typical commercial wireless sensor node. The figure shows that about 80 % of the power consumption is in the radio and baseband

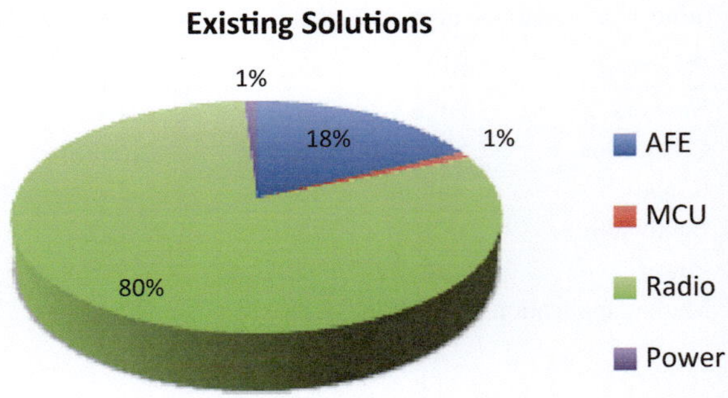

Fig. 5 Power breakdown in wireless sensor nodes

functions. It is therefore imperative that we first try and optimize the power in the modem sub-system.

There are two definitions of power for the modem; the active/ON/peak power and the average power. The active power of a modem is the power required to transmit or receive a burst of data. In a Time Division Duplex (TDD) system, only the transmit section or the receive section is turned on during the packet transmission and reception. Thus, TDD systems are by design more suited for ultra-low power radios. Other advantages of TDD over Frequency Duplex Division (FDD) systems include not needing a diplexer and not needing to maintain a very low cross talk that enables the reception of <-85 dBm signals while transmitting >0 dBm signals. Both of these advantages result in a lower design complexity and a lower power than comparable FDD solutions. For these reasons, standards such as 802.15.4/Zigbee, 802.15.6 MBAN, 802.11ah and BLE tend to favor TDD protocols.

The peak/ON current of the radio is primarily dependent on the radio architecture and the PHY layer specifications—such as center frequency, channelization, bandwidth, the adjacent and alternate adjacent channel requirements, sensitivity level for the receiver, the transmit power and modulation schemes used. Thus, the peak power can be minimized by careful circuit design techniques and by the choice of various PHY/MAC parameters in the standards, discussed elsewhere in this book.

Importantly, the peak/ON current that a radio is allowed to consume can be dictated by the choice of battery technology used for the application. Table 2 shows the capacity and corresponding peak current capability of several representative battery types used in wireless applications and the peak current that can be drawn from typical batteries. It can be seen from this table that peak current needs to be low to enable new applications that may use paper battery operation (e.g. disposable medical BAN) as the capacity of these batteries as well as the discharge currents are low.

Typically, in systems optimizing for low average power, the radio is duty cycled. The average current draw (power) to a first degree of approximation can simply be written as

$$\left(i_{ON}T_{ON} + i_{Sleep}T_{Sleep}\right) \Big/ \left(T_{ON} + T_{Sleep}\right).$$

Table 2 Battery technologies and current capacity

Shape	Chem.	Size	Typical capacity (mAh)	Nom. voltage	Min. voltage	Maximum current
AA	ZnC	50 × 14 mm	400–900	1.5 V	0.9 V	500 mA (continuous)
CR2032	Lithium	20 × 3.2 mm	210	3 V	2 V	15 mA (pulse)
L(S)R1154	$ZnMnO_2$ Alk.	11 × 5.4 mm	150–200	1.5 V	0.9 V	
AAA	$ZnMnO_2$ Alk.	44.5 × 10.5 mm	1,200	1.5 V	0.9 V	250 mA (continuous)
Paper		39 × 39 mm	18	1.5 V	0.9 V	15 mA (1 ms)

The average power therefore depends on the peak power, the sleep power as well as the radio duty cycle. The radio power optimization should therefore look at both active power optimization and low sleep current optimization.

The radio duty cycle is in turn dependent on the ratio of the data rate that needs to be communicated to the peak data rate at which the radio is capable of transmitting in addition to current channel conditions.

The sleep current of a radio is dominated by the sleep clock, which periodically wakes up the radio to synchronize to the network. Typically, the radio wakes up a certain time before its transmit opportunity (TXOP) or the time reception of network information is due. More information regarding sleep clocks can be found in chapter "Synchronization Clocks for Ultra-Low Power Wireless Networks".

Radios can also be kicked out of the network because of inactivity. Usually when this happens, the radio has to re-acquire and re-authenticate, which is usually a long and energy-expensive process. Enhancements are built into ultra low power MAC protocols such as BLE, 15.6 BAN, 802.11ah etc. to enable long sleep times for ultra-low-power operation in deeply duty-cycled systems.

There have been a number of proprietary radio implementations [3–5] that focus on minimizing either the average power or the peak power of the radio. Proprietary radios are usually developed for narrow use cases and for those use cases achieve very low power. These radios in general are difficult to extend to other use cases. Also, since many of the proprietary radios are developed by a single company, interoperability, lack of secondary sources for manufacturers, lack of a developed eco-system, cost of maintaining and upgrading the specification cause significant roadblocks for wide adoption and tend to restrict the commercial use. Standards based radios on the other hand are designed for a set of use cases with the objective of alleviating the issues stated above.

In general, there is a fear that standards based radios, because of their flexibility, have a much higher power penalty compared to proprietary systems. This, is not true when the standard is designed with power optimization kept in view and, as will be shown in this chapter, when the architecture is designed with power optimization as a major criterion.

In this section we derive the principles of implementing an ultra-low power, multi-standard modem, by performing a case study to implement two standards—viz. the IEEE 802.15.6 medical BAN standard and the BLE standard.

Let us first examine the high level requirements for the 802.15.6 BAN and BLE standard.

From Table 3, it can be seen that both the 15.6 BAN and the BLE have similar bandwidth requirements and are defined for the 2.4 GHz bands. So for the 2.4 GHz band design, it makes sense to re-use the same architecture for the radio portion of the modem, including the ADC for the receiver.

The 15.6 BAN and the 802.11ah standards are also defined at the 816/900 MHz ISM bands so in the following, we will also consider requirements on the VCO/PLL to see if the same design can be used for both the 2,400 MHz and the 816 MHz designs.

Table 3 Physical layer parameters for BLE and 802.15.6 BAN standards

	BAN	BLE
Channelization	2402 + n MHz, 0 < n < 80	2402 + 2n MHz, 0 < n < 40
Bandwidth	1 MHz	1 MHz
Modulation	Pi/2 DPSK, pi/4 QPSK	GFSK
Filtering	RRC, beta = 0.2	Gaussian, BT = 0.5
Transmit power	−4 (3 m) to 0 dBm	0 dBm (10 m)
TX EVM	17 dB @ highest data rate	N/A
PPM tolerance	+/− 20 PPM	+/− 50 PPM
Turnaround time	50 μs	150 μs
ACI	9 dB @ 1 MHz rate	17 dB @ 2 MHz
Alternate ACI	N/A	27 dB
Blocker	48 dB	48 dB
Sensitivity	−94 dBm	−94 dBm
Preamble length	64 symbols	8 symbols

The following subsections discuss the requirements for the overall radio architecture, followed by specific requirements and corresponding trade-offs in the main radio blocks: the transmitter, receiver and the MAC.

2.1 *Modem System Architecture*

Almost all the present generation low-power RF transceiver solutions in the market are being offered in combination with a low-power microcontroller subsystem (MCUSS) including application peripherals. The radio, PHY/modem and the low level Link-Layer/MAC is typically implemented as a self-contained slave subsystem connected to a general purpose MCUSS over a bus interface. Typically the software protocol stack along with the applications code runs on the main MCUSS. Figure 6 shows the architecture of a wireless sensor node with MCUSS for the BAN standard.

2.2 *Transmitter Architecture Requirements*

On the transmit side, the data to be transferred is first packetized by the MAC at the appropriate TXOP and is then sent over to the PHY. The bits are scrambled and encoded and the appropriate PHY header and preamble for packet detection and synchronization is added. The data is then modulated using the modulation scheme allowed in the standard, pulse shaped at an oversampled rate and sent to the DAC. The samples at the output of the DAC are further filtered, up-converted to the desired transmit frequency, amplified by the power amplifier and transmitted by the antenna. The processing for BLE is similar. The differences between BAN and BLE transmitters are

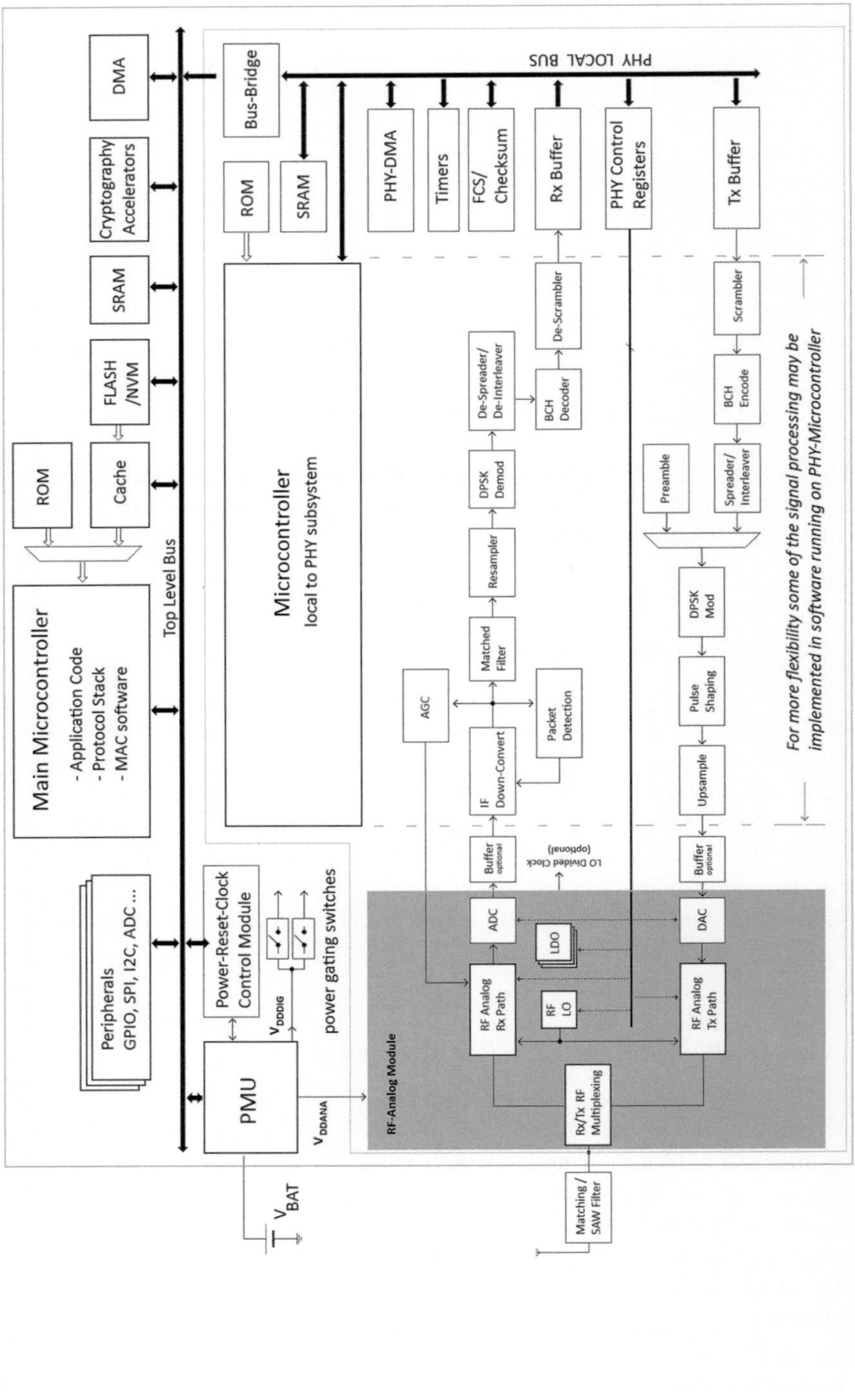

Fig. 6 Architecture of a typical wireless micro-controller system-on-chip

- There is no BCH encoder/decoder for BLE;
- The SRRC matched filter for BAN is replaced by a low pass channel select filter;
- There is no spreading/interleaving that is required in BLE to achieve the low data rates of BAN;
- The DPSK modulation and pulse shaping is replaced by up-sampling, GFSK pulse shaping and then phase to amplitude conversion.

For constant envelop signals (e.g., BLE), a direct TX synthesized architecture using an All-Digital PLL is typically a popular choice. But for schemes having amplitude modulation and/or higher signal bandwidth (BT, WLAN etc.), the ADPLL based implementation becomes complicated, leading to a higher power consumption and additionally, suffering from LO pulling issues.

Note that 15.6 BAN uses mandatory modulations of pi/2 D-BPSK and pi/4 DQPSK. Though, BPSK and QPSK are constant envelope modulations, the square root raised cosine pulse shaping filter, defined in the standard introduces a peak-to-average ratio greater than 1. It is quite possible to use a polar type of architecture for 15.6 BAN—however, these types of architectures require higher bandwidth for the phase and the amplitude modulation paths with tight synchronization requirements between the paths, resulting in a higher power implementation. This implies that for the 15.6 BAN at least, we would need to support a regular I/Q architecture with amplitude modulation capabilities for the transmit side.

A separate loop modulator type transmitter path is certainly possible for the BLE portion. However, this increases the area and thus the cost of the final solution. Thus, for dual-mode or multi-mode transceivers, where at least one of the standards needs amplitude modulation, some compromise is established in power in order to have a common transceiver path for all the modes.

Some of the major optimizations to be considered in the transmitter design are the DAC sampling rate and resolution, low IF vs. direct conversion, filtering required for image suppression, LO architecture and the PA architecture (including the antenna matching). These topics will be dealt in more detail later in Sect. 3.3.

2.3 Receiver Architecture Requirements

On the receiver side, the signal received on the antenna is first passed through the low noise amplifier to get to the appropriate range. The signal is then down-converted with an LO at the appropriate frequency to either zero IF (intermediate frequency) or an appropriately low IF. After some more gain to get to the right dynamic range for the ADC, along with filtering to meet the adjacent channel and blocker specifications set by the standards or by the industry for a competitive radio, the signal is sampled. The signal may be further filtered with a digital filter to remove any residual ACI/blockers. Automatic Gain Control (AGC) for setting the right front end gain, packet detection and synchronization to correct the frequency and sample error between the transmitted and received signal are performed. The signal can then be down-sampled further and demodulation, error correction and descrambling are performed and the packet is transmitted to the MAC. The MAC checks the CRC and if the CRC passes, depending on the packet type (control/data) performs the appropriate functions and transfers the data to the host for further processing.

The major design tradeoffs in the RX architecture depends on the dynamic range of the receiver as well as the blocker specification. The dynamic range (maximum desired signal to minimum desired signal) and the blocker tolerance affect the filtering partition (digital vs. analog), the AGC scheme and the ADC sampling rate. The 802.15.6 BAN standard has a preamble size of 63 bits for packet detection and initial synchronization. On the other hand, the training sequence for AGC and frequency offset estimation is quite short. As a result, the AGC must have a fast convergence. At the same time, there are not enough number of symbols to estimate the power accurately—hence, there should be enough headroom. Due to the blockers, one can also consider wideband and narrowband power measurements to set the AGC to its optimal level. This topic will be dealt in more detail in Sect. 3.2.

From the baseband design perspective, most of the receive power is spent in filtering/down-conversion, preamble detection and error correction. Thus, careful design attention should be given to these functions. Multi-rate filtering techniques can be utilized for the best power performance of the post ADC decimation (down-sampling) filters. Power consumption in the preamble detect function for 802.15.6 BAN is driven primarily by correlation some of these techniques are further discussed in Sect. 4.3.

2.4 MAC Design Requirements

The lowest power is obtained by hardening the MAC path through a gate level hardware implementation. However, by fully hardening it, flexibility is compromised. Thus, the architecture should consist of an appropriate ultra low power processor with hardware acceleration for specific computationally intensive operations. It is important to figure out which features are run in hardware and which in software, while at the same time, thus preserving the flexibility. The criterion to be used over here is area vs. power. Usually, time critical and power hungry tasks that will not change much are mapped to hardware (examples include CRC computation, security engines such as the encryption and decryption engines etc.).

Power efficient hardware DMA's should be considered to move the packets from the PHY boundary to the MAC and vice versa. One needs to consider whether the packet for transmission/reception is shifted out/in into a buffer for the PHY to be transmitted/received without the MAC processor intervention (area) vs processor intervention (power).

The choice of whether to use an RTOS (real time operating system) or not is critical in an ultra-low power system. While using an RTOS can significantly reduce the development and debug time, it can add a significant overhead ($\sim$25–30 %) in terms of code size (memory) and processing. Multiple RTOS's (open source/proprietary) are available for the user to evaluate e.g. TI RTOS, Contiki, Free RTOS, TinyOS etc. [6–9].

Both the PHY and the MAC sub-blocks should have power gating where individual blocks can be turned off when not in use in order to minimize sleep mode power. For example while the CRC check is being performed, some of the receive blocks can be turned off as by this time, the packet would have been received. Similarly, the MAC processing can be turned off during the packet transmission or reception when the blocks are not needed. This implies that the design of the blocks have to be done so that they can be rapidly turned on and configured to the right parameter set and that the modem state machine is designed carefully. Power sequence design in this regard is further discussed in Sect. 4.5.

3 Transceivers: RF and Analog Techniques

This section deals with the practical aspects of the RF and analog design techniques for ultra-low power transceivers, specifically focusing on implementing these transceivers to comply with the IEEE 802.15.6 BAN and BLE standards.

Many strategies and consequent architectures have been developed to implement low power transceivers. These are dependent on factors such as communication standards to be implemented, cost targets, choice of silicon process technology, signal properties like frequency of operation, bandwidth, peak-to-rms ratio etc. and finally competitive performance targets. We will look at some of these aspects in detail.

Standards based radios are typically targeted at ISM bands, which range all the way from 433 MHz to 5.8 GHz. If the implementation is for the 433 MHz band, typically a tuned circuit implementation is not suitable, requiring bulky and expensive off-chip components. A wide variety of baseband circuit techniques can be applied in order to offset the unavailability of "cheap" on-chip tuned circuits. For 800–900 MHz bands, a front-end tuned element for impedance transformation and matching, followed by inductor-less circuits can be implemented. However, for 2.4 GHz and higher bands, where LC tanks can be built with smaller area and baseband amplifiers are not power efficient, heavy use of tuned circuits can be made to obtain power optimized transceivers.

Process technology plays a key part in achieving cost, power and performance metrics optimally. From a cost perspective, IoT applications, especially those for BLE and BAN standards do not mandate moving to the latest available CMOS technologies. This is due to many reasons—(a) the small amount of digital circuitry needed does not mandate the choice of fine geometry CMOS (b) fine geometry CMOS technologies, inherently tend to be much leakier compared to the larger ones (c) they tend to be non linearly more expensive (d) most of the IoT applications need to have embedded non-volatile memory (NVM), which tends to not be available in these finer geometry processes. It is therefore not uncommon to see many commercial products (circa 2015) for BLE and similar IoT applications to be in 130–65 nm processes.

Signal characteristics like bandwidth and peak-to-rms ratio play a key role in architecture choice. Higher bandwidth signals (e.g.: 20 MHz or higher for WiFi) with a large variation in amplitude like WiFi need more elaborate circuitry and tend to consume high power. Low power short distance standards like BLE and BAN, which support a modest (1–2 Mbps max) data rate, typically use lower bandwidth (1 MHz) signals with non amplitude modulated (only phase modulated) signals. As discussed in Sect. 2.2, however, multi-mode transceivers do dictate some compromise to take into account different modulation types.

In addition, designs also need to take into account signal path loss, which is both distance and frequency dependent. An antenna placed on or inside a human body, for example, is heavily influenced by the human body absorption coefficient as described in chapter "Channel Modeling for Wireless Body Area Networks". To accommodate the extra losses due to this absorption, commercially available radios tend to exceed the specifications in the IEEE 802.15.6 standard by 5–10 dB.

3.1 General Design Techniques for Design Robustness

Robustness of the design for commercial and production environments is of critical importance. This depends to a large extent on the choice of architecture. This section looks at a few important aspects pertaining to designing a commercially robust RF transceiver.

3.1.1 Blocker Tolerance

The radio spectrum today is quite crowded, with multiple users and standards co-existing. For example, a low power radio like BLE, which works in the 2.4 GHz ISM band needs to be tolerant to WiFi signals in the same ISM band. The WiFi signal can be as high as −30 dBm at the antenna, while a BLE signal at sensitivity levels could be as low as −93 dBm. It is important to design the radio to be sufficiently linear to tolerate these blockers. Typically either direct conversion or low IF architectures excel in this area. On the other hand, for example, 2-step or sliding IF receivers tend to have issues of down-converting blockers that lie on the carrier +/−first LO frequencies.

3.1.2 VCO Pulling

If the PA and the VCO frequencies are harmonically related, then they can interact with each other and as a result drift from the original frequency, depending on the output signal power. This can severely limit the performance (EVM) of the Radio. The interaction can happen via (a) on chip supply/ground coupling, capacitive signal coupling, coupling through the substrate etc. (b) on package through bond-wire mutual inductance, lead-frame and pin coupling or (c) on PCB. VCO pulling is a highly difficult phenomenon to debug and fix. Therefore, it is much preferable to mitigate it at the design stage. One most popular method for this is to use an offset LO scheme, where the VCO and the PA are not harmonically related.

3.1.3 Digital Coexistence

The RF transceivers are typically part of an SOC which include processors, memory, I/Os, DC–DC converters etc. These create a lot of high frequency noise which can affect radio performance. As discussed in the context of VCO pulling, there are a lot of mechanisms for signals to couple within the chip. This situation can be tackled at three different levels: (1) at the source or aggressor (2) at the coupling mechanisms and (3) at the victim. It becomes progressively difficult from one level to the next. Hence the best strategy is to look at the SOC comprehensively for co-existence planning.

At the source level there are several strategies that can be adopted. The digital circuits are analyzed to find switching hot spots. If these frequencies are aggressors that couple, then simple techniques such as controlling the supply/ground bounce via appropriate decoupling capacitors, clock dithering to spread the clock related spurs or even re-architecting the digital subsystem can be implemented.

At the chip level, the overall SOC power management and clock distribution are typically the primary paths for coupling. Closely placed bond-wires can couple at a −20 dB level. On the other hand, orthogonal bond-wires couple at a much lower level. In compact wireless IoT SoC's, RF/analog and digital circuits are both on the same silicon substrate. While it is important that substrate coupling be looked at carefully, a simple disciplined approach of keeping guard rings and some physical separation of the domains typically mitigates the issue.

At the victim side, robust architectures, fully differential circuits, frequency planning to avoid digital switching harmonics and power management separation etc. are often implemented.

3.1.4 Antenna VSWR

The effective antenna impedance can vary a lot—for example a 1:10 VSWR range, based on environmental conditions and position. System designers usually compromise by limiting the VSWR variation tolerance to 1:2 or 1:3. However, another approach that is adopted is to design circuits which can adapt the output impedance of the PA. This helps prevent the system from being over-designed and thus has an impact on the overall power consumption of the solution.

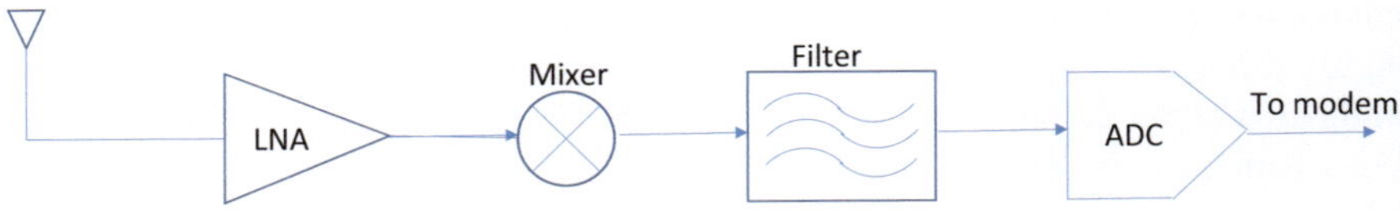

Fig. 7 A typical low power wireless receiver

3.1.5 Temperature Tolerance and Process Variation

Typically IoT systems on chip have one or many temperature sensors on chip. The radio can take appropriate action (changing bias current/voltage, changing gain steps etc.) based on this temperature information.

In addition, silicon process technology can have statistical variations over the production cycle. For example transconductance, resistivity, capacitance per unit area etc. can vary by a large percentage. Instead of designing for worst case conditions, it is better to adapt the design according to the process corner. Process technology information, therefore, can be made available as factory calibration (coded through fuses or non volatile memory etc.). For example, RC filters can be trimmed based on frequency measurement of an on-chip RC oscillator etc.

3.2 Receiver Architectures

The receiver aims at processing the desired signal to ensure at least a minimum SNR is available at the modem decoder. The interface to the antenna is therefore a low noise amplifier (LNA), which helps to increase the signal to a level, which can be digitized and processed without adding much noise to the signal. For ease of signal processing and reducing the bandwidth implication on the ADC, a frequency down-converter (mixer) is inserted between LNA and ADC. To reduce the ADC dynamic range an analog filter is used between the mixer and the ADC. A typical receiver chain looks as shown in Fig. 7.

3.2.1 Specifications

The range of a transceiver is limited by the output power of the transmitter and the sensitivity of the receiver. Standards typically set the minimum required sensitivity of a received signal at a given bit error rate. For example, the reference sensitivity specified for BAN (IEEE 802.15.6) over a 1 Mbps data rate is −83 dBm. However, most typical commercial solutions meet −85 dBm or better sensitivity for enhanced range. The front-end LNA is the most sensitive block in the receiver chain from a noise figure perspective. This implies that the LNA would typically be the highest current consumption block. Also, for robust and reliable co-existence performances, commercial solutions target high interference specifications (typically 40–45 dB). The Adjacent channel interferers typically define the linearity requirement for IF section (analog filter order and the ADC dynamic range).

3.2.2 Receiver Front-End General Requirements

It is quite common to have external 'band select' filters (which are typically either SAW or BAW filters) in order to reduce emission and jammer constraints on the transceiver. Unfortunately, the sharper transition SAW filters also introduce an in-band loss, which impacts the power consumption of LNA and PA (these are typically the major contributor to the overall power consumption of the transceiver chains). One typically has to go through a judicious compromise between the external SAW filter (and its insertion loss and cost) vs. the impact on the power consumption of the LNA. Also, for cost reasons, the receivers are designed as single-ended circuits without additional baluns either on or off chip. For a simpler implementation of the RF front-end, the pin for Rx and Tx is shared so that no separate on-board switches and controls are needed for Rx and Tx operation [10]. In such a case, the off-impedance of PA needs be accounted for while matching the LNA input impedance and vice versa. The combined matching optimization of LNA and PA may cause the LNA to be tuned away from its best noise-current trade-off. The matching components for the LNA are typically integrated providing a compact solution. This integration can be challenging in low power cost packages, which tend to have longer bond wires. For this reason, common gate LNA architectures are often used as they provide a good trade-off under the constraints of combined port integrated matching.

Direct conversion receivers [11] are often used in transceiver designs for their simplistic architecture and reduced mixing spurs/noise. For transceivers designed at 400 MHz frequency range it is optimal to have VCO running 2×/4× the LO frequency. For transceivers at 2.4 GHz, VCO running at twice the LO frequency can become prohibitively power expensive. Instead, sliding-IF receivers [12] can avoid a VCO running at twice the LO frequency and hence save power for VCO/LO section. A typical sliding IF architecture is shown in Fig. 8.

The double conversion mixer in a sliding IF architecture may not meet the noise performance of a single down converter but is effective in reducing power consumption in the LO. The choice of the two LO frequencies in this case needs to be carefully done as dual conversion folds many more frequencies in-band and so increases the risk of jammer and noise de-sensing the receiver. The harmonically related LO_1 and LO_2, in the case of a sliding IF architecture, avoid the need for two different VCOs.

3.2.3 LNA and Mixer

Current-reuse architectures [12], as shown in Fig. 9, have been a popular technique for implementing LNAs. The design challenge in these architectures, however, is to

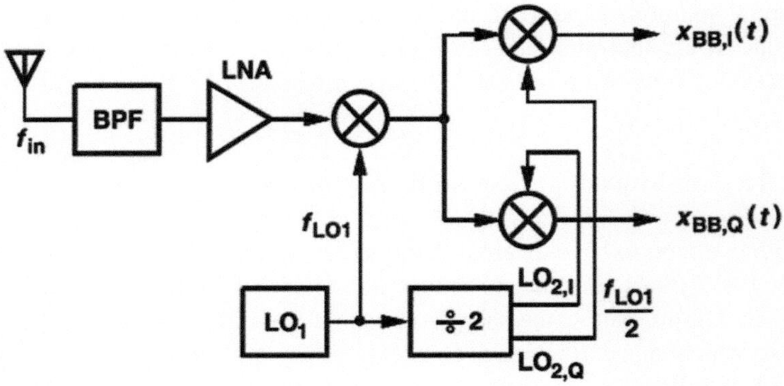

Fig. 8 Sliding IF receiver architecture

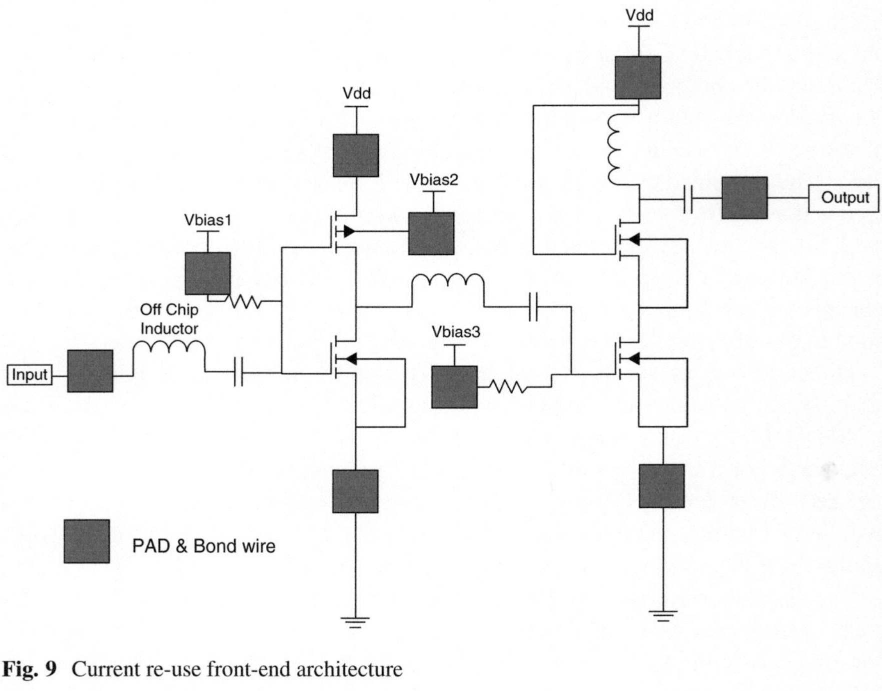

Fig. 9 Current re-use front-end architecture

lower the operating voltage as far down as possible for lower power consumption while maintaining the required gain, noise figure and linearity.

It is essential to have programmability on the LNA gain as it provides further flexibility for noise, linearity and power consumption.

Another method to solve the linearity issue is by sensing the signal/jammer level at the LNA input/output and feeding this information as another input into the AGC in addition to the energy detection from the ADC output. With a greater visibility to the receiver chain, the AGC has the ability to make decisions to preserve the linearity of the chain for a wider dynamic range at the expense of reduction in

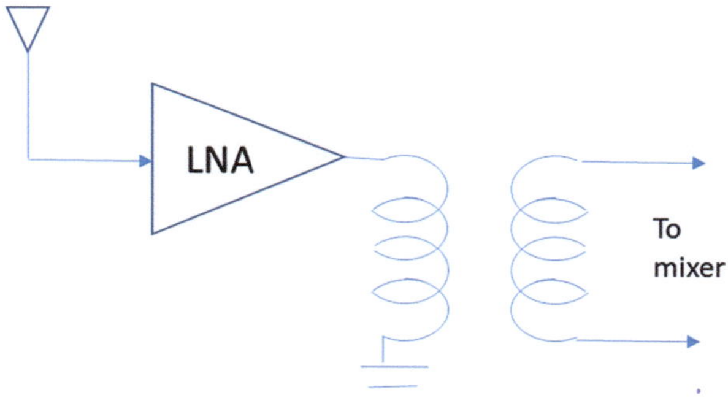

Fig. 10 Representative noise immune front-ends

the sensitivity of the receiver in the presence of jammers. Implementing gain steps in the LNA helps to alleviate this problem but results in the addition of parasitics and increases in the noise figure of the LNA, if not implemented carefully. This extra noise typically gets missed out in a lot of published works that do not get commercialized.

The LNA outputs typically have a tuned network to help in reducing the jammer level seen by next stage. The passive tuned network can be utilized to generate a differential signal, which is needed to have immunity against common mode noise on the SoC as shown in Fig. 10.

The mixer design is typically a Gm input stage with passive switches using 25 % duty LO to provide a good trade-off for noise and gain. 25 % duty cycle is easily available for low frequency transceiver implementations (say 400 M) but for operation at 2.4 GHz more involved techniques are needed to generate the 25 % duty cycle LO clocks which turn out to be power hungry and are mostly used only in high performance receiver chains.

3.2.4 IF and ADC

The IF section of the receiver is typically constrained by the filtering, gain, noise and the dynamic range requirements for the receiver. As discussed in Sect. 3.2.1, commercially available receivers target a higher (40–45 dB) specification for the jammer performance. The in-band IP1 dB of the receiver is limited by the IF and ADC sections. This mandates the need for a variable gain amplifier (VGA) section, which can also function as a filter for close-in jammers. Finer geometry CMOS processes which force lower supply voltages pose a greater challenge for meeting linearity requirements in the IF section. Closed loop structures for VGA [12] require good overall gain, thus implying a cascade of stages and hence greater area and power consumption. Open loop structures [10] provide low power solution but pose challenges on gain, in-band droop, out-of-band attenuation from process variation perspective which need to be corrected/tolerated by the modem. Digitally assisted calibrations like AGC, pole calibration/IQ mismatch for IF, DC offset are typically needed to make the radio robust to process and temperature variations. A detailed discussion of these calibrations is outside the scope for this chapter.

The dynamic range requirements for the A/D converter in the low power transceivers are driven by the specification of close-in jammers as well as the overall power consumption budget. For example, a chain designed for GFSK modulation (BLE) to tolerate 40 dB adjacent channel jammer will need better than 53 dB ADC SNR. Comparatively, an ASK modulation receiver designed for a 10 dB adjacent channel performance requires only a 10 dB SNR in the ADC.

The choice of IF, ADC and digital filters go hand-in-hand and should be seen as an overall power optimization problem. SAR ADC's need good anti-aliasing filters but can be implemented with low power digital logic. Sigma Delta modulators can be designed with relaxed analog filtering but need high frequency decimation filters, which tend to be power hungry. Deep-Submicron technologies (say 90 nm and lower) can offer advantages of low power (and area) for digital logic and hence allow an oversampling data converter at an overall lower power. Depending on the modulation (FSK, QAM etc.) the ADC and demodulator specifications can be really reduced for power consumption [11].

3.2.5 Synthesizer

The crystal selection [12] is a compromise between crystal size and oscillator power requirement, whose typical power is approximately 200 μA. Also, the frequency of crystal is chosen so that the digital PHY and MAC can directly use the divided clock without any extra synthesizers (e.g.: 24 MHz).

In a traditional analog PLL architecture, the VCO and the feedback divider dominate the power consumption. For the feedback divider section, a counter based approach gives lower power. For the VCO the spot phase noise requirements come from the adjacent channel interferer specification. To optimize the power of the VCO and the divider sections, frequency planning and process technology node considerations are important. Faster technology nodes (lower channel length) result in lower divider/buffer power but as explained in the previous sections may not be the suitable choice due to other factors (cost, leakage). Divider/buffer power increases linearly with frequency. Also, the receiver architecture and the VCO frequency planning are tied together. To achieve LC-VCO power optimization for a required spot phase noise profile, an optimal frequency choice comes from the quality factor of the inductor and the available transistor gain.

Polar architecture [13] being a popular choice for the transmitter implementation for GMSK modulations (only phase modulations), a two-point modulation PLL as shown in Fig. 11 is adopted which allows lower power consumption than the mixer based architectures.

Another aspect to consider in commercially robust ultra low power ICs is the integrated power management. Designing analog circuits at a low supply can increase the susceptibility of the design to noise and spurs from supply. This may increase the specification on the LDO circuits, which need to operate under low headroom conditions. Also, the overall number of LDO's to improve isolations

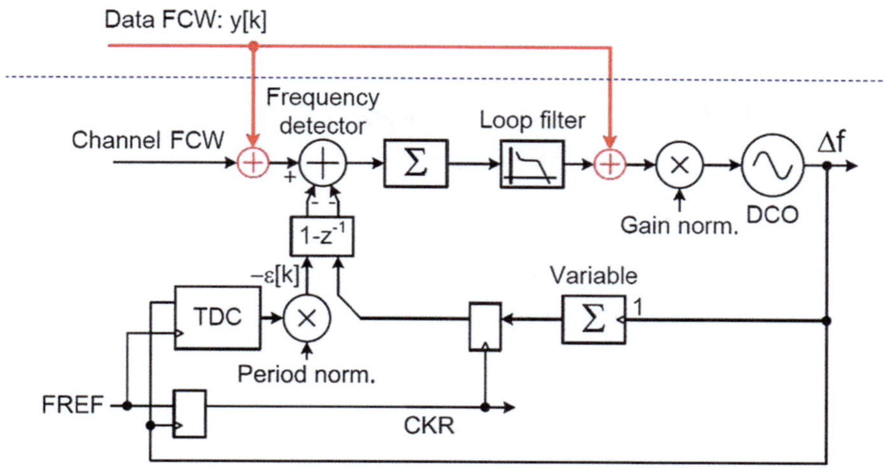

Fig. 11 A two point modulation PLL

between the blocks (VCO to PA, LNA to VCO etc.) increases as a result. For example, certain VCO/DCO architectures may provide good phase noise at low current but have higher gain from the supply path as well. Designing power management circuits for such cases might pose tighter constraints on the required VCO PSRR resulting in an overall suboptimal solution. For circuits requiring a higher bias point without drawing much current, a higher supply voltage can be considered (e.g. cascode gate bias of LNA). Multiple supply requirements for various blocks increase the area of the solution, but circuit performances can be achieved with low power consumption. The LDOs' no-load condition currents are important they remain ON during RX/TX transitions. This is especially true for BLE with tighter turn-around time requirements ($\sim$100 μs).

3.3 *Transmitter Architectures*

For digital modulated signals, the complex baseband signal can be expressed with its quadrature components *I(t)* and *Q(t)* as: $s(t) = I(t) + jQ(t)$. The aim of the transmitter is to take this signal and translate and amplify it to the desired carrier frequency band and power level. This process needs to be linear enough so as to meet output error vector magnitude (EVM) and transmission mask requirements. Also the system needs to be able to generate enough power under various output load (antenna VSWR) conditions.

Transmitter design is highly dependent on the output power needing to be delivered. In typical IoT applications, the commonly required output power is in the order of 0 dBm (1 mW) or less and for the most part, never more than +6 dBm (4 mW). Even at these modest wattage levels, the overall power consumption can be dominated by the power amplifier (PA). Low-power transmitters are thus typically

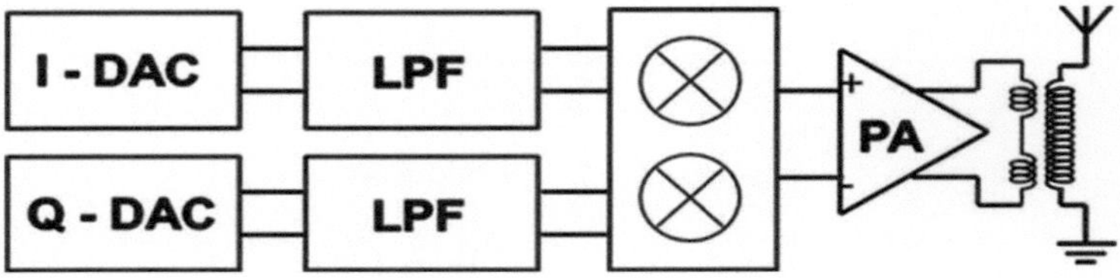

Fig. 12 A cartesian implementation of a wireless transmitter

designed to make use of high efficiency (higher than 50 %) class C or saturated PAs. The next few sub-sections will look at different transmit architectures suitable to low-power applications.

A large variety of architectures are available to choose from. A careful analysis of the pros and cons of each architecture can be used as a design guide for an optimal implementation. Overall the transmit architectures can be categorized as either being more digital intensive vs. more analog intensive. The digital intensive circuits are more suitable for deep-submicron (typically 65 nm and below) designs as they make use of large number of high speed switching logic. In these cases, finer geometry processes tend to allow for more aggressive power reduction. On the other hand, analog intensive circuits are preferable for coarser geometry process choices (130 or 180 nm).

3.3.1 I/Q or Cartesian

The most common implementation is a *direct up-conversion* transmitter. These are typically used in *high bandwidth high performance* systems. In this scheme, the quadrature (I and Q) digital signal is converted to analog using a DAC, filtered (typically active low pass filters with sharp cutoff to meet the emission requirements) and then up-converted using I/Q mixers. The RF signal is then amplified using several stages including the PA and then passed through an external band pass filter to drive the antenna as shown in Fig. 12. A recent example can be seen in reference from Texas Instruments [14], which implements a WLAN a/b/g/n transmitter in 45 nm technology. Such a scheme is usually power hungry but provides the best performance. This scheme can easily handle a large signal peak-to-rms ratio and can achieve very high linearity. Another advantage of this architecture is that it can be implemented efficiently in any process that can support the basic RF amplification. It also tracks well with process scaling. The I/Q configuration also helps in mismatch, leakage corrections using baseband techniques.

The overall power consumption of this scheme can be reduced significantly if the application bandwidth and performance requirements are lessened. For example, for a BLE/BAN transmitter, where the signal bandwidth is less than 1 MHz, a highly oversampled DAC can reduce the post filtering requirement to a point where simple passive R-C filters can be used, as shown in Fig. 13. The mixing operation need not be done in a single step. Instead, a 2 step mixer can reduce the overall system power, in turn also helping alleviate LO pulling.

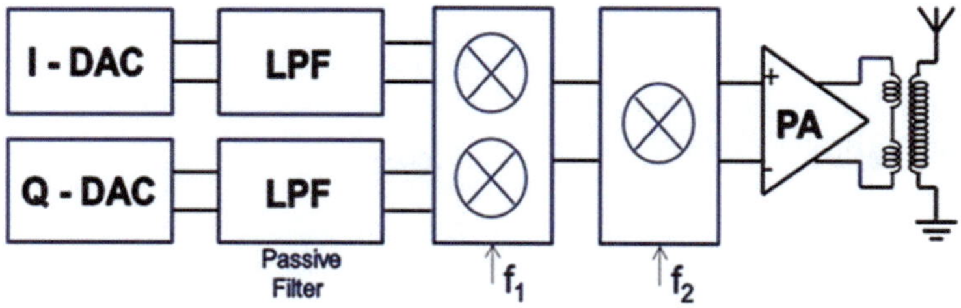

Fig. 13 A 2-stage mixer based transmitter chain

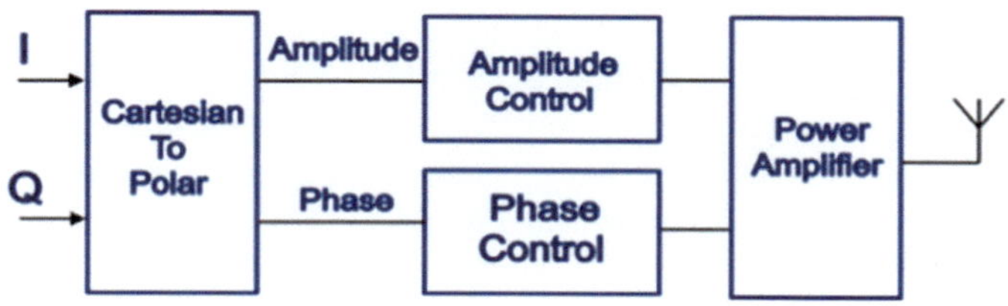

Fig. 14 Polar transmitter architecture

In this scheme, the mixer frequencies f_1 and f_2 are chosen such that $f_{carrier} = f_1 + f_2$. A popular choice is to use a low frequency (say $f_1 = 1/3* f_{carrier}$) quadrature local oscillator (LO) to convert the baseband signal to an intermediate frequency. This is followed by a single mixer clocked at a higher frequency (say $f_2 = 2/3* f_{carrier}$) to achieve the final carrier frequency. This scheme also reduces the power consumption due to 4 clock buffers driving the quadrature mixer running at the lower frequency with only 2 clock buffers required for the higher frequency mixer. The frequency choices are mostly done keeping the over application and SOC requirements in mind. Note that the frequencies f_1 and f_2 typically leak as spurious emissions at the output, which may need additional filtering in the overall path. This scheme can be quite power efficient even for 180 nm or 130 nm process nodes.

3.3.2 R/θ or Polar

Polar transmitters are a popular and power efficient choice for low bandwidth systems like EDGE, Bluetooth, BLE, BAN and others. In this scheme the complex baseband signal $s(t) = I(t) + jQ(t)$ is converted into amplitude $R(t) = sqrt(I(t)^2 + Q(t)^2)$ and phase $\theta = arctan(Q(t)/I(t))$ signals (cartesian $I + jQ$ to polar $Rexp(j\theta)$ format). The phase information directly modulates the VCO, which in turn drives the PA. The amplitude information is fed to the PA through a separate path. The advantage of this architecture, shown in Fig. 14 is that the PA can be operated in saturation, vastly improving its power efficiency.

There are several issues with polar schemes that restrict its use to low bandwidth standards. One of the issues is that the transformation is non-linear ($R(t) = sqrt(I(t)^2 + Q(t)^2)$) and thus the signal bandwidth of the individual components (R and θ) more than doubles [12]. This tends to be challenging to handle (in terms of linearity and power consumption) for high signal bandwidth systems like WLAN. The other issue is that delicate delay matching is needed between the phase and amplitude path before they are combined back at the PA, which becomes quite challenging for higher bandwidth systems, thereby needing elaborate pre-distortion algorithms to compensate.

The simplest polar implementation is possible where the system only uses phase modulated signals with constant amplitude, as in BLE. In such conditions, direct synthesis of RF is possible. Here the VCO control voltage is modulated with the baseband signal. Since the closed loop PLL tries to suppress the low frequency component of the modulation, a copy of the modulation is applied prior to the low pass filter of the PLL as shown in Fig. 11. This restores the low frequency component. This scheme is popularly known as the 2-point modulation (since the modulation is applied at 2 points of the PLL).

An example can be found in a recent work from Texas Instruments [13]. Here a Digital PLL is used to implement the 2-point modulation scheme explained above. The advantage is that the 2 point modulations can be completely matched (since implementation is digital) and the implementation is usually compact. Such a digitally-intensive scheme is particularly useful for deep submicron technologies like 65 nm (the choice of the reference).

For coarser geometry processes like 130 nm, a more analog intensive architecture is preferable, as found in an implementation from IMSE [15]. In this case, an analog PLL is employed, thus minimizing the high speed switching circuitry and resulting in lower power. In this scheme, a PLL is used to lock an LC VCO to the required career frequency. At the time of transmission, the PLL is opened and the resulting free running VCO tank capacitor is directly modulated with the baseband signal. Such a scheme can be prone to pulling/noise issues as the VCO is kept free running for the duration of transmit.

The output of the VCO thus obtained is already a modulated signal at the carrier frequency. This is directly fed to the power amplifier, which amplifies the signal to the required power level and drives the antenna.

The implementation becomes more complex for signals with both amplitude and phase. As before, phase component modulates the VCO and the output of the VCO drives the power amplifier stage. The complexity arises in processing the amplitude information and combining the information at the PA to generate the final modulated RF signal. Although there are many schemes for combining the amplitude, it can be generally categorized into two different classes. A popular scheme is to modulate the supply of the PA with the amplitude information (drain modulation). For example, this can be accomplished by controlling the output of a linear regulator, which supplies the PA. An example of such a scheme can be seen in a recent design from Toumaz [12]. Another frequently used scheme is to control the gain of the PA by turning ON/OFF some segments of the PA based on the amplitude signal. The gain control of such a segmented PA is often done with a delta-sigma loop to spread out spurs and meet the transmit mask. An example can be seen from IMEC in [16].

3.3.3 External Filter and Antenna

The antenna design and external filter design are crucial for low power. This can also keep the external BOM count low. For example, based on the application requirement, an omni-directional antenna may not be needed. A directional antenna with some gain can greatly reduce the power delivery and sensitivity requirements. For low power delivery, an aggressive impedance transformation is done to translate the 50 Ω antenna impedance to as high value as possible at the chip pin limited by the voltage swing acceptable. Also, given noise and linearity requirements, the antenna can be designed to a non 50 Ω impedance, which can lead to lower power.

A good external filter can greatly attenuate blockers thus reducing the linearity requirements. This is especially useful for sub-GHz designs, where the available bandwidth is narrow.

3.3.4 Summary of Transmitter Design

To conclude, as we discussed, there are many possible architectures and schemes for implementing a commercial standards compliant wireless transmitter. The choice mostly depends on cost and performance constraints. It is not always necessary to go to the latest process technology to build a low power transmitter. One can choose a flavor of cartesian or polar architecture based on the silicon process and performance goals. For standards which employ binary phase modulation like BLE, polar architectures usually provide most power efficient implementations. For standards like BAN, where there is amplitude variation due to signal processing, either cartesian or polar modulation can be considered. Also, for finer geometry process choice of 65 nm and below, more digital intensive architectures are attractive.

4 Low Power and Robust Digital Architectures

In this section, we will look at the digital architecture techniques for low power IoT system implementation. This involves a discussion of the micro-architecture techniques for the baseband modem, the MAC as well as techniques related to hardware/software partitioning. We will also look at system level energy consumption optimization techniques, including MAC based periodic fast wake-up for overall battery optimization.

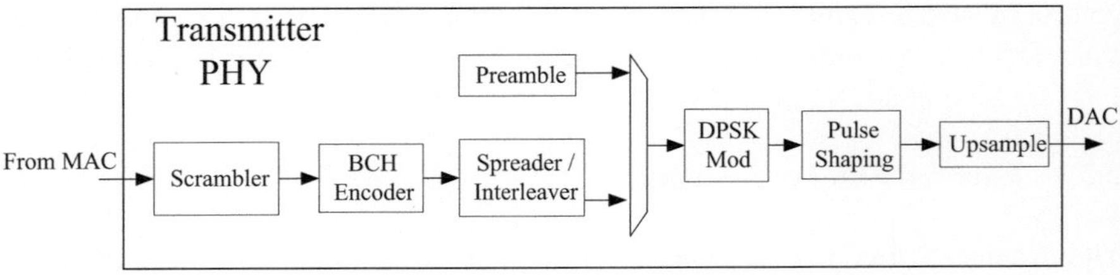

Fig. 15 Transmitter physical layer architecture

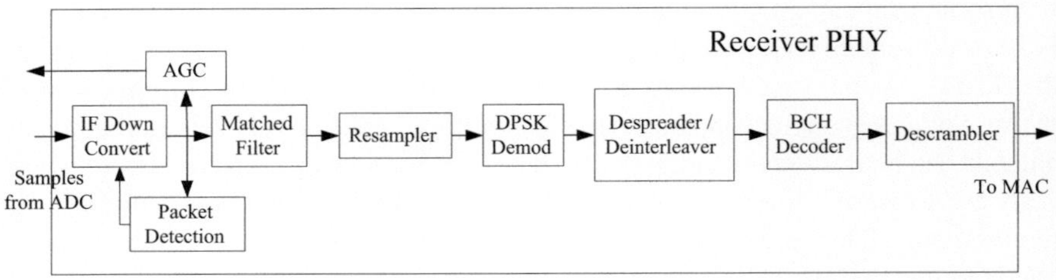

Fig. 16 Receiver physical layer architecture

4.1 Baseband Phy Architecture

In this section we will concentrate on the different architectural choices for the baseband PHY portion of both the transmitter and the receiver. Since the BAN and BLE standards are similar in bandwidth and processing requirements, the front end (AGC/filtering/sample rate conversion) processing is valid for both technologies. The backend processing (from demodulation to CRC checking) between the two technologies is kept separate to avoid unnecessary control complexity. However, the typical area penalty for such an implementation is low due to the low precision requirements.

Figures 15 and 16 show a typical transmit and receive baseband processing for the 802.15.6 BAN transceiver respectively. For the remainder of the architecture discussion, we limit ourselves to BAN. The difference between the BAN and BLE baseband modems were discussed in Sects. 2.2 and 2.3 of this chapter.

4.2 Transmit Processing

A majority of the transmitter baseband blocks are processed at the symbol rate and at a single bit precision. Therefore, the key drivers of power savings are the DAC sampling rate and the DAC precision. These two parameters set the precision and processing rate for the pulse shaping and up-sampling blocks.

Higher DAC sample rates allow for relaxed analog filtering in the transmit path, while increasing the complexity/power of the DAC and the pulse-shaping/up-sampling filters. In addition, LO leak-through requirements need to be taken into account to meet a specified transmit mask. Finally, the DAC precision, sets the noise floor of the transmitter and affects the transmit mask. An analysis of the BAN and BLE transmit systems keeping the relevant mask and power consumption requirements shows that an 8× to 12× oversampling of the symbol rate and a DAC precision of 6–8 bits are required.

4.3 Receive Processing

The receiver baseband modem processing can be divided into three major domains:

- Front end processing, which is a function of the ADC sampling rate
- Packet detection—which includes correlation to the preamble to find the preamble, initial synchronization and
- Data path processing after down sampling.

The receiver front end processing converts the low IF signal to base band, does the channel selection and required matched filtering and down samples the receiver to the required rate.

The front end filtering and down-sampling blocks are always on during the signal reception. Therefore, these need to be optimized for power. The power and performance of the analog filter, ADC and the digital filter determine the required adjacent channel interference (ACI) and blocker suppression. Also, a low IF design avoids the DC due to LO leakage and reduces some of the complexity requirements on the mixer but increases the complexity and power for the analog filter, ADC and the digital sections that follow.

Given the blocker specifications for BAN and BLE and the fact that a low IF architecture can avoid the DC due to LO leakage and 1/f noise of the amplifiers, an analysis shows that a sample rate of about 4× to 8× of the bandwidth is required for processing.

Also, the receiver AGC algorithm optimizes the number of ADC bits and the precision of the filters, while still meeting the dynamic range requirements of the input signal. The AGC algorithm sets the gain level based on both the desired signal as well as the broadband receive signal strength indicator (RSSI) that includes the residual blocker. With such an AGC scheme, an 8 bit ADC is sufficient to obtain the necessary performance. The AGC logic is also switched off as soon as packet detection is complete to reduce the power consumption.

To further optimize the front-end digital power, it is common to combine decimation filtering and matched filtering blocks. Any of the well-known techniques of multi-rate filtering [17] can be used to compute only the desired samples and thereby reduce the power. Filter symmetry can also be exploited to further reduce the number of multiplications and thus save power. Techniques such as canonical signed digit (CSD) are also used to design the filter coefficients such that power expensive multiplications are eliminated.

A second place where significant power saving can be achieved is the packet detection and initial synchronization block. Packet detection is usually performed using self-correlation or correlation with a known sequence that is transmitted to facilitate packet detection. In BAN a 63 symbol sequence is transmitted for this purpose. Packet detection performance requires that the correlation be done at least 2× the data rate for the preamble. For BAN, this turns out to be about 250 Ksps while for BLE, it is 2 Msps. Performing correlation using full multipliers turns out to be power hungry as well as area intensive. Also, the multipliers cannot be repurposed for some other task after packet detection due to latency constraints. Full precision correlation makes the threshold setting for packet detection AGC dependent and increases the complexity.

Many low complexity/power correlation schemes have been presented in literature [18, 19]. In [19], the output of the differential detector is quantized in the phase domain. This converts the multiplications for correlations into simple additions on the unit circle. After the phase addition, the result is next mapped back into the I/Q domain. These partial results are then added to obtain the correlation metric. Avoiding expensive multipliers results in low power architecture. Since the amplitude information is neglected, the correlation threshold does not depend on the signal level received and thus the AGC information is not required.

While a comprehensive discussion of low-power BCH error correction is beyond the scope of this chapter, techniques described in [20], where divisions are avoided in computing the error locator polynomial and in [21] for further power reduction in the PHY are effective.

Figures 17 and 18 show the effectiveness of the PHY data path processing for a 15.6 BAN type modem using some of the optimizations described. It can be seen that the active power for both transmit and receive for the digital baseband can be optimized to a level that the most significant power contributors are limited to the front end RF analog and the processor.

4.4 PHY Sequencer Design

The PHY and radio in any wireless system require timing critical event handling and sequencing. Thus, it is common to have a dedicated controller alongside the signal processing blocks. A programmable sequence controller is desirable since it allows the radio and PHY to be adapted to multiple protocols and standards as well as allowing for over-the-air firmware upgrades.

It has become common place to use a dedicated MCU, ROM and a small amount of SRAM for the PHY controller implementation. To ensure a deterministic schedule, only the radio related interrupts and events are handled by this core. In some implementations, the lower level MAC function (e.g. BLE Link Layer) implemented in firmware may also run on this MCU core. Some of the more

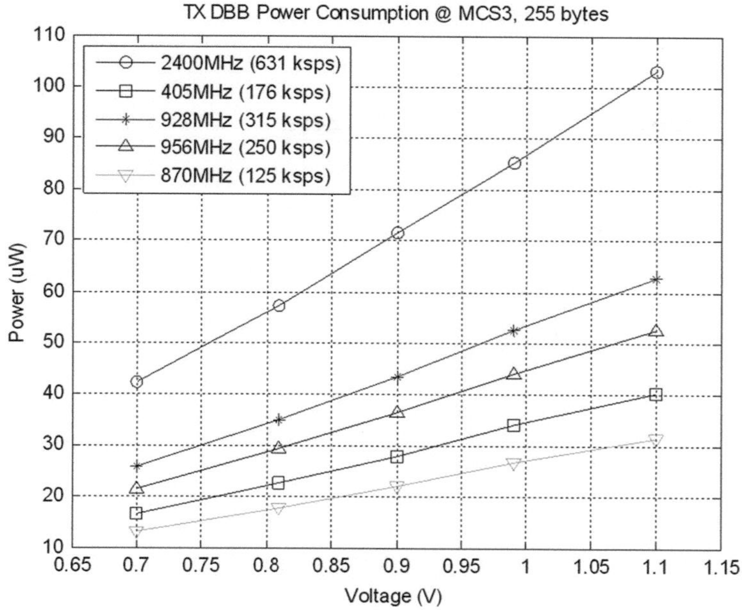

Fig. 17 Transmitter data path power consumption for a 802.15.6 modem

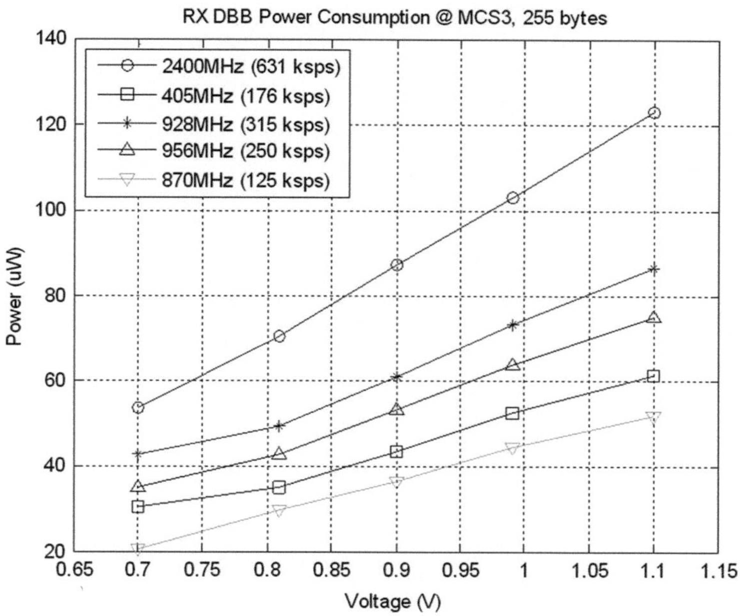

Fig. 18 Receiver data path power consumption for a 802.15.6 modem

Radio Analog	3	mW	75%
PHY Accelerator HW (digital signal chain)	0.15	mW	4%
Lower-MAC Accelerator HW	0.15	mW	4%
Micro-controller	0.12	mW	3%
SoC Active Overhead (incl. PMU quiescent current, Clock Osc, PLL, Clock Tree, Interconnect)	0.6	mW	15%
Total	4.02	mW	100%

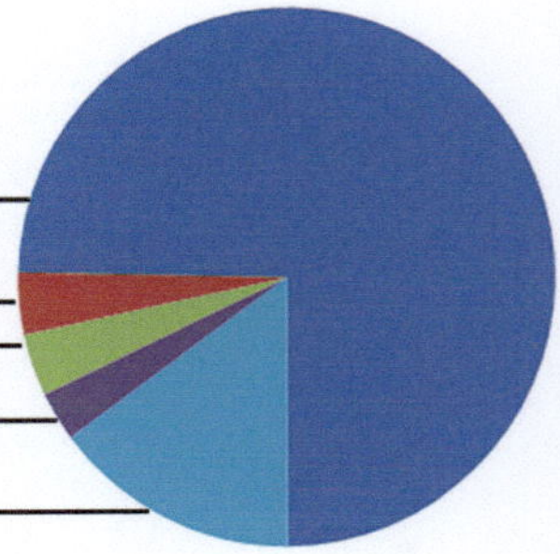

Fig. 19 Active power breakdown for a state-of-the-art BLE implementation

commonly used microcontroller cores for this purpose include ARM Cortex™ M0+, Synopsys ARC™ etc.

In an efficient implementation, the PHY-controller enters an active state before the radio is enabled and enters clock-gated sleep shortly after the radio is disabled. Multiple timers are used to wake-up and trigger sequence of events leading to Rx and Tx. Since the controller must be active while the radio is enabled its power is typically bundled along with the Rx and Tx power consumption.

For low data rate standards (e.g.: Zigbee), a significant portion of the Rx/Tx signal processing may be implemented in software—enabling greater flexibility. This approach saves silicon area (ROM code is much denser that logic) but incurs some power penalty. The MIPS requirement for implementing a standards compliant PHY on an MCU ranges from 4 MIPS for basic sequencing tasks to above 24 MIPS when significant baseband signal processing and link-layer management are also included. An efficient commercial microcontroller core like ARM Cortex™ M0+ in 65 nm consumes around 25 μW/MHz at 0.9 V when continuously executing from ROM or RAM (including the clock tree and memory read power in silicon). This contributes between 100 and 600 μW to the overall power consumption.

Figure 19 below shows the distribution of active power consumption in receive mode for a state of the art ultra-low power BLE implementation. The transmit case is similar for low output power.

Some of the key techniques for optimizing the logic and controller power consumption in PHY are briefly listed below.

- Optimizing and grouping analog and radio control registers.
- Proper partitioning between retention and non-retention states for wake-up.
- Minimizing dynamic configurations and tuning requirements of the radio and analog baseband, thus using a simple finite state machine instead of a programmable controller.
- Optimizing the hardware–software partition such that the MIPS requirement on the MCU is minimized, leading to a lower operating frequency.

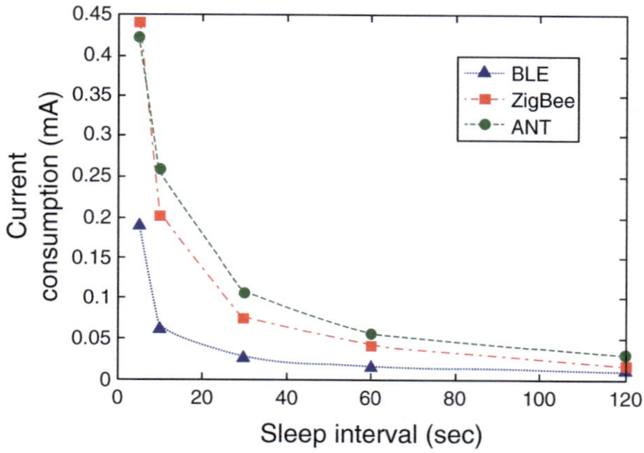

Fig. 20 Current consumption as a function of "active" duty cycle

- Minimizing access from an external or on-chip Flash memory, since any read/execute command current draw typically ranges from 175 to 400 μW/MHz, depending on process technology.
- Proper code block partitioning with appropriate clock gating for each instance of these code blocks.
- Incorporating efficient power management techniques to generate the logic and memory drive voltages as well as utilizing known techniques such as forward body biasing.
- Incorporating efficient clock and power gating techniques . . .

4.5 System Level Energy Consumption

In a majority of IoT applications, the wireless sensor node is in deep sleep for a majority of the duration, waking up either periodically based on the real-time clock (RTC) or asynchronously based on any sensor event.

Figure 20 shows current consumption scaling with duty-cycling, using data based on existing commercial SoCs, in the cyclic sleep scenario, in which a short-range and low-power wireless sensor node periodically sends a data packet to a remote 'hub' with intervening sleep intervals. The plot shows three standard protocols BLE, Zigbee and ANT. As can be seen, severe duty cycling achieves a very low average power (30 μW at 120 s interval for BLE).

The amount of time a radio stays awake depends on the protocol and impacts the average power consumption. This can be conceptually represented by plotting the effective duty cycle of the radio as a function of the sleep interval, as shown in Fig. 21.

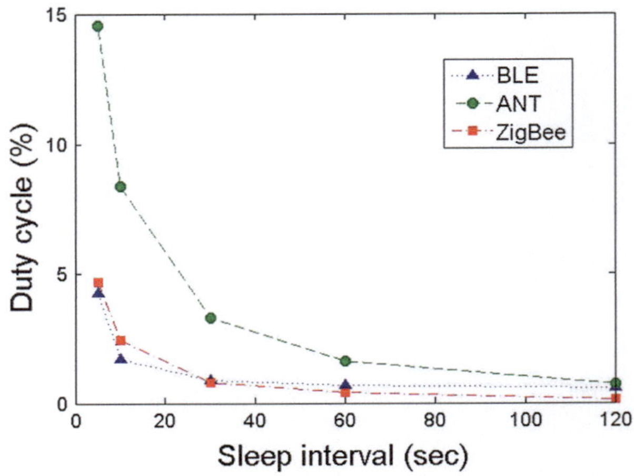

Fig. 21 Radio duty factor vs. sleep interval

Under long sleep interval conditions, a sensor node typically gets dis-associated from the hub. For example in BLE nodes, the maximum time between two received data-packet PDUs before the connection is considered lost is the range of 100 ms to 32 s (max supervision timeout). Hence a new connection has to be re-established for each cycle. The wake time is largely governed by the fact that re-establishing a connection requires several packet exchanges between the node and the hub. Upon waking the node starts advertising its presence, while the master listens, and data is only sent after master and slave are synchronized. Most of the power consumption differences amongst protocols can be attributed to the time taken for this activity. Figure 22 shows this phenomenon (configured for shorter timeout interval). In proprietary protocols and network implementations, this overhead is optimized by trading flexibility and security in favor of lower power consumption.

Figure 23 depicts oscilloscope plots showing the voltage drop on a high-side shunt resistor as a proxy for sensor node activity and current consumption. The voltage spikes represent the node being awake and active and the plateau between the spikes is when the node is asleep. In each case one data packet is transmitted every 5 s. However several packets get exchanged between the hub and the node as can be seen from the expanded view of one transmission event (smaller boxed graphs at top-right).

The following example is based on an actual BLE stack. It is running on a Texas Instruments TI CC2541 device and illustrates the time profile of active power during steady state data exchange in "Connection" mode (i.e. no time-out and no re-handshake) (Fig. 24).

The states within a "connection event" are:

- **MCU wakeup**—the MCU wakes up on sleep timer expiry. It may have to reboot and reinstate the hardware configuration from saved context memory.

Fig. 22 Wakeup characteristics for various protocols

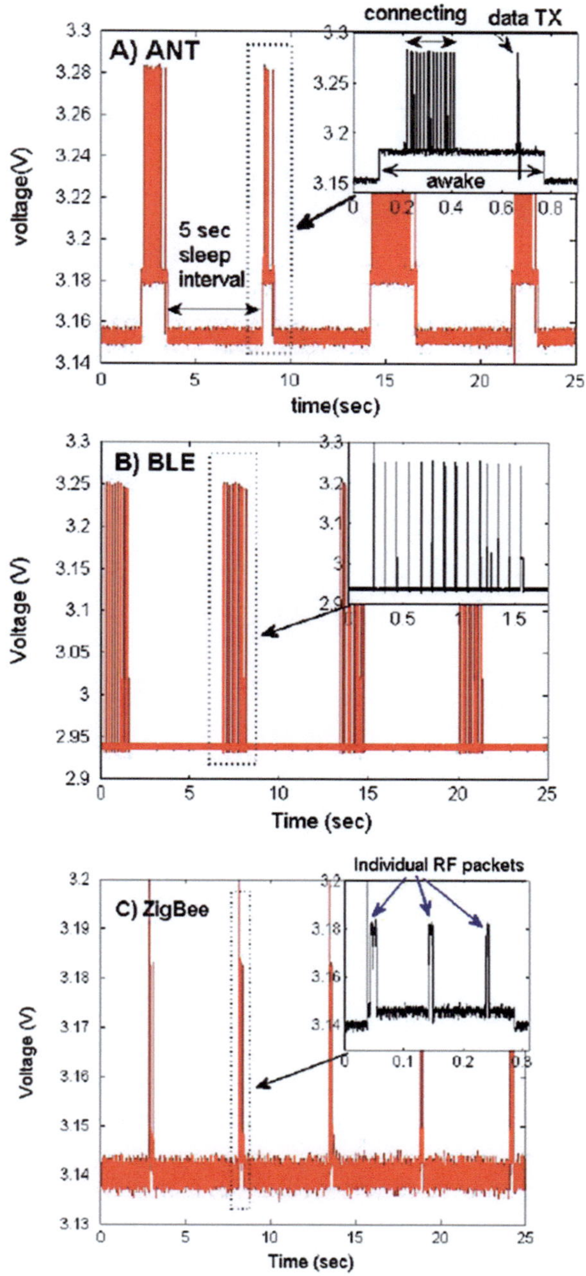

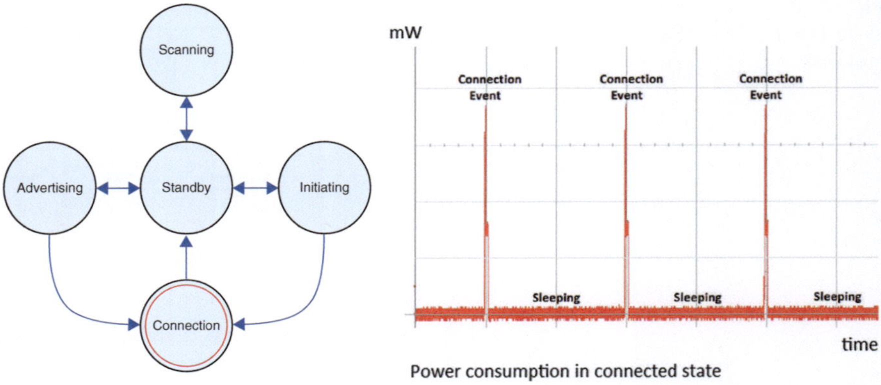

Fig. 23 Representative BLE power profile across events

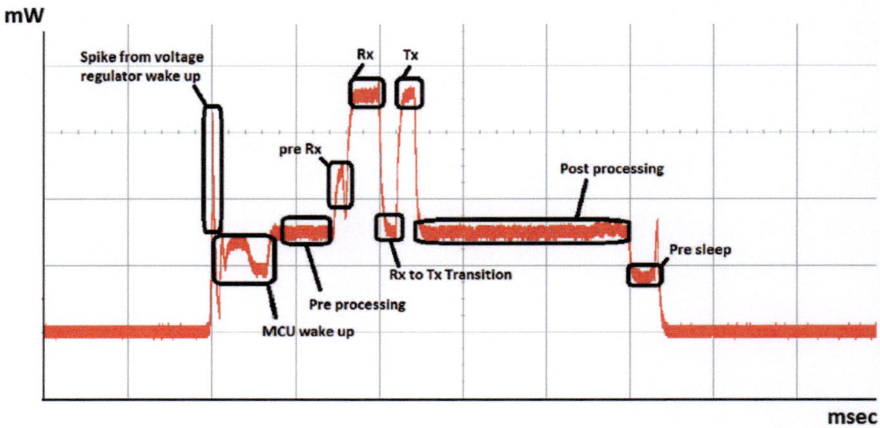

Fig. 24 An expanded view of the BLE power profile

- **Pre-processing**—the BLE protocol stack prepares the radio for sending and receiving data.
- **Pre-Rx**—the radio turns on in preparation of Rx and Tx.
- **Rx**—the radio receiver listens for a packet from the master.
- **Rx-to-Tx transition**—the receiver stops, and the radio prepares to transmit a packet to the master.
- **Tx**—the radio transmits a packet to the master.
- **Post-processing**—the BLE protocol stack processes the received packet and sets up the sleep timer in preparation for the next connection event.
- **Pre-Sleep**—the BLE protocol stack prepares to go into sleep mode.

In a standard protocol compliant sensor node, the portion of consumption other than the radio and PHY can be quite significant—about 71 % as seen from Fig. 25 (based on TI CC2541).

State	uS	mA (at 3V)	uJoule	Percentage
Wake Up	400	6	7.2	10%
Pre-processing	315	7.4	7.0	10%
Pre-Rx	80	11	2.6	4%
Rx	275	17.5	14.4	20%
Rx-to-Tx	105	7.4	2.3	3%
Tx	115	17.5	6.0	9%
Post-Processing	1325	7.4	29.4	41%
Pre-Sleep	160	4.1	2.0	3%
Total Energy (uJ)			71.0	

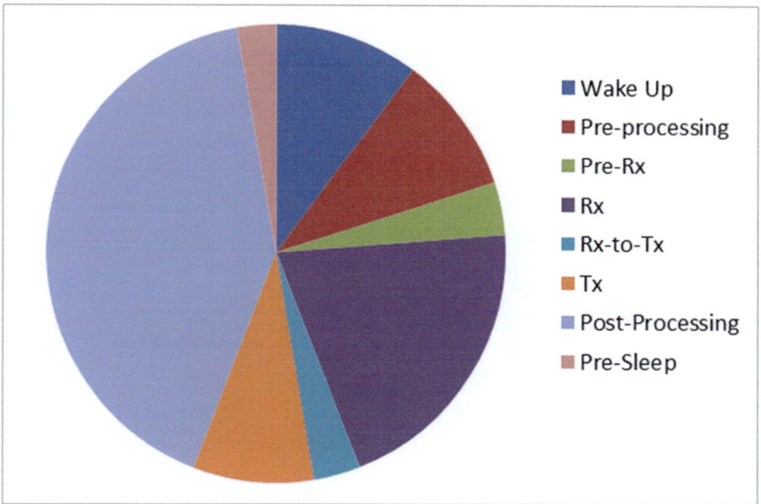

Fig. 25 Energy consumption breakup during each activity slot

As the sleep interval increases, the contribution of the active currents towards the effective average current becomes less dominant. The deep-sleep current of the SoC then becomes the dominant factor. Figure 26 shows this behavior. Hence, significant improvement in battery life requires optimization at the SoC level. While a detailed discussion of this optimization is beyond the scope of the present chapter, a few key techniques can be highlighted as below.

4.6 SoC Design Techniques for Minimizing Energy Consumption

As indicated in Sect. 4.5, the next few topics relate to the highlights of design techniques for an ultra-low power SoC both under deep-sleep and active mode

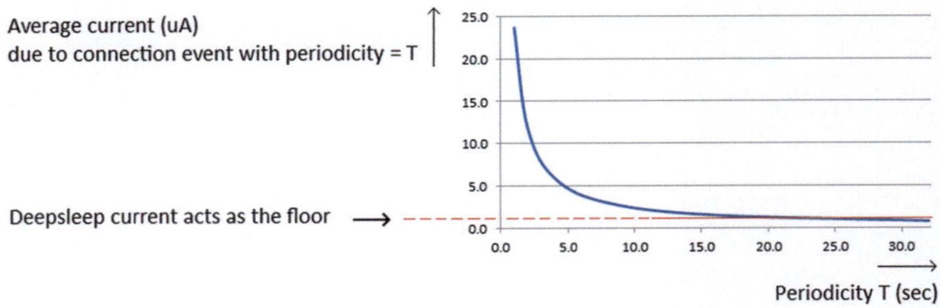

Fig. 26 Current profile across "duty cycled" BLE

conditions. The criteria for each of these states are different and we highlight them accordingly.

4.6.1 Deep Sleep Mode

The typical state of wireless SoC subsystems during "deep sleep" mode is as follows.

- Processor, Peripheral Logic, MAC and PHY Digital are power gated
- Radio Analog is power gated (DC–DC or LDO turned OFF)
- PMU for digital and analog cores is partially power gated
- 32 kHz XOSC/RCOSC is left to continue to run off the battery
- Always-ON Digital, Wake Controller continue to run off 32 kHz
- Minimal amount of logic and SRAM in state retention mode

Brownout/Battery Monitors are kept optionally ON, depending on the system criteria

Most IoT applications need some small amount of SRAM in retention mode. Depending on the amount of memory and the process, the retention current requirement can vary between 0.3 μA to several tens of μA. For very small retention currents it may be more economical to use a duty-cycled Bandgap reference followed by a Sample-and-Hold buffer and an LDO or a simple source-follower to generate the SRAM array supply. For designs with larger retention current requirements, a DC–DC regulator can be used in low duty factor mode with a relaxed ripple requirement.

4.6.2 Active Mode

For a given data-transform or data-movement task, dynamic energy optimization boils down to minimizing the following product:

Dynamic Energy Consumption = (Clock ***Cycles*** *Required To Complete the Task) * (Clock* ***Frequency****) * (Logic and SRAM* ***"Gate" Count*** *Being Clocked) * (Average* ***Cpd*** *per "Gate") * (****Voltage****)^2*

A number of design and architecture parameters need to be looked at in order to reduce the overall active mode or dynamic energy consumption.

Processor Core

In a standard protocol compliant sensor node, the main processor needs to execute a lot of software code—(a) application code (b) protocol stack (c) profiles (e.g.: BLE) and supplicants (e.g.: WiFi) (d) security and encryption related code. An efficient and fast microcontroller core with high execution throughput per μA rating reduces the MHz requirement as well as the amount of time the SoC needs to be in active state. Furthermore, an optimized protocol stack helps lower the peak MIPS requirement.

NVM & Cache

As described earlier, the read/execute current of on-chip parallel flash memory is quite high. Therefore, the microcontroller should run most of the protocol code from either the on-chip ROM or SRAM. Alternatively a cache can be used between the flash and the microcontroller core. A simple N-way associative cache greatly reduces the number of stead state flash accesses. In some applications, the indeterminate nature of a cache may be undesirable. In those cases, the most frequently executed and timing critical routines of the protocol stack may be copied to and execute from SRAM.

SRAM Bank Granularity

Smaller memory blocks help reduce the SRAM dynamic power per read/write (or equivalently Cpd) and total delay (sum of setup time and clock to output delay). Hence, the SRAM bank should consist of smaller sized instances with clock gating for each instance. A granular memory bank also enables software greater control over the amount of memory retained, thereby avoiding deep-sleep current penalty for smaller applications.

Efficient Clock Gating

Maximizing clock gating efficiency and minimizing the clock tree power when most/all blocks are gated is a well known and highly effective technique. A clock distribution scheme with a star-topology is typically used—with clocks coming to each block gated at the root.

Figure 27 below shows an example of fine grained clock gating—where every functional block in the MCU subsystem is provided its own dedicated gated clock from the global clock manager module. In addition, the clock to a functional block may be dynamically gated based on requests coming from the block itself. This way when only a few blocks needs to be clocked, the rest of the blocks are gated and there is no un-used clock tree toggle. This results in lower active in low power active modes. This style of gating is typically used in ULP microcontrollers.

However, this has two drawbacks. Firstly, when many blocks are active at the same time, there is some duplication in the clock tree buffering (especially towards the root). Secondly, when on-chip-variation (OCV) methodology is followed for static timing closure, the early divergence of the clock paths leads to a higher setup and hold timing margin requirement for data-paths between blocks, thus increasing the number of delay buffers and gate upsizes (for setup timing fix)—which increases the overall post-layout logic gate-count and power consumption. So the most power optimal solution merges some of the gated clocks.

Voltage Down-Scaling & Forward Body Biasing

Reducing the supply voltage also reduces the maximum operating speed of the logic gates. This reduction in speed is particularly severe for “weak” devices.

Forward Body Bias (FBB) during active mode can be used to increase the speed of the devices even at low supply voltages. FBB also minimizes the impact of device variability and hence alleviates the worst case margin requirement for timing closure. Lower design margin improves the logic area and hence power.

Figure 28 shows that the highest frequency of an example ring oscillator is 16 MHz at the minimum V_{dd} of 0.6 V and 327 MHz at a V_{dd} of 1.2 V. However with a 0.4 V FBB applied to both the NMOS and PMOS devices increases the oscillator frequency to 46 MHz at the same minimum V_{dd} of 0.6 V. Put another way, the operating frequency scales by a factor of (46.6/327) or 0.142. The “μW per MHz” value scales by a factor of $(0.6/1.2)^2$ or 0.25 and the total power scales by 0.25×0.142 or 0.0355. As an example,

Near-threshold/Sub-threshold Design

If there was no lower limit on the required MIPS and latency, one could scale down the logic supply voltage and the μW/MHz value further. This would involve operating the design at a voltage that is either near or somewhat below Vt of the transistors. This may be an interesting option, provided the system level MIPS requirement is much below 1 MIPS. For example [22] reports a 27.2 pJ/cycle (i.e. 27.2 μW/MHz) MCU core in 65 nm at 0.5 V with maximum operating frequency of 434 kHz at 25 °C.

In practice, the application and the protocol sets a lower bound for the MCU clock frequency. At very low MIPS the SoC needs to stay ON for longer period to finish data processing and movements. This is not optimal from an overall energy

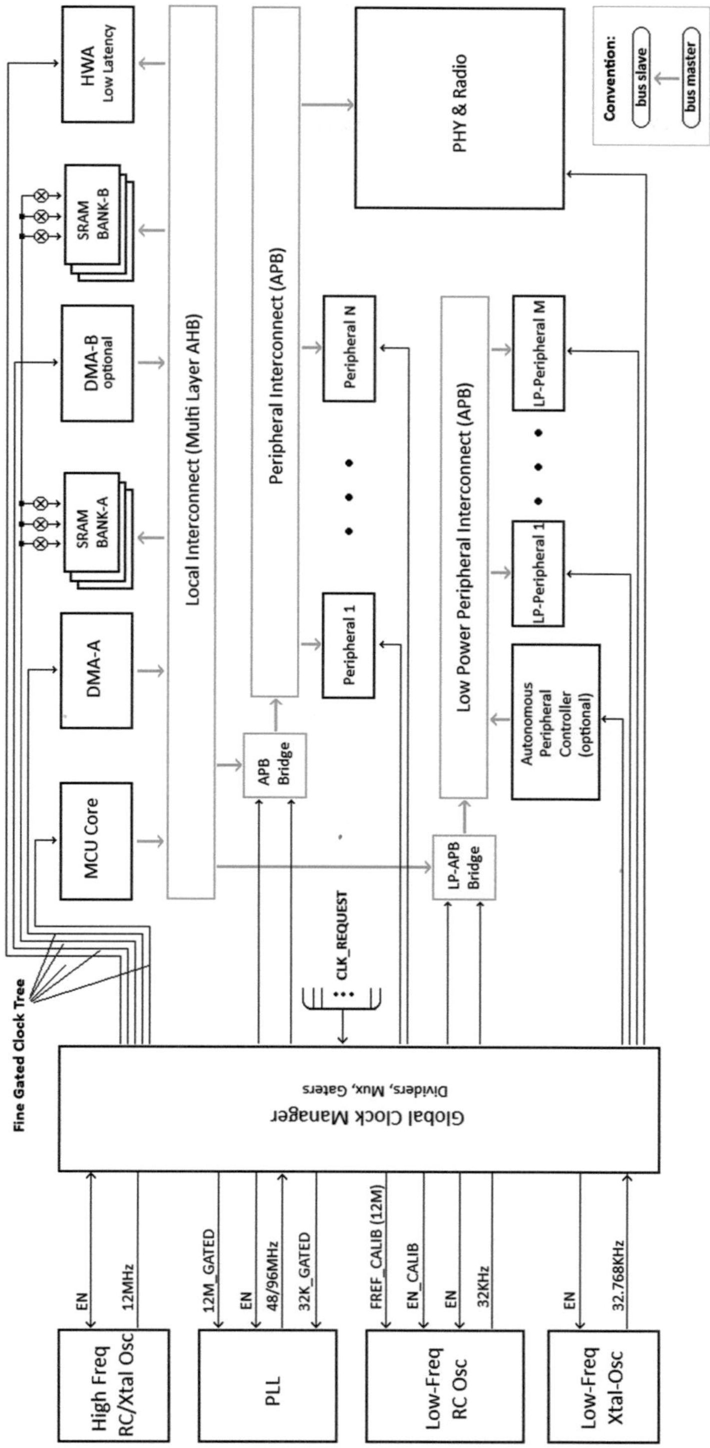

Fig. 27 SoC level clock gating techniques for low power

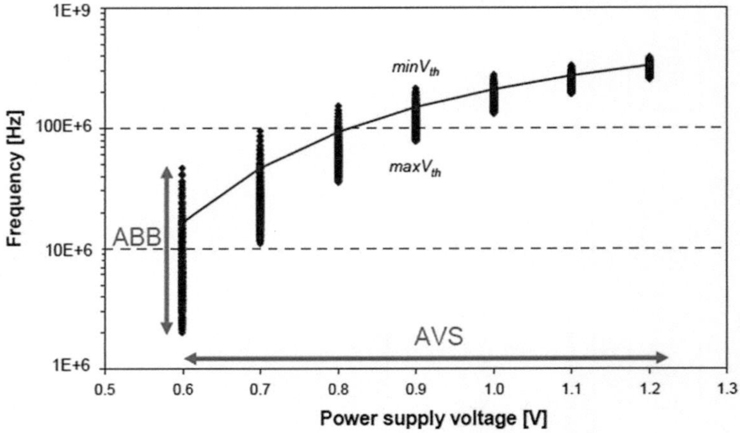

Fig. 28 Body-bias techniques vs. frequency of operation

consumption stand-point. Because of these reasons sub-threshold implementation is not suited for wireless-microcontrollers.

Retention Flop vs Save-Restore

If deep-sleep/active transitions are very frequent, then it is more economical from an energy standpoint to use retention flip-flops. However for more infrequent deep-sleep scenarios, a save-restore mechanism could be more attractive. In this scheme the contents of the flip-flops in a design are shifted in and out of a retention-SRAM. A save/restore mechanism using the scan chains typically has lesser area overhead compared to replacing the flip-flops with retention flip-flops (typically 25% layout overhead). This is because the area per bit is smaller in the case of an, compared to the per bit overhead inside a retention flip-flop. This scheme is shown in Fig. 29 below. It should be noted that the parameter values shown are for illustration only.

Faster Power Management Transitions

In traditional designs the power-reset-clock module (PRCM) and its state machines are run at 32 kHz. This slow clock is choosen since it is always running and available prior to the digital logic coming out of reset. However this is not an energy efficient implementation.

Designing the system for faster sequencing and running the PRCM at a faster clock frequency helps minimize state transition latencies between active and deep-sleep modes. The frequency of this clock can be made a function of the PM mode. In lower power modes, this can be slowed down to 32 kHz or lower.

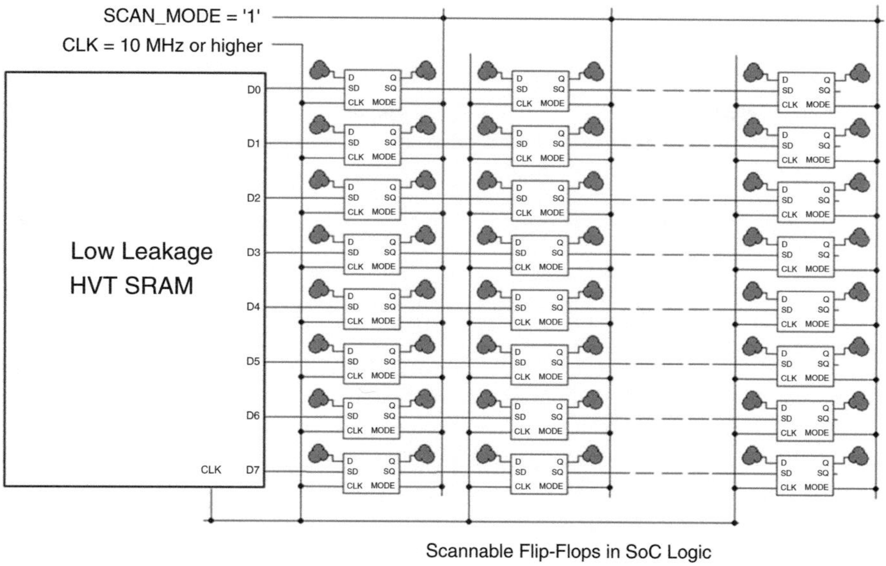

Fig. 29 An example save/restore technique for low power applications

4.7 *MAC Mechanisms*

Most of the techniques we have discussed so far try to optimize the peak power consumption of the radio. A reduction in peak power consumption, automatically leads to an average power consumption reduction. Another way to reduce the average power consumption is duty cycling to make sure that the radio is not switched ON unnecessarily. For e.g., if the time between two connection events is long enough, the protocol can decide to put itself into a deep sleep mode to conserve current. In some cases it can shift to a low power mode where some computations can be at a slower clock rate.

The choice of the protocol modes in BAN such as scheduled traffic vs. contention based traffic has a huge impact on the overall power saving. In case of scheduled traffic, the radios can wake up just in time to transmit the data. The choice of these modes also depends on the application. The MAC architecture should be generic enough to allow for such optimizations.

Synchronization, with the network for maintaining the timing impacts the average power as the system has to decide how often the how often to wake-up the radio for synchronization. Having precise references that drift very slowly compared to each other implies that the radio can sleep for a longer time without synchronization. Recent works on low drift BAW resonators [23] have pushed this area in the right direction to help alleviate this issue. The BLE and BAN protocols allow for the system to go into deep-sleep and re-synch by increasing the inactivity time before the radio has to be re-authenticated.

Low power wakeup radios can potentially ease a lot of protocol constraints by operating in a low power mode and wakeup the rest of the radio when there is a data packet. A number of wakeup radios have been developed [24, 25]. However, issues of false alarm performance at a good sensitivity have to be resolved before they become widely adopted. Integration of the radio into the protocol is currently proprietary and standard mechanisms have to be developed for interoperability and system wide power reduction. With higher false alarms, there will be little or no advantage. Also an out of band wakeup radio will impact the area and cost.

Other ways of minimizing the ON time include beacon suppression and reducing number of channels to search over (in case of frequency hopping) in order to find the timing information. BLE addresses this issue by having the node search within the three advertisement channels to do initial synchronization [26]. In case the packets are small, most of the transmit power could be in transmitting the preamble, as the preamble length is designed to perform synchronization for the worst channel conditions. A good strategy to conserve average power in this case would be to aggregate the packets and burst them out in a single sequence. In such a case, the architecture has to facilitate aggregation of packets, fragmentation and managing the queues at a low power penalty.

In a wakeup and sleep type of application, a huge amount of power currently is burnt in the wakeup time the processor/host. Thus, the radio/processor architecture should facilitate ultra-fast wakeup from deep sleep.

The battery life of the system can also be increased if one improves on the provided MAC mechanisms, schedules the traffic to wakeup only for data and minimize the radio wakeup for beacons and other network information traffic that does not change significantly over time [27].

4.8 Extensions for IEEE 802.11ah

In this section, we discuss a new standard defined in IEEE 802.11 called 802.11ah. This standard was developed to support the wireless sensor use case. The standard is also defined in the 900 MHz bands. Table 4 gives the mandatory parameters for the sensor node side of 802.11ah.

As discussed previously in this chapter, the major factors affecting peak power are (1) processing bandwidth/symbol rate, (2) complex modulation and coding and (3) band of operation.

(1) Effect of Bandwidth on Power: One can see from Table 4 that the mandatory processing bandwidth of the 802.11ah system is a maximum of 2 MHz. This bandwidth is twice that of BLE. Thus, the ADC and the front end digital sections might have to operate at a higher sampling rate requiring a little more power for these blocks. Also, 802.11ah requires an FFT/IFFT in the receive/transmit paths, which add to the digital power. While the encoding complexity for BAN might be of the same order of magnitude as that for

Table 4 Parameters for IEEE 802.11ah

	11ah	Comments
Frequency band	900 M	BAN/BLE operate at 2.4 GHz
BW	1–2 M	BLE bandwidth is 1 MHz
FFT size	64/32 (OFDM operation)	No FFT required in BAN/BLE
Modulations	BPSK/QPSK	Higher modulations are optional for the STA side
Max PHY rate	2.1 Mbps (2 MHz)0.97 Mbps (1 MHz)	For the mandatory modes. Higher data rates are possible but might require higher power.
Coding	Viterbi	BCH for BAN and none for BLE
Scalability (STA's in 1 BSS)	Up to 8191	Hierarchical AID structure for .11ah. Much higher than BAN/BLE
Network access	Partially managed	
Path loss @ 1 m	31.5 dB	8.5 dB advantage @ 900 MHz1
MAC throughput	New enhancements such as short MAC headers and NDP ACKs control frames	

802.11ah, the decoding for BCH can be much simpler than 802.11ah. However, if one considers that there is no frequency selectivity for the typical sensor network applications that 802.11ah might be deployed in, for a minimal performance loss one can obtain a low complexity Viterbi decoder that will increase the decoder power only slightly. As discussed earlier, the digital baseband PHY power for BAN/BLE is not very significant. Thus, the increase in power due to the increased bandwidth should affect the overall SoC active power only slightly.

(2) Effect of modulation: 802.11ah mandatory modes support only BPSK and QPSK modulations. Given the low SNR required for these modulation schemes, the effects of peak to average ratio and receive/transmit linearity requirements should only be a few dB tighter over BAN and BLE. This requirement is much easier to meet compared to 802.11a/g where the modulation can be as high as 64QAM. Note that, in general, one needs an additional 3 dB/bit in SNR for supporting higher modulations. This implies that 64QAM requires an SNR 12 dB higher compared to QPSK. This in turn means that the radio performance requirements are at least 12 dB tighter and thus, a much higher radio power. Therefore, limiting the modulation to BPSK or QPSK, combined with a 900 MHz operation, should not increase the radio power even though 802.11ah incorporates OFDM.

(3) Effect of band of operation on power: There are multiple advantages of operating in the 900 MHz band versus the 2.4 GHz band. To start with, there is a 9 dB advantage in propagation loss compared to 2.4 GHz This implies that, to get the same range as BLE or BAN in 2.4 GHz, one can operate at a 9 dB lower transmit power, significantly reducing the overall power consumption in the transmitter. A better way is to split this 9 dB budget appropriately between the transmitter and the receiver. The receiver budget can be used towards a lower complexity analog RF circuits with a lower power but with increased sensitivity degradation while still maintaining the same range as BLE/802.15.6 BAN. Some of this budget can also be used to degrade performance of digital modules such as the Viterbi decoder for a lower power implementation. A second advantage is of 900 MHz over a 2.4 GHz operation is that, due to the lower center frequency, power is lower in the VCO and the dividers. The phase noise performance is also better. The PA/LNA complexity/power is also lower at 900 MHz. Since the PA/LNA and VCO/divider can be a significant portion of the radio power, 900 MHz operation gives 802.11ah a power advantage over BLE or 802.15.6 BAN operating at 2.4 GHz.

In order to reduce the average power, the 802.11ah standards also introduce new power savings modes. For example, the sensor node can inform the AP that it is going to sleep and the packets to the sensor node can be buffered at the AP and delivered when the sensor node wakes up. Also, the sensor nodes can wakeup only for receiving the beacon that could have the relevant information for the sensor node. There are also additional new channel access mechanisms that are defined in the MAC that can allow the sensor node to receive and transmit at certain times in a contention free manner. Thus, further reducing power. The standard also improves on the MAC throughput using short MAC headers and by improving on frame exchanges by reducing ACK overheads [28].

Given the above advantages, using the techniques/tradeoffs discussed in this chapter, it is completely feasible to build a radio for 802.11ah that is in a similar realm of power as BLE/BAN radios, both in the active as well as the average power.

References

1. Gartner, Forecast: The Internet of Things. Worldwide 2013, published 18 November, 2013
2. IHS Technology, Industrial Internet of Things – 2014 Edition
3. X. Huang et al., A 0 dBm 10 Mbps 2.4 GHz ultra-low power ASK/OOK transmitter with digital pulse shaping, in *Radio Frequency Integrated Circuits Symposium (RFIC)*, May 2010, pp. 263–266
4. Z. Qi, K. Xiaofei, W. Nanjian, An ultra-low-power RF transceiver for WBANs in medical applications. J. Semicond. **1**(6), 200–201 (2011)
5. J.M. Rabaey et al., *PicoRadios for wireless sensor networks: the next challenge in ultra-low power design* (ISSCC, 2002)
6. http://www.ti.com/tool/ti-rtos
7. http://www.contiki-os.org

8. http://www.freertos.org
9. http://www.tinyos.net
10. J. Bae et al., A 490 uW fully MICS compatible FSK transceiver for implantable devices, in *2009 Symposium on VLSI Circuits Digest of Technical Papers*, pp. 36–37
11. S. Wu, B. Razavi, A 900-MHz/1.8-GHz CMOS receiver for dual-band applications. IEEE J. Solid-State Circuits **33**, 2178–2185 (1998)
12. M. Vidojkovic et al., A 0.33 nJ/b IEEE802.15.6/proprietary-MICS/ISM band transceiver with scalable data-rate from 11 kb/s to 4.5 Mb/s for medical applications, in *ISSCC*, 2014, pp. 170–172
13. S. Chakraborty et al., An ultra low power reconfigurable multi-standard transceiver using fully digital PLL, in *Proc. Symp. VLSI Circuits*, June 2013, pp. 148–149
14. R. Kumar et al., A fully integrated 2 × 2 b/g and 1 × 2 a-band MIMO WLAN SoC in 45 nm CMOS for multi-radio IC, in *ISSCC*, 2013
15. J. Masuch et al., A 1.1 mW RX −81.4 dBm sensitivity CMOS transceiver for Bluetooth low energy. IEEE Trans. Microw. Theory Tech. **61**, 1660–1673 (2013)
16. Y.H. Liu et al., A 2.7 nJ/bit multi-standard 2.3/2.4 GHz polar transmitter for wireless sensor networks, ISSCC Dig. Tech. Papers, February 2012, pp. 448–450
17. R.E. Crochiere, L.R. Rabiner, *Multirate Digital Signal Processing*, Prentice-Hall Inc., Englewood Cliffs, New Jersey 07632 (Prentice Hall, 1983)
18. T. Ha, S. Lee, J. Jim, Low-complexity correlation system for timing synchronization in IEEE802.11a wireless LANs, in *Proceedings of Radio and Wireless Conference*, 2003
19. J.C. Roh, A. Batra, S. Hosur, Packet detection and coarse symbol timing for rotated differential M-ary PSK modulated preamble signal, US Patent 8,630,374
20. H.-S. Kim, S.-J. Lee, M. Goel, Method, device, and digital circuitry for providing a closed-form solution to a scaled error locator polynomial used in BCH decoding, US Patent 8,392,806
21. P. Reviriego, C. Argyrides, J.A. Maestro, Efficient error detection in Double Error Correction BCH codes for memory applications. Microelectron. Reliab. **52**(7), 1528–1530 (2012)
22. J. Kwong, Y.K. Ramadass, N. Verma, A.P. Chandrakasan, A 65 nm sub-Vt microcontroller with integrated SRAM and switched capacitor DC–DC converter. IEEE J. Solid-State Circuits **44**(1), 115–126 (2009)
23. R. Tabrizian et al., A 27 MHz temperature compensated MEMS oscillator with sub-ppm instability, in *IEEE 25th Int'l Conf. on Micro-Electro Mechanical Systems (MEMS)*, 29th January - 2nd February 2012, pp. 23–26
24. N. Fletcher, J.M. Rabaey, *Ultra-Low Power Wakeup Receivers for Wireless Sensor Networks* (EECS Department, University of California Berkeley, 2008)
25. X. Huang, S. Rampu, X. Wang, G. Dolmans, H. de Groot, A 2.4 GHz/915 MHz 51 μW wake-up receiver with offset and noise suppression, in *IEEE Solid-State Circuits Conference*, February 2010
26. The Bluetooth 4.0 specification: https://www.bluetooth.org/en-us/specification/adopted-specifications
27. A. Xhafa, B. Campbell, S. Hosur, Towards a perpetual wireless sensor node, in *IEEE 2013 Sensors Proceedings*
28. W. Sun, M. Choi, S. Choi, IEEE 802.11ah: a long range 802.11 WLAN at sub 1 GHz. J. ICT Standardization **1**(1), 83–108 (2013)
29. K.-H. Chen, H.-P. Ma, A low power ZigBee baseband processor, in *Proceedings 2008 International SoC Conference*, 24–25 November 2008, pp. 140–143
30. C.-C. Wangt et al., A 6.57 mW ZigBee transceiver for 868/915 MHz band (ISCAS, 2006), p. 45
31. IEEE Standard for Local and Metropolitan Area Networks – Part 15.6: Wireless Body Area Networks, 2012
32. http://www.arm.com/files/pdf/dspconceptsm4presentation.pdf
33. http://www.ti.com/lit/an/swra347a/swra347a.pdf
34. A. Dementyev, S. Hodges, S. Taylor, J. Smith, Power consumption analysis of Bluetooth low energy, ZigBee and ANT sensor nodes in a cyclic sleep scenario, *IEEE International Wireless Symposium*, Beijing, China (2013). doi 10.1109/IEEE-WS.2013.6616827

35. M. Meijer, J.P. de Gyvez, Technological boundaries of voltage and frequency scaling for power performance tuning, in *Adaptive Techniques for Dynamic Processor Optimization, Springer Series on Integrated Circuits and Systems* (2008), pp. 25–47
36. C.-M. Hsu et al., The low power MICS band biotelemetry architecture and its LNA design for implantable applications, in *Solid-State Circuits Conference, 2006, ASSCC 2006* (IEEE Asian), pp. 435–438
37. F. Wang et al., Wideband envelope elimination and restoration power amplifier with high efficiency wide band envelope amplifier for WLAN 802.11g applications, in *Proc. IEEE Int'l Microwave Symp.*, 2005, pp. 645–648
38. IMS Report on Consumer and Wearable Applications – August 2012

Synchronization Clocks for Ultra-Low Power Wireless Networks

Danielle Griffith

Abstract An effective method to reduce radio power consumption in applications whose average data rates fall below the maximum data rate capabilities of the underlying radio is to duty-cycle the radio front ends between bursts of data transmission. Doing so in a network, however, requires careful synchronization amongst cooperating radios, often requiring ultra-low-power, yet precise and stable clocks. Such synchronization clocks can be implemented as low frequency crystal oscillators, temperature compensated crystal oscillators, MEMS oscillators, or integrated oscillators. Each of these options has advantages and disadvantages. It is important to understand both the expected system duty cycle and temperature variation as well as the required form factor, cost, and power consumption to know which clock source is most appropriate for the application. This chapter discusses these trade-offs along with several design examples.

Keywords Low power • Wireless • Oscillators • Timers • Clocks • Crystal oscillator • Synchronization • TCXO • MEMS • Bluetooth Low Energy • Zigbee • Sleep timer • Wake up timer • Wake-up receiver • Real time clock • RC oscillator

1 Introduction

In recent years, wireless connectivity has been added to many objects that previously did not have it, such as thermostats, watches, utility meters, lights, shoes, and even toothbrushes. This wireless connectivity, combined with low cost sensors, has enabled new applications to be developed that promise to have significant impact on many parts of daily life. Examples of these applications include home automation, physical activity and health tracking, precision agriculture, and monitoring of infrastructure such as bridges, dams, and water quality.

D. Griffith (✉)
Texas Instruments Dallas, TX
e-mail: danielle.griffith@ti.com

P.P. Mercier, A.P. Chandrakasan (eds.), *Ultra-Low-Power Short-Range Radios*,
Integrated Circuits and Systems, DOI 10.1007/978-3-319-14714-7_7

Low cost and low power consumption are critical for these wireless sensor networks to become widespread. The cost target will vary by application, but must be low enough that adding wireless capability has a negligible contribution to the total system cost. For sensor networks using a primary (non-rechargeable) battery, power consumption should be low enough that the battery can power the node for the lifetime of the device, which might be days, months, or even years. If the device uses a secondary (rechargeable) battery, the time between recharge cycles needs to be as high as possible. If a very long operating life is needed and physical location prevents the node from being accessed easily to change or charge the battery, the node may be self-powered by harvesting energy sources such as solar or motional energy as discussed in Chapter "Energy Harvesting Opportunities for Low-Power Radios". The power source used, such as type of battery or energy harvesting method, will place limits on both the wireless node average and peak power consumption.

2 Low Power Operating Modes

Various operating modes for synchronizing data transmission are possible for low power radios. Depending on the mode implemented, the average system power will vary over several orders of magnitude. Several of these modes are described below.

- Continuous operation: In this mode, the radio is always on and capable of transmitting or receiving continuously. There is little latency in data transfer through the network. The average power consumption for each node will be a 5–50 mW as long as the transmitted power is limited. If a coin, or button, cell battery is used as a power source, its lifetime will be measured in hours. This mode is not compatible with energy harvesting power sources.
- Unslotted Carrier Sense Multiple Access with Collision Avoidance (CSMA/CA): In this implementation, power consumption is quite asymmetrical. Each network has a coordinator device which controls and maintains the network. The coordinator device is always able to receive, and therefore draws significant power. The end nodes, however, spend most of the time in sleep, and wake only when triggered by an external event, such as a sensor detecting an environmental change. When the end node wakes due to a detection event, it first briefly listens to the channel first to see if it is open. When it is, the node sends data to the coordinator device, waits for an acknowledgement, and then returns to sleep until the next trigger event occurs. This implementation is used in Zigbee for non-beaconed enabled networks.
- Wake on Radio (WOR): In this mode, the receiver periodically wakes itself up from sleep and measures the received signal strength indicator (RSSI). The RSSI trigger point must be set low enough to not miss a valid transmission, but high enough that false detections rarely occur. If a sufficiently large signal level is detected, the radio checks the received signal for the correct preamble

and correlation code. When the correct code is detected, the radio remains awake to receive the data packets which follow, and if required, transmit an acknowledgement or data. If the receiver detects either a low RSSI value or the incorrect correlation code when it wakes, it returns to sleep mode until the next predefined wake-up time. The wake-up timer is implemented with a low power, low frequency oscillator inside of each wireless node. The frequency stability of this oscillator determines the required preamble length used by the transmitter. If this oscillator's frequency varies due to supply voltage, temperature, or other environmental changes, the receiver could be woken up earlier or later than it ideally should be. To compensate for possible sleep time variations, the transmitter needs to use a longer preamble to ensure that the receiver will detect the transmission. Therefore, the WOR method allows low power consumption in the wireless node due to duty-cycled operation, but it shifts the power consumption burden to the transmitter when a long transmission time is required. The transmission time is usually limited by radio regulations, which means that there will be a limit to the achievable duty cycle. Also, all nearby receivers will wake during the preamble, even if the correlation code later determines that the data is meant for only one of them. This limits how low the average power for each node can be in some use cases. Average power consumption in this mode can vary significantly, depending on sleep time, sleep power, wake-up timer accuracy, RX and TX power. Latency will depend on the sleep time.

- Always-on wake-up receiver (WRX): In this implementation, a secondary receiver, called a wake-up receiver, is added. It has low power consumption and degraded sensitivity compared to the primary receiver. The WRX operates continuously, typically with power consumption in the range of 1–50 μW. When this receiver detects the presence of a signal that is determined to have the appropriate correlation code, it wakes the main receiver and prepares it to receive the data. The correlation must be done so that the primary radio is only enabled to receive the appropriate data, and not whenever a transmission is detected in the frequency band. Average power consumption is reduced compared to a continuously operating receiver because the main receiver spends more time in sleep, but network latency will be longer. Also, extra silicon area is required to implement the WRX. Detailed design examples of WRXs are discussed in Chapter "Ultra-Low Power Wake-Up Radios".
- Duty-cycled wake-up receiver: Rather than operating the WRX continuously, in this implementation the WRX is periodically turned on to sample the incoming RF signal. Because the WRX is duty-cycled, it can have higher power consumption and achieve better sensitivity with the same average power as used in the always-on wake-up receiver described above. The improved sensitivity comes at the cost of increased network latency, typically in the range of milliseconds. Radio regulations often limit the total transmission time, which must be considered when implementing the transmitter duty cycle and wake up message length.
- Synchronized transceivers: Synchronization is used in Zigbee beacon-enabled networks as well as Bluetooth Low Energy (BLE) networks. In Zigbee networks,

a master node sends periodic time synchronization beacons. All nodes wake up to receive these beacons and use the information to adjust their internal wake-up timer to correct for any timing inaccuracies that have accumulated since the last beacon. In Bluetooth Low Energy networks, a master, or controller, node controls slave, or peripheral, nodes, including the frequency hopping sequence, connection interval, and sleep time. In both of these standards, as well as others using synchronization, each node spends most of the time in sleep, and only wakes to transfer data intermittently. Radio wake-up is controlled and synchronized by a wake-up timer present in each node. The internal oscillators used to create the wake-up timers will differ slightly between each node. To account for this, a guard time is added to the time each radio is active to ensure that the appropriate receiver and transmitter operation overlaps in time. After the data is received, resynchronization is done to adjust for any accumulated time drift between the nodes. Then the nodes go back into sleep mode until the next time the wake-up timers enable the radios. Intermittent data transfer allows low average power consumption in the range of 10 μW. The sleep current, wake-up timer accuracy, and duty cycle (ratio of active mode time to total time) are significant factors in determining the average power consumption. The time to transition between sleep and active mode can also add to the average power. System latency is determined by how frequently each node wakes up to send or receive data.

For all of these systems, the radio's active power consumption while receiving and transmitting data is important. However, that is not always the dominant factor in determining average system power consumption. For systems using wake-up receivers, the receiver power consumption can determine the average power consumption. For WOR and synchronized transceivers, average power consumption will also depend on sleep power and wake-up timer accuracy.

3 Average Power Consumption

In many applications, it is sufficient to collect sensor data, wake up the radio to send or receive the data, and then return the sensor node to sleep until it is time for the next data transfer. Each end device in the network therefore spends most of the time in sleep, waking only after milliseconds, seconds, or even hours, to send or receive data. Coordinator or master nodes, however, may be awake more frequently or even continuously. When the intermittent data transfer is implemented with the optimal operating mode for a given use case, the resulting average power consumption of the end nodes can be very low, allowing a battery life with a coin cell on the order of months or years.

3.1 Power Source Considerations

Although average power consumption is an important consideration, there are other factors that guide ultra-low-power radio and timing circuit design. For example, while the rated capacity can be 220 mAh for the commonly used CR2032 battery, this capacity can be degraded if the peak current drawn is higher than what the battery is meant to provide (typically in the range of 15–30 mA). The battery must be chosen such that the peak power needed when the radio and MCU are on can be provided without exceeding the maximum current sourcing capability of the battery.

A self-powered solution can be created with a power management IC, a thin film battery or large capacitor for energy storage, and an energy harvesting device, such a solar cell or piezoelectric material for converting vibration to voltage. Self-powered solutions impose additional constraints on average power consumption, data transmission frequency, and duty cycle beyond what exists in battery powered solutions. For example, a solar cell can support different values of average system power consumption depending on the ambient light level. Therefore, in low light conditions, a high duty cycle for data transmission will be unsustainable. Also, the maximum energy available for any given data transmission event will be limited to what is available in the energy storage device. The data transmission and node synchronization must be implemented to account for an uncertain energy supply available for each node.

3.2 Synchronized Transceiver Power Profile

Figure 1 shows a simplified diagram of wireless node power consumption versus time, assuming synchronized transceivers are used. Most of the time, T_S, the device is in sleep mode consuming power P_S. Each node has its own real time clock (RTC) which implements the wake-up function. The RTC is one of the few circuits operating continuously, even when the node is asleep. Therefore, its power consumption adds directly to the system average power consumption.

At the end of the sleep time, the wake-up timer starts up the rest of the node in preparation to send or receive data. First, the node enters the pre-processing phase. During this time, radio startup and initialization occurs. This includes turning on any necessary voltage regulators, enabling the crystal oscillator used as a reference clock by the radio PLL, locking and calibrating the PLL, and enabling the radio circuits. This period occurs in time T_W and at power P_W.

Once the radio is operational, the receiver waits for the data to arrive. Due to possible inaccuracies in the RTC between the receiving and transmitting node, a guard time, T_G is introduced. A less accurate RTC requires a longer guard time to ensure that the data is received. During this time, the radio is on and consumes approximately the same power as in the active mode, P_A. During the active time period T_A, data packets are transferred. In the reverse case, when the node is

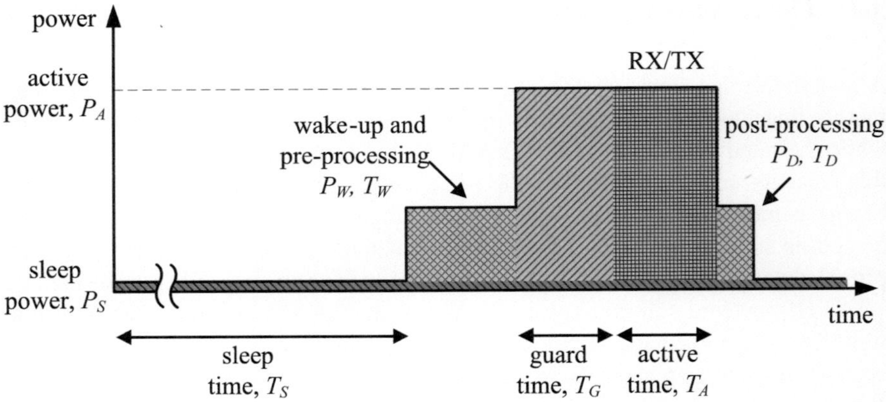

Fig. 1 Wireless node power consumption versus time, showing sleep, wake up, pre-processing, guard, active, and post-processing time

transmitting data, a preamble is sent by the transmitter during time T_G to ensure that the receiving node is able to receive the data in spite of any possible timing inaccuracies.

After the data transmission is complete, the node enters the post-processing phase for time T_D with power consumption P_D. All circuitry not needed in sleep mode is disabled. The received data packet is processed and the wake-up timer is resynchronized to the other node in preparation for the next connection event.

The duty cycle D is defined as the time spent transmitting or receiving data over the total cycle time.

$$D = \frac{T_A}{T_S + T_W + T_G + T_A + T_D}$$

For example, the device might spend 200 ms in sleep mode, 2 ms in the wake-up and pre-processing mode, 100 μs in the guard time waiting for data to arrive, 1 ms transmitting or receiving data, and 2 ms in the post-processing mode. In this case, the duty cycle is approximately 0.5 %.

3.3 Guard Time and Real Time Clock Accuracy

For cellular applications, each handset receives information indicating the precise current time from the base station. This information is used to calibrate the handset low power RTC, creating a very accurate wake-up timer. In some low power wireless networks, however, there may be no external source available to provide the time or to calibrate the node's real time clock. Therefore, each node in the network must have its own accurate RTC used to synchronize data transmissions with other nodes.

Typically the RTC can be made more accurate by increasing its power consumption. However, it is on constantly, so increasing its power consumption adds directly to the node's average power consumption.

The guard time, T_G, is a function of the sleep time and the RTC frequency variation, V_S. This frequency variation can be caused by changes in environmental conditions, supply voltage, and long term frequency variation, quantified by Allan variance [1]. If the real time clocks in two radios have a 500 parts-per-million (ppm) frequency difference, then during the sleep time T_S, the time base can drift apart by $T_S{\cdot}500{\cdot}10^{-6}$ s. Therefore, the guard time must be set to $T_G = T_S{\cdot}V_S$ to ensure that the data is not missed due to clock drift between the sending and receiving nodes. If the sleep time is long, error between two nodes can accumulate, requiring a long guard time. Also, some standards require the wake-up timer to meet certain accuracy requirements. For example, to maintain a Bluetooth Low Energy link, the wake-up timer must have a $<\pm 500$ ppm frequency accuracy [2].

3.4 Average Power Curves for Synchronized Transceivers

The average power consumption, P_{sync}, for a wireless node using synchronized transceivers is

$$P_{sync} = P_S + \frac{P_W T_W + P_A\,(T_S A_S + T_A) + P_D T_D}{T_S + T_W + T_S A_S + T_A + T_D}$$

where the wake-up timer frequency variation is V_S and the guard time has been defined as $T_G = T_S{\cdot}V_S$. This is the area under the curve in Fig. 1, divided by the total time period.

In Fig. 2, the average node power, P_{sync}, is plotted versus duty cycle D for various real time clock variation V_S from 1 to 10,000 ppm (or 1 %). For this example, the following values are used:

- sleep power $P_S = 2\ \mu$W
- receiver power $P_A = P_G = 10$ mW, packet length $T_A = 200\ \mu$s
- wake-up and pre-processing power $P_W = 2$ mW and time $T_W = 500\ \mu$s
- post-processing power $P_D = 2$ mW and time $T_D = 300\ \mu$s.

These values can change depending on amount of data transferred, wireless standard implemented, and specific transceiver performance, but they are typical enough to show the system tradeoffs involved.

At low duty cycles with a very stable real time clock as the wake-up timer, corresponding to point A on the plot, the average power consumption becomes approximately equal to the sleep power, P_S. The power needed for wake-up, pre-processing, post-processing, and data transmission is averaged out over a long period and becomes insignificant. At this point on the graph, the required guard time T_G is also small due to the highly stable wake-up timer. Note that the average

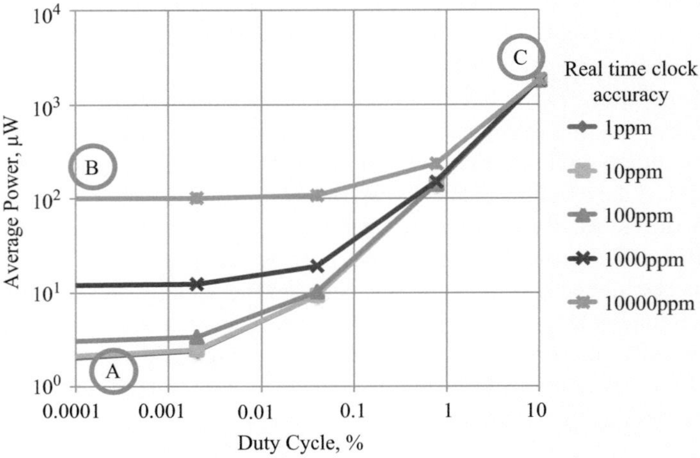

Fig. 2 Example of average node power consumption versus data duty cycle for various RTC accuracies

power using a wake-up timer with 1 and 10 ppm frequency stability is nearly the same. This is because the power consumption is limited by the sleep power which was 2 µW for this example. If, however, the sleep power were lower, it would be possible to see a power reduction when using a 1 ppm-stable clock instead of a 10 ppm-stable clock. In reality, the sleep power consumption and the wake-up timer stability are not independent. It is typically easier to implement a more stable oscillator using higher power consumption, either through calibration or other design techniques.

If instead the wake-up timer has 1 %, or 10,000 ppm, variation in frequency with environmental changes, the average power consumption increases dramatically, even at low duty cycles. This corresponds to point B on the graph in Fig. 2. Here, the average power is dominated by the power consumed by the radio during the long guard time required to maintain synchronization. It is worth noting that this wake-up timer frequency variation is too large to support some low power connectivity standards. Bluetooth Low Energy, for example, requires wake-up timer frequency variation of $<\pm 500$ ppm to allow a connection. For example, if the sleep time is 1 s and the wake-up timer has a frequency variation of 0.05 %, or 500 ppm, the guard time needed is 500 µs. In other words, the radio must wake up 500 µs early and wait for data to arrive. A BLE data transmission might be only 200 µs long, so in this example, the guard time is 2.5 times longer than the data packet. Therefore, the energy used to receive the data is 2.5 times larger than it would need to be if the wake-up timer had a perfectly accurate and stable frequency.

Point C on the graph corresponds to much higher duty cycle. In this case, the power consumption is dominated by the wake-up, pre-processing, post-processing, and active power. Sleep power and wake-up timer stability make little impact on the average power at higher duty cycles. Instead, if operation is expected in this area of the graph, time to wake up and shut down the radio should be reduced for best power efficiency. Using the power consumption values in this example,

$$P_W T_W + P_D T_D = 1.6\mu\text{J}$$

$$P_A T_A = 2\mu\text{J}$$

Therefore, 1.6 μJ are used to wake up and shutdown the radio, while 2 μJ are used to actually receive or transmit the data. Wake-up includes enabling voltage regulators and bias circuits, enabling the crystal oscillator used as the PLL reference clock, calibrating and locking the PLL, enabling the radio, and preparing to send or receive data. In post-processing, the wake-up timer is recalibrated, received data is processed, regulators and the radio are disabled, and the device prepares to enter sleep. To minimize average power consumption in higher duty cycle cases, it is beneficial to have regulators and a crystal oscillator that can be enabled quickly.

As shown in this example, when comparing power consumption among various radios, it is not sufficient to look only at the active power consumption of the radio. Other factors, such as sleep power and wake-up timer frequency stability, can often dominate.

3.5 Comparison to a Wake-Up Radio

If a wake-up receiver is used instead of synchronized transceivers, the average power consumption equation changes slightly. Now, there is no need for a guard time. Instead of sleep power P_S, a wake-up receiver is operating during sleep. Defining the wake-up receiver power as P_{WRX}, the node average power, P_{WU}, is

$$P_{WU} = P_{WRX} + \frac{P_W T_W + P_A T_A + P_D T_D}{T_S + T_W + T_A + T_D}$$

For networks where low power is critical, the logical question is which approach allows the lowest possible power consumption given achievable wake-up receiver power consumption and wake-up timer frequency stability. The point at which the two approaches have the same average power can be solved for by setting $P_{WU} = P_{sync}$ and making two assumptions. First, it is assumed that the wake-up timer variation V_S is much less than 1. Second, it is assumed that the sleep time is a large portion of the total cycle time, or $T_S \gg T_W + T_A + T_D + T_G$. These assumptions will be valid is nearly all duty-cycled low power radio uses cases. Using these assumptions, the average node power using a WRX equals that obtained by using a synchronized receiver when the following equation is satisfied.

$$P_{WRX} = P_S + P_A V_S$$

For example, if a highly accurate wake-up timer (low V_S) is available for the synchronized receiver, the WRX power must be lower to achieve the same average power consumption for a node using a wake-up receiver. Using the values from the example in the previous section of $P_S = 2\ \mu W$, $P_A = 10$ mW, and $V_S = 500$ ppm, $P_S + P_A V_S = 7\mu W$. If the wake-up receiver uses less than 7 μW, lower average power can be obtained using it. Otherwise, a synchronized receiver approach will provide lower power. Other than impact on average power consumption, wake-up receivers have other advantages and disadvantages. The first disadvantage is increased cost since more silicon area is required to implement a second receiver. The second disadvantage is that the WRX sensitivity is lower than that of the main receiver. Therefore it is possible that weak signals will not be detected, although this may not be an issue with networks requiring only a very short range. The first advantage of a wake-up receiver is that it can allow lower latency in the system in some cases. Synchronized receivers with low connection interval can have latency on the order of a few seconds, while the latency of a WRX is usually in the range of milliseconds. The second advantage is that it removes the requirement for the node to resynchronize its wake-up timer to other nodes after each connection event, which simplifies the networking protocol. For many low power connectivity standards, wake-on radio or synchronized receivers are used, but for applications with low latency requirements, wake up receivers are beneficial.

4 Real Time Clock Sources

A real time clock is a circuit that keeps track of the current time. Often the RTC frequency is 32.768 kHz. At this frequency, 1 s is 2^{15} clock cycles. This frequency is convenient to use with simple counter circuits. Because it is a low frequency relative to the capabilities of modern CMOS, it is possible to generate the clock with low power consumption. This is critical because this clock needs to operate at all times in synchronized systems, even when the device is in sleep, and adds directly to the average power consumption. Real time clocks to generate a wake-up timer can be implemented in multiple ways, with associated advantages and disadvantages. Four of these methods are low frequency crystal oscillators, temperature compensated crystal oscillators, MEMS oscillators, and integrated oscillators.

4.1 Crystal Oscillator (XO)

One of the most common RTC implementations uses a low frequency crystal oscillator. A crystal oscillator is created by using active circuitry to create and sustain a resonance in a piezoelectric material, usually quartz. Crystal oscillators are a mature technology. The first was created and patented in 1917 [3]. It consisted of a vacuum tube amplifier and a sodium-potassium tartrate crystal. In 1921 the first oscillator using a quartz crystal was built by Walter Guyton Cady [4]. For modern crystal oscillators used in wireless nodes, the transistors used to create the active circuitry can be either included inside of the same package as the quartz crystal, or they can be implemented on the same silicon die as the rest of the radio with only the crystal resonator external to the radio die.

There are two primary advantages of a crystal oscillator for a wake-up timer. First, frequency stability V_S can be in the range of ± 20 to ± 100 ppm over a wide temperature range such as −40 to 90 °C, enabled by the low resonant frequency drift of the quartz material itself. This stable frequency allows for short guard times T_G, which decreases the synchronized node's average power consumption. The second advantage is low oscillator power consumption, typically well under 1 μW. Both of these advantages are due to the quartz resonator having a very high quality factor, or Q. This means little energy is lost in the resonator, and the active transistors need to provide little additional power to sustain the oscillation. A high quality factor also means that it is difficult to pull the frequency away from the desired resonance frequency due to any environmental changes.

There are three primary disadvantages to crystal oscillators. The first is size. The quartz resonator itself is large and is packaged in metal case. The 3.2×2.5 mm surface mount device (SMD) size is common. A smaller size of 2.0×1.6 mm is often available, but typically with poorer performance and at higher cost. The radio itself could come in a package size that is only 4×4 mm, so the crystal, including the tuning capacitors needed, can take up a significant area compared to the radio. Using a crystal oscillator thus limits the minimum form factor possible for the wireless node. The second disadvantage of a crystal oscillator is that the quartz crystals are sensitive to shock and vibration. In harsh environments, the crystal oscillator can have spurious tones caused by vibration or even fail if it experiences shock. The third disadvantage is cost. The crystal is another component needed to create the complete wireless node in addition to the radio. In some applications, the total cost needs to be extremely low and therefore the crystal is undesirable.

4.1.1 Crystal Oscillator Implementation

A conceptual schematic of a crystal oscillator is shown in Fig. 3. The quartz crystal can be modeled by four lumped elements, R_m, L_m, C_m, and C_o. The parameter C_o is the parallel capacitance across the crystal electrodes. This value includes parasitics from the crystal packaging, and for 32.768 kHz SMD crystals is usually in the range of 2 pF. The remaining three parameters model the acoustic properties of the crystal. The parameter L_m is the motional inductance, C_m is the motional capacitance, and R_m motional resistance, which models the loss in the crystal.

Crystals are manufactured such that they resonate at the precisely desired frequency when they are loaded with a specific load capacitance, C_L. If the oscillator circuitry loads the crystal with more capacitance than the specified C_L, the oscillation frequency will be too low. Likewise, if the effective C_L is lower than the specified C_L, the oscillation frequency will be too high. In Fig. 3, two capacitors are shown, each of value $2C_L$. These two capacitors create an effective load capacitance of C_L due to being in series through an AC ground. If the crystal oscillator is implemented with the negative resistance on the same integrated circuit as the radio, the capacitors can be implemented either with the radio in silicon, or implemented with surface mount devices directly on the printed circuit board (PCB). Integrating them in silicon allows a smaller footprint compared to using capacitors mounted directly on the PCB. However, integrated capacitors may have lower quality factor, which will slightly increase power consumption, and also will vary more with changes in environmental conditions, which will shift the oscillation frequency.

Finally, the active circuitry has a transconductance, g_m, which creates the "negative resistance", R_n, to overcome losses in the crystal and sustain the oscillation. The associated bias circuitry is shown as R_{fb} in this figure.

4.1.2 Design Considerations

Several parameters must be considered to optimize the crystal oscillator design and make the desired tradeoffs between performance and low power consumption. First, to maintain the oscillation, the active circuitry must supply enough power to the crystal to overcome losses in the crystal. Therefore, the negative resistance from the oscillator, R_n, must be greater than the equivalent series resistance, *ESR*, of the crystal.

$$R_n = \frac{-g_m}{(2\pi f)^2 (2C_L)^2}$$

$$ESR = R_m \left(1 + \frac{C_o}{C_L}\right)^2$$

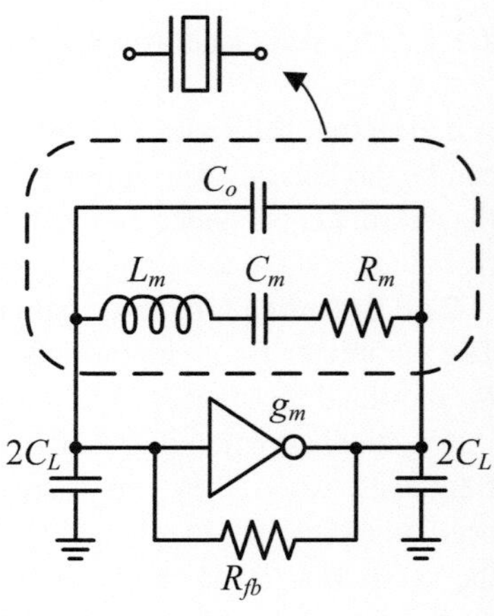

Fig. 3 Conceptual diagram of a crystal oscillator

If $R_n > ESR$, the oscillation will be sustained. However, changes in temperature, voltage, or manufacturing variation among units will cause these parameters to change. Therefore, to ensure reliable startup, the negative resistance must be much larger than the ESR, typically $R_n > 5{\cdot}ESR$. Biasing the active circuitry to constantly have $R_n > 5{\cdot}ESR$, however, is not power efficient. Variable bias is often used, in combination with an amplitude control loop, either analog [5] or digital [6], to allow robust startup and operation with the lowest power consumption. When the oscillator is first enabled, the amplitude is zero, so bias current is set to the maximum value such that $R_n > 5{\cdot}ESR$. Then, once the oscillation amplitude is sufficiently large, the bias current is reduced until the minimum current is used to sustain the oscillation.

The next design consideration impacts the choice of crystal. Crystals are available in various values of C_L for the same frequency. For example, 32.768 kHz crystals may have a load capacitance specified of 6 or 12 pF. As can be seen in the equation for R_n, a 2× larger load capacitance causes a 4× reduction in negative resistance. Therefore, the power consumption needs to be significantly higher to achieve the same negative resistance and sustain the oscillation with a 12 pF C_L crystal as compared to a 6 pF C_L crystal. At first glance, it may seem that choosing a 6 pF C_L crystal will always result in the lowest power consumption. This is true for the oscillator itself, but may not allow the lowest possible system average power consumption. Crystal oscillator frequency stability is given by

$$\frac{\Delta f}{f_s} = \frac{C_m}{2\left(C_L + C_o\right)}$$

where Δf is the frequency shift and f_s, the series resonance frequency, is

$$f_s = \frac{1}{2\pi\sqrt{L_m C_m}}$$

A larger value of C_L makes the resonance frequency f_s less sensitive to changes in parasitic capacitance, which can be modeled as changes in C_o. These changes can occur due to either an environmental change, such as temperature, or manufacturing variations between nodes. If a larger value of C_L is used, the oscillator itself will consume more power due to the tradeoff between frequency stability and oscillator power consumption. However, if the resulting wake-up timer has better frequency stability, the guard time can be shorter, which will reduce the node average power consumption. An understanding of the expected node duty cycle and achievable wake-up timer frequency stability is needed to make the best choice on crystal load capacitance.

4.1.3 Further Power Consumption Improvements

Although the first crystal oscillator was made approximately 100 years ago, new architectures are still being proposed for lower power. One approach, proposed by Hsiao in 2014 [7], involves supplying energy to the crystal during a only short pulse, a fraction of each clock cycle, rather than having a continually biased oscillator core. A second approach, proposed by Iguchi in 2013 [8], is to periodically turn off the oscillator core, and allow the amplitude envelope to decay until it hits a threshold, then turn the oscillator back on to recharge the crystal. In both cases, the power to the oscillator is duty cycled to reduce the average power consumption, whether during each clock cycle or over hundreds of clock cycles. The high quality factor of the crystal, and therefore slow amplitude decay, allows these techniques to work.

4.1.4 Temperature Compensated Crystal Oscillator (TCXO)

As was shown in Fig. 2, in use cases with extremely low duty cycles, power consumption can be dominated by the guard time required if the real time clock is not extremely accurate. A typical 32.768 kHz crystal oscillator can have ±20 to ±100 ppm frequency variation over a wide temperature range. In the use case with a low duty cycle where ambient temperature is expected to vary widely, a viable solution is use a temperature compensated crystal oscillator, or TCXO.

The uncompensated crystal resonance frequency variation over temperature is determined by the crystal cut angle, and is fairly predictable. Figure 4 shows the typical inverted quadratic frequency versus temperature curve for a tuning fork 32.768 kHz crystal. Using this curve, the crystal oscillator frequency stability over temperature can be improved by using a temperature sensor and performing periodic calibration to correct for the frequency variation.

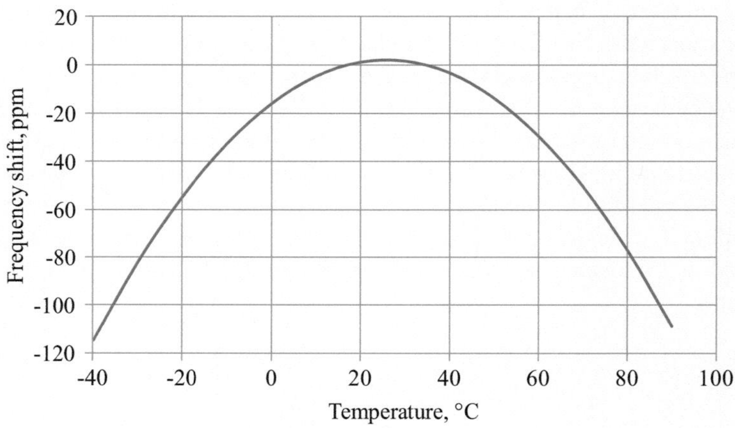

Fig. 4 Typical frequency shift for a 32.768 kHz tuning fork crystal versus temperature

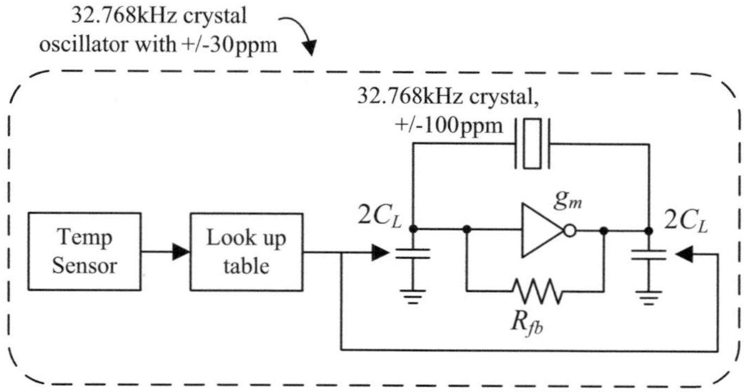

Fig. 5 Block diagram for crystal oscillator with compensation applied to the capacitor array to achieve lower frequency variation with temperature

In Fig. 5, a block diagram shows one method of doing this compensation. Assuming the load capacitors are on the same IC as the oscillator, and not discrete components on the PCB, they can be implemented as an array and adjusted to compensate for the temperature variation. As the temperature either decreases or increases from room temperature, the frequency drops. The temperature sensor measures the temperature, uses a look-up table, and switches out the appropriate number of capacitors from the array to increase the frequency enough to compensate for the drop due to temperature. The effective frequency stability is shown as ±30 ppm, but in reality will be limited by the temperature sensor accuracy and resolution, any temperature gradients between the temperature sensor and the crystal, and how closely the crystal frequency behavior with temperature matches what is programmed into the look up table.

In Fig. 6, a digital approach is used for compensating the frequency. In this case, rather than adjusting the capacitor array to change the oscillation frequency, a temperature correction code is provided to the digital real time clock module. For example, if a 1 s sleep time is desired, at room temperature the RTC module will send a wake-up signal after 32,768 clock cycles. If the temperature increases such that the look up table determines the frequency is 30 ppm low, the RTC module sends a wake-up signal after 32,767 clock cycles. This implementation is useful in situations where the oscillator load capacitors are components mounted on the PCB and therefore can't be adjusted.

Finally, the oscillator and temperature compensation functionality can be implemented in the same package as the quartz crystal. This allows more accurate compensation because the temperature gradient between the resonator and the temperature sensor is less. Frequency stability can be ±3 ppm. The power consumption, physical size, and cost for a TCXO are higher than that of a crystal oscillator, but in some use cases the average system power can be reduced.

In summary, with a crystal oscillator it is possible to generate a wake-up timer with very low power consumption (nanowatts to a few microwatts) and excellent

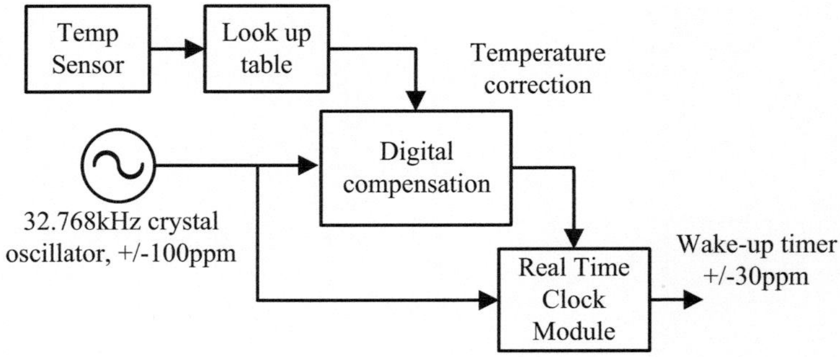

Fig. 6 Block diagram for crystal oscillator digital compensation to achieve lower frequency variation with temperature

frequency stability of much less than 100 ppm variation over a wide temperature range. Both of these are important for wake-up timers, especially in applications with a very low duty cycle. However, crystal size can limit the minimum size and cost of a wireless node. Some emerging Internet of Things (IoT) applications require smaller physical size than can be achieved with an external crystal.

4.2 MEMS Oscillator

MEMS (micro electro-mechanical system) resonators are an alternative to quartz crystal resonators. MEMS resonators are electromechanical structures that are designed to vibrate at a resonant frequency. They can be implemented with various methods of transducing the mechanical motion to electrical signals. Piezoelectricity is a commonly used transduction method used for higher frequency MEMS resonators. This is the same phenomenon as exists in the quartz crystals described earlier because quartz is also a piezoelectric material. The maximum frequency of a quartz crystal is limited because it is not possible to deposit quartz in thin film form to produce resonators at frequencies much above 100 MHz. However, MEMS resonators can achieve much higher resonant frequencies through the use of certain piezoelectric materials. For example, aluminum nitride (AlN) has evolved as the best suited thin film piezoelectric material for high frequency resonators.

For low frequency resonators suitable for use in wake-up timers, MEMS resonators often use electrostatic transduction. In this implementation, an air gap in the resonator creates a capacitance that changes as the resonator mass vibrates, creating electrostatic forces. A signal is driven into the resonator on a drive or input electrode to excite the resonance and the resulting signal appears at the sense or output electrode. Regardless of which method is used to fabricate the resonator, the oscillator is formed by coupling the resonator to an amplifier that generates negative resistance to sustain the oscillation, just as is done with a quartz crystal.

Unlike quartz crystals, MEMS resonators require temperature compensation. MEMS resonators have a temperature coefficient of frequency (TCF) that is strongly negative, usually −10 to −50 ppm/°C. This is because many materials soften and expand as temperature increases. From −40 to 85 °C, an uncompensated MEMS resonator will have between 1,250 and 6,250 ppm frequency variation compared to around ±20 to ±100 ppm for a quartz crystal. Mechanical, or passive, temperature compensation can be implemented by adding a material with a positive TCF such as SiO_2 to the MEMS resonator structure. The disadvantage of this is that the resonator quality factor is degraded, increasing the power consumption required to sustain the oscillation. Active temperature compensation is another option. Similar to the method described earlier, a temperature sensor is added to the system and the oscillator tuning capacitance is adjusted to counteract the temperature change. Active compensation will also increase the average power consumption, depending on how often it must be done.

MEMS resonators are smaller than quartz crystal resonators. This allows them to be placed in smaller form factor packages than crystals, or packaged together with the radio IC. Lower parasitics due to smaller size helps to reduce the power consumption needed for the sustaining amplifier. Also, the MEMS resonators do not require load capacitors, unless needed for temperature compensation, which also reduces the power consumption. Finally, MEMS resonators tend to be more resistant to shock and vibration than quartz crystals. At this point, MEMS resonators tend to be a higher cost solution for wake-up timers than a quartz crystal, but the technology is rapidly improving. MEMS resonators are a realistic solution when the application requires a very small form factor. A more detailed discussion of high-frequency MEMS resonators for RF carrier generation applications is discussed in Chapter “Architectures for Ultra-Low-Power Multi-Channel Resonator-Based Wireless Transceivers”.

4.3 Integrated Oscillator

As seen in Fig. 2, the wireless node average power consumption is less dependent on real time clock frequency stability at higher duty cycles. In the example in Sect. 3.4, at duty cycles of around 1 % and above, the total power consumption is limited by wake-up, pre-processing, active, and post-processing power, and the required guard time has little impact on power consumption. In this use case, it is possible to use a real clock source with worse frequency stability than a crystal oscillator without impacting the average power consumption significantly. The stability requirement is limited only by what is required for implementing the wireless network protocol.

A fully integrated oscillator implemented on the same silicon as the radio is a lower cost alternative to a quartz crystal oscillator, TCXO, or MEMS oscillator because no separate resonator must be purchased. The physical wireless node size is also lower, which can enable new applications. The power consumption of an integrated oscillator is very low, either similar to a crystal oscillator or slightly lower. It is also easier to use, because unlike a crystal or MEMS oscillator, there is no question about which resonator is compatible with the oscillator circuitry. The main disadvantage of an integrated oscillator is the degraded frequency stability, which means the frequency varies more in response to changes in temperature and supply voltage.

4.3.1 Example Architectures

Oscillators implemented using a higher quality factor resonance will have a more stable frequency in the presence of environment changes and also consume less power. Therefore in high frequency (above 1 GHz) integrated oscillators, the resonant circuit is implemented with an inductor-capacitor (LC) tank circuit. However, for low frequency integrated oscillators used as wake-up timers, the inductor value needed would be far too large to use this approach. Ring oscillators composed of CMOS inverters or current-starved inverters are a low power approach, but the frequency change with supply and temperature tends to be large. For example, a CMOS inverter based ring oscillator will have frequency variation of far more than $\pm 1\ \%$ over voltage and temperature. If the frequency variation were $\pm 1\ \%$, then to meet the $\pm 0.05\ \%$ frequency stability required by Bluetooth Low Energy, compensation to achieve $20\times$ stability improvement would be needed. Assuming a linear TCF, compensation must be performed after each 5 °C temperature change. As the frequency variation with temperature increases, the demands placed on the temperature sensor or other compensation circuits increases. More frequent compensation will also increase the average power consumption.

Because of these issues, the most stable low frequency integrated oscillators are RC-based. Several different topologies exist. Two possible ones are shown in Fig. 7. In Fig. 7a, a current I is used to charge a capacitor C. The voltage across the capacitor is compared to the same current across a resistor and the capacitor is reset when the voltage reaches $I{\cdot}R$. To the first order, the frequency of this oscillator is $f = 1/RC$. The frequency stability will be set by the resistor and capacitor temperature coefficients. If zero or near-zero temperature coefficient passives are used, the frequency stability will be limited by variations in the current mismatch, comparator offset, and comparator delay. These variations can be caused by noise or changes in temperature or supply.

In Fig. 7b, an oscillator is implemented using inverters rather than a comparator. In this oscillator, the capacitor is charged through a voltage range equal to twice the supply voltage. To the first order, the oscillator frequency is $f = 1/(2.2{\cdot}RC)$ and again depends on the resistor and capacitor temperature coefficients. The frequency also varies with changes in inverter delay, which can be caused by noise, or changes in supply voltage and temperature.

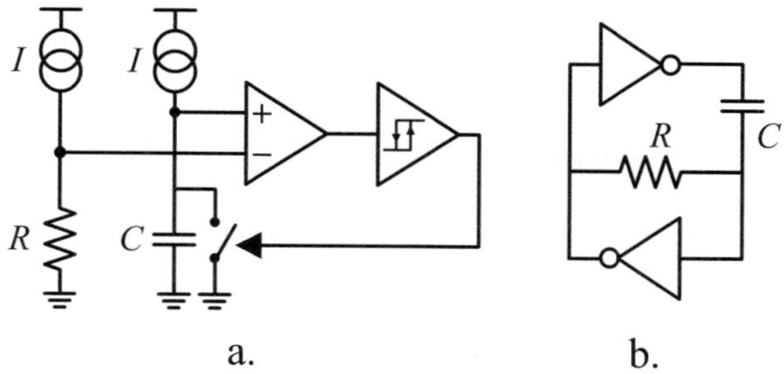

Fig. 7 Two RC oscillator architectures

4.3.2 Design Considerations

As seen from the equations in the previous section, the integrated oscillator frequency stability is directly impacted by the temperature coefficients of passive components. Metal fringe capacitors are linear with little variation over temperature, but integrated resistors can often have a strong temperature coefficient in a CMOS process. To compensate for this, the resistor can be implemented with a combination of two resistors of different temperature coefficients, one positive and one negative, scaled so that the sum is approximately zero. If that is not possible, a resistor external to the CMOS die can also be used. This slightly increases the physical size and cost, but significantly less so than using a crystal.

4.3.3 Further Improvements

Integrated oscillators with frequency stability better than ±1 % over temperature and voltage changes have been published in recent literature [9–13]. Techniques to achieve a stable frequency include feed-forward correction [9], self-chopping [10], and basing the frequency on the electron mobility in a MOS transistor [11]. In [12], the frequency stability for the oscillator in Fig. 7a has been improved 7× by A. Paidimarri by adding chopping to cancel offsets and mismatches. The improved architecture is shown in Fig. 8. Frequency stability was measured to be ±0.1 % from 0 to 90 °C and ±0.25 % from −40 to 90 °C. The remaining sources of frequency variation are any temperature coefficient of the resistor and delay variation in the inverter with temperature or supply changes.

In [13] and in Fig. 9, the oscillator from Fig. 7b has been improved by connecting the oscillator inverters to a locally regulated supply which tracks the inverter parameters. Doing so causes the inverter delay, and therefore the frequency, to vary less with supply and temperature changes. The oscillator is also low power because the inverters are biased in the sub-threshold region. Unlike the oscillator in Fig. 7a or Fig. 8, the capacitor is charged in both the positive and negative direction, from $-VDD_{LOCAL}/2$ to $3\cdot VDD_{LOCAL}/2$, with halves the capacitance area needed for a given frequency, minimizing the oscillator silicon area and cost. Any noise or offsets appearing on the *INV1* input impacts the rising and falling voltage on the capacitor almost equally but with opposite sign so the effect is largely cancelled and the frequency remains stable.

Recent literature shows that it is possible to achieve very low power oscillators, on the order of 100–300 nW at 32 kHz with frequency variation of ~1,000–2,000 ppm (0.1–0.2 %) over a moderate-to-wide temperature range. Both of these are parameters are critical to use integrated oscillators as wake-up timers. The frequency stability achievable by RC oscillators makes them most useful for wireless applications with moderate-to-high duty cycles (0.1–1 %). Finally, having a fully integrated oscillator is useful for applications that need small physical size or ultra low cost. Figure 10 is a conceptual drawing of the wake-up timer clock source choice as a function of the duty cycle and application temperature variation. Crystal oscillators will provide the lowest average system power when temperature variation is high or the duty cycle is very low. In the opposite extremes, using an integrated oscillator will make little impact in the average power consumption while allowing lower system cost and form factor.

4.3.4 Long Term Stability

Even if an oscillator has been designed with a very stable frequency over temperature and voltages changes, the performance may not be sufficient to meet some wireless connectivity standards due to noise. The oscillator frequency drift around the nominal frequency must be measured to understand its suitability for use as a wake-up timer. Allan variance [1] is a statistical measure of the relative frequency instability of an oscillator and is defined as one half of the time average of the squares of the differences between successive readings of the frequency deviation sampled over the averaging period τ.

$$\sigma_y{}^2(\tau) = \frac{1}{2}\left\langle \left(\overline{y}_{n+1} - \overline{y}_n\right)^2 \right\rangle$$

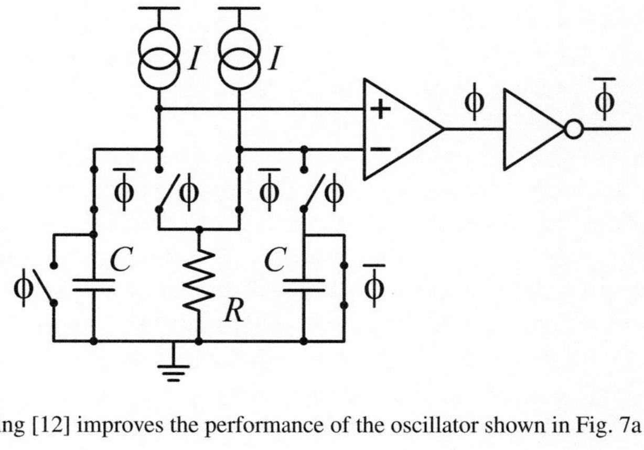

Fig. 8 Chopping [12] improves the performance of the oscillator shown in Fig. 7a

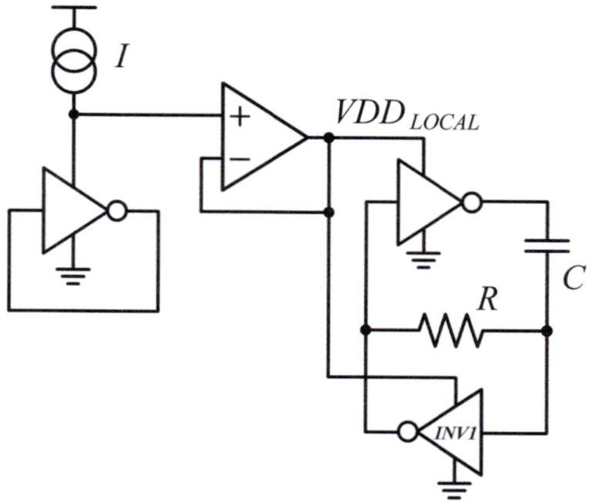

Fig. 9 A local supply [13] that tracks inverter parameters improves the oscillator shown in Fig. 7b

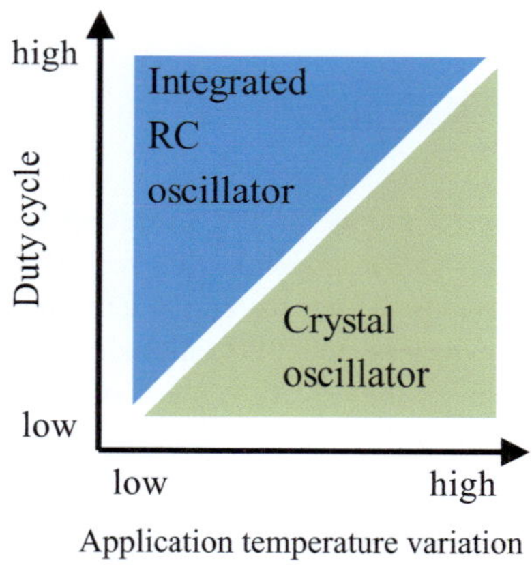

Fig. 10 Conceptual drawing of wake-up timer clock source choice

The Allan deviation, σ_y, is the square root of the Allan variance. Allan deviation is plotted on a log scale versus the averaging time τ. Low Allan deviation means that the oscillator drift is predictable, while high Allan deviation means the oscillator varies a lot around the nominal frequency.

In Fig. 11, Allan deviation is plotted for two oscillators. Curve A is for a crystal oscillator at 32.768 kHz, consuming approximately 300 nW. Allan deviation was measured up to averaging times of 60 s. The deviation is less than 0.01 ppm and the floor has not yet been reached. Curve B is an RC oscillator at 33 kHz with approximately 200 nW power consumption. The Allan deviation floor is around 4 ppm and is reached by averaging times of 20 s. The crystal oscillator achieves much lower Allan deviation allowing better performance as a wake-up timer, at the cost of an external crystal.

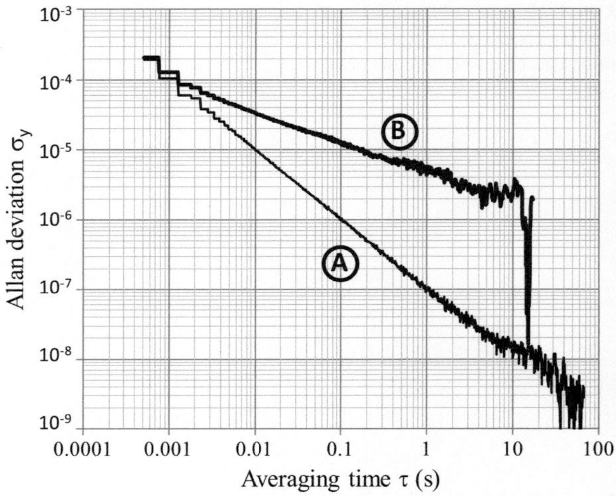

Fig. 11 Allan deviation of a crystal oscillator and an RC oscillator

5 Summary

Improvements in the power consumption, form factor, and cost of integrated radios have enabled wireless connectivity to be added to many applications that previous did not have it. Several methods are possible to lower power consumption of wireless networks, including aggressive duty-cycling of the radio front-ends. Popular duty-cycling methods include waking the primary radio up periodically to check for the presence of an RF signal, using a wake-up radio to detect when the correct RF signal is present and then waking the main radio, and finally using synchronized data transmission. For a wake-on radio or synchronized data transmission approach, each node needs wake-up timer to determine the correct time to enable the radio.

The wake-up timer can be implemented with either a crystal oscillator, temperature compensated oscillator, MEMS oscillator, or a fully integrated oscillator. Crystal oscillators will always give better frequency stability than integrated RC oscillators, but integrated oscillators are rapidly improving, allowing smaller form factor designs supporting new applications. However, for some applications, using an integrated RC oscillator as the RTC source makes little difference in average system power. An RC oscillator for the RTC is a reasonable choice for space-constrained, cost-sensitive applications with a high duty cycle or moderate duty cycle and low temperature variation.

References

1. D. Allan, Statistics of atomic frequency standards. Proc. IEEE **54**(2), 221–230 (1966)
2. Bluetooth specification version 4.0, vol. 6, Part A. www.bluetooth.org
3. A.M. Nicholson, Generating and transmitting electric currents, U.S. Patent 2,212,845, granted 27 August 1940
4. V.E. Bottom, A history of the quartz crystal industry in the USA, in *IEEE Proc. 35th Frequency Control Symp.*, 1981, pp. 3–12
5. E. Vittoz et al., High-performance crystal oscillator circuits: theory and application. IEEE J. Solid-State Circuits **23**(3), 774–783 (1988)
6. J. Lin, A low-phase-noise 0.004-ppm/step DCXO with guaranteed monotonicity in the 90-nm CMOS process. IEEE J. Solid-State Circuits **40**(12), 2726–2734 (2005)
7. K.-J. Hsiao, A 1.89 nW/0.15 V self-charged XO for real-time clock generation, in *2014 IEEE International Solid-State Circuits Conference Digest of Technical Papers (ISSCC)*, February 2014, pp. 298–299
8. S. Iguchi et al., 93 % power reduction by automatic self power gating (ASPG) and multistage inverter for negative resistance (MINR) in 0.7 V, 9.2 μW, 39 MHz crystal oscillator, in *2013 IEEE Symposium on VLSI Circuits (VLSIC)*, June 2013, pp. C142–C143
9. T. Tokairin et al., A 280 nW, 100 kHz, 1-cycle start-up time, on-chip CMOS relaxation oscillator employing a feedforward period control scheme, in *2012 Symposium on VLSI Circuits (VLSIC)*, June 2012, pp. 16–17
10. K.-J. Hsiao, A 32.4 ppm/°C 3.2–1.6 V self-chopped relaxation oscillator with adaptive supply generation, in *2012 Symposium on VLSI Circuits (VLSIC)*, June 2012, pp. 14–15
11. F. Sebastiano et al., A low-voltage mobility-based frequency reference for crystal-less ULP radios. IEEE J. Solid-State Circuits **44**(7), 2002–2009 (2009)
12. A. Paidimarri et al., A 120 nW 18.5 kHz RC oscillator with comparator offset cancellation for ±0.25 % temperature stability, in *2013 IEEE International Solid-State Circuits Conference Digest of Technical Papers (ISSCC)*, February 2012, pp. 184–185
13. D. Griffith et al., A 190 nW 33 kHz RC oscillator with ±0.21 % temperature stability and 4 ppm long term stability, in *2014 IEEE International Solid-State Circuits Conference Digest of Technical Papers (ISSCC)*, February 2014, pp. 300–301

Pulsed Ultra-Wideband Transceivers

Patrick P. Mercier*, Denis C. Daly*, Fred S. Lee, David D. Wentzloff, and Anantha P. Chandrakasan

Abstract Ultra-wideband (UWB) radios offer tremendous promise in terms of achievable data rates due to the large capacity afforded by their inherently large occupied bandwidth. While achieving ultra-high data rates may have been one of the original intents of UWB radios, pulsed-UWB radios have another potential advantage over their narrowband counterparts: energy. By exploiting the large available bandwidth in conjunction with non-coherent signaling, low-complexity and ultra-energy-efficient transmitters can be designed using all-digital architectures that do not require the use of a PLL. Similarly, energy-detecting receivers can receive pulses with low energy-per-bit at high data rates and can be rapidly duty-cycled to minimize overall power consumption. This chapter outlines the main challenges in UWB design, while discussing several representative receiver and transmitter implementations in detail.

Keywords Ultra-wide band • UWB • IR-UWB • Pulsed radios • Low-power radios

---

*Author contributed equally.

P.P. Mercier (✉)
University of California San Diego, La Jolla, CA 92093, USA
e-mail: pmercier@ucsd.edu

D.C. Daly
Maxim Integrated, North Chelmsford, MA 01863, USA
e-mail: denis.daly@maximintegrated.com

F.S. Lee
Google[x], Mountain View, CA 94043, USA

D.D. Wentzloff
University of Michigan, Ann Arbor, MI 48109, USA

A.P. Chandrakasan
Massachusetts Institute of Technology, Cambridge, MA 02139, USA

P.P. Mercier, A.P. Chandrakasan (eds.), *Ultra-Low-Power Short-Range Radios*, Integrated Circuits and Systems, DOI 10.1007/978-3-319-14714-7_8

1 Introduction

1.1 Background

The world-wide use of portable electronics has never been more prevalent than in today's society. To stay competitive, it is increasingly important to design portable electronics to have high performance, small form factors, and long battery life. As a result, much research and development efforts have been spent maximizing the performance and integration of portable electronics in power-constrained environments. However, for applications such as wireless sensor networks, medical monitoring, and asset tracking, the ultimate goal of maximizing performance is superseded by minimizing energy consumption, area, and/or cost [8, 41]. In these types of energy-starved applications, the radio-frequency (RF) circuits typically dominate the overall energy budget. Thus, in order to maximize battery lifetime or minimize the required amount of energy harvesting, new and innovative design techniques are required to reduce the RF circuitry energy burden. This chapter describes ultra-wideband (UWB) circuits and their applicability for low power RF applications, focusing primarily on impulse radio ultra-wideband (IR-UWB).

UWB communication was first demonstrated at the end of the nineteenth century by Marconi with spark gap transmitters, but by the end of the twentieth century, UWB communication was primarily used for only niche military and radar applications; instead, narrowband communication was the dominant wireless communication scheme. However, in 2002 the United States Federal Communications Commission (FCC) issued a First Order and Report permitting the development and operation of UWB systems for communication, measurement, imaging and vehicular radar, which reinvigorated academic and industrial UWB development efforts [19, 20]. The FCC established emission limits of −41.3 dBm/MHz in three different frequency bands: below 960 MHz, 3.1-to-10.6 GHz and 22-to-29 GHz. The 22-to-29 GHz band is intended for vehicular radar systems whereas the other two bands can also be used for communication, measurement and imaging systems.

Naturally, these bands overlap with other FCC band allocations, resulting in direct interference between UWB devices and competing narrowband devices. The FCC only permits this by limiting the *average* emissions of UWB radios to be less than the Part 15 radiation limit for consumer electronic devices (−41.3 dBm/MHz). In other words, UWB devices operate below ambient noise levels. This brings up two important questions. First, if the average signal power is below ambient noise, how does a receiver correctly receive and demodulate UWB communications? To answer this, keep in mind that the FCC regulates the *average* radiated output power; since an IR-UWB signal outputs narrow pulses followed by periods of zero radiated output power, the *instantaneous*, or *peak* radiated output power can be large—often well above the noise floor.[1] In other words, UWB signals are,

[1]The FCC also limits the peak power of a UWB signal to be less than 0 dBm in a 50 MHz resolution bandwidth, though this is often well above the noise floor.

instantaneously, not necessarily completely buried in noise (though to be fair, signal-to-noise ratios (SNRs) can often be low). The second question is: why bother using UWB in the first place? The answer to this question lies with Shannon: a channel's theoretical capacity (assuming additive white Gaussian noise (AWGN)) is equal to $C = B \times log_2(1+SNR)$, and thus a larger bandwidth will give much larger capacity. UWB circuit and system designers can thus leverage this additional capacity in order to achieve ultra-high data rates; alternatively, they can trade-off the available capacity to employ modulation schemes that may be spectrally inefficient, yet result in circuits or architectures that reduce power dramatically. In addition, the fine time resolution enabled by short pulses can enable accurate measurement of distances between objects (i.e., ranging).

According to the FCC, a signal is considered ultra-wideband if it has a -10 dB bandwidth that exceeds the lesser of 20 % of its own center frequency or 500 MHz. Ultra-wideband is commissioned to be an overlay technology, such that it does not disrupt the operation of narrowband devices operating in the same frequency span. Due to concerns about interference with low SNR devices such as the global positioning system (GPS), the average power spectral density (PSD) limit is further reduced in other frequency ranges, as shown in Fig. 1.

The resulting power limit constrains high data rate communication to a range of approximately 1-to-10 m, which is appropriate for wireless personal area network (WPAN) and body-area network (BAN) applications. It is possible, however, to trade-off data rate and/or spectral efficiency for increased transmit distance and/or energy efficiency, which opens up UWB to other low power applications. For instance, applications such as miniaturized flying vehicles require communication distances upwards of 100 m, while minimizing both energy consumption and weight due to limited payload carrying capacities [6]. As discussed throughout this chapter, leveraging the wide available bandwidth of UWB signaling can lead to the possibility of achieving small, energy efficient radios.

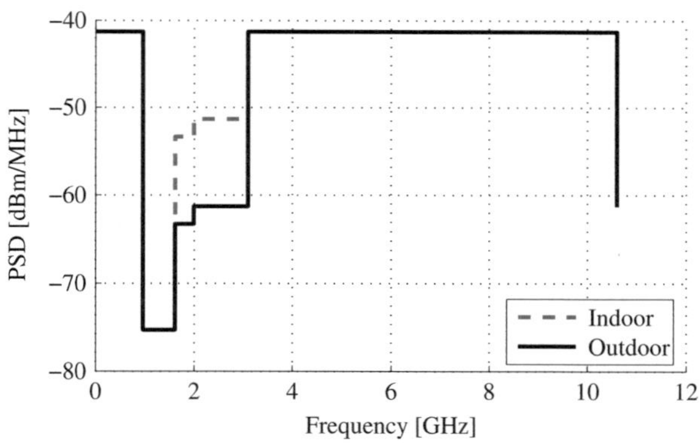

Fig. 1 FCC mask restricting power spectral densities from 0-to-10.6 GHz

1.2 UWB Standards, Proposals, and Communication Schemes

Since the 2002 FCC report did not restrict UWB signaling to any particular scheme, circuit and systems designers have the freedom to choose any type of implementation, provided the spectral masks are met. As a result, several very different techniques were proposed for standardization. One of the early standardization efforts was initiated by the Institute for Electrical and Electronics Engineers (IEEE) 802.15.3a task group, which attempted to add a UWB physical (PHY) layer to the 802.15.3 high-rate WPAN standard. After much deliberation, the task group consolidated the many submitted proposals into two separate proposals: one relying on orthogonal frequency-division multiplexing (OFDM), and the other relying on a form of IR-UWB communication called direct sequence UWB (DS-UWB). Unfortunately the parties could not agree to further consolidation and the 802.15.3a task group was disbanded in 2006 and each technology sought standardization and development elsewhere. The OFDM proposal was adopted by the WiMedia Alliance for high-rate communication and eventually standardized as ECMA-368 and ECMA-369 in 2008. Several industry products compatible with this standard were released but ultimately little commercial success was achieved. The DS-UWB proposal was adapted for low-rate communication as part of the IEEE 802.15.4a amendment. 802.15.4a compliant UWB radios have been released by industry, but long term commercial success of the standard is too early to be determined.

Despite the limited commercial success to-date of products compliant with the EMCA and IEEE UWB standards and amendments, there has been ongoing research and development into using UWB technology in applications that require either energy efficiency, high data rates, localization, difficult-to-intercept communication, or some combination therein. For example, UWB systems have found utility in applications ranging from automotive radar [42], respiration monitoring [63], medical implants, RFID tags [1], and secure military communications. Given the relatively low output power limits, UWB technology appears most differentiated in short range links where either precision ranging is required or unique features of UWB signaling can be exploited in the circuit domain to reduce power consumption, cost, or area. The majority of these applications employ impulse-based communication (IR-UWB).

1.3 IR-UWB

A promising approach for implementing low-power UWB communication involves a time-domain IR-UWB approach [56]. With this technique, pulses of very short duration (on the order of 200 ps to 2 ns) are used to create inherently wideband signals capable of both transmitting digital data and providing ranging and localization information [26]. These wideband signals can be generated to lie directly in the band of interest, or can be generated at baseband and subsequently mixed-up to RF frequencies. Since the radiated pulse power is relatively low due to FCC regulations, IR-UWB receivers must operate at very low signal-to-noise ratios. Correlation and comparison operations are typically required to separate signal information from noise, even at low-to-medium transmission distances.

Compared to narrowband signals, IR-UWB signals are more amenable to be processed in the time domain rather than the frequency domain, which allows for different transceiver architectures with the potential for reduction in cost, area, or power. Due to the wide bandwidth of UWB signals, they can be efficiently amplified and processed with wide-bandwidth, low Q resonant or non-resonant circuits, which can be easily integrated on-chip with minimal area [55]. IR-UWB signaling is highly compatible with digital architectures, and very simple digital pulse transmitters consisting of only digital logic and delay elements have been successfully demonstrated [54].

This chapter describes several IR-UWB transceiver implementations in detail while also highlighting other implementations to provide an overview of the current state-of-the art. Section 2 focuses on IR-UWB receiver implementations, describing a 3-to-5 GHz noncoherent receiver for insect motion control applications and a 9.8 GHz noncoherent receiver for ultra-low power cubic-mm sensor nodes. Section 3 focuses on IR-UWB transmitter implementations, beginning with an overview and classification of architectures, followed by detailed descriptions of an all-digital 3-to-5 GHz transmitter with pulse shaping [29, 31].

1.4 Coherency

Before diving into architectural details, it is first necessary to make an important note about modulation schemes. There are two fundamentally different ways to demodulate data in carrier-based communication systems: coherent versus non-coherent demodulation. Coherent receivers typically lock the incoming carrier phase with a locally generated (and very accurate) carrier or pilot tone, whereas non-coherent receivers discard phase information. For example, consider the non-coherent receiver shown in Fig. 2. The incoming signal is amplified, squared, then integrated over a set window of time. The squaring and integrating operation does

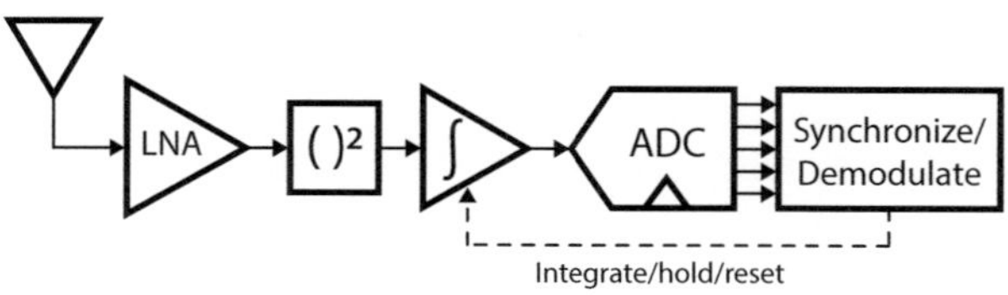

Fig. 2 Block diagram of a non-coherent energy-detecting receiver

not consider phase, and is in fact equivalent to finding the *energy* of a signal in a given window of time. For this reason, this type of non-coherent receiver is called an energy-detecting receiver.

Non-coherent systems have a lower effective data rate for a given bit error rate (BER) compared to the coherent case, since the loss of phase information reduces the number of potential signaling dimensions by one. This limits the types of modulation that can be used, and may result in decreased symbol distances in the constellation diagram.

However, since phase information is discarded, non-coherent systems do not require phase alignment between the transmitter and receiver. Thus, non-coherent receivers are only sensitive to variations in the transmitted *frequency*. If the fractional bandwidth of a system is large (as is the case for UWB), then the *absolute* transmitted frequency accuracy can be relatively low compared to narrowband systems. For example, the IEEE 802.15.4a standard specifies an RF frequency accuracy requirement of ± 20 ppm for coherent signaling, whereas noncoherent UWB signaling can tolerate RF frequency accuracies over $\pm 1{,}000$ ppm. For this reason, non-coherent systems can employ simple architectures with relaxed frequency requirements, and often do not require the use of phase-locked loops (PLLs) or cordic blocks. Thus, noncoherent signaling is frequently used in systems where minimizing power consumption is the main priority over spectral efficiency or wireless range.

2 IR-UWB Receiver Design

2.1 System and Architecture Level Considerations

Due to the differences between narrowband and UWB signals, UWB receivers frequently are implemented with different architectures and circuits than traditional narrowband receivers. For instance, to achieve ultra-low power operation, it is useful for a UWB receiver to be able to quickly turn on and off multiple times during a packet between individual pulses or bits. This behavior differs compared to narrowband receivers which typically remain on at all times while receiving a packet.

Synchronization is also a key challenge of UWB receiver design because pulses are often transmitted with large gaps in between them, multi-path must be carefully considered, and extremely precise synchronization is required for ranging. Included in 802.15.4a is a packet structure and frame format for the UWB PHY. The frame consists of a synchronization header, a start frame delimiter (SFD), a packet header and a data field. The synchronization header provides time for the for the receiver to detect a signal, realize automatic gain control (AGC), synchronize with the transmitter, and implement frequency tracking and several other functions. Embedded in the synchronization header are length 31 or length 127 ternary codes which are repeatedly sent by the transmitter. The UWB PHY specifies forward error correction to be implemented with an outer Reed-Solomon systematic block code and an inner half-rate systematic convolutional code [40]. An interesting characteristic of the UWB PHY is that both coherent and noncoherent signaling are supported. With noncoherent signaling, the receiver can only demodulate the pulse-position modulation (PPM) modulated data and not the binary phase-shift keying (BPSK) modulated data. Thus, the overall data rate is lowered, but simpler, energy-detection receiver architectures are supported.

2.2 Receiver Performance Metrics

To evaluate the performance of receivers, several performance metrics are used including power consumption, data rate and sensitivity. For UWB radios, which frequently operate at fast instantaneous data rates but low duty cycles, it is important to differentiate between peak power consumption and average consumption when duty cycled. A key metric used to compare the energy efficiency of radios is energy per bit, corresponding to the energy required to send or receive a bit of information. Low power radios typically consume less than 5 nJ/bit. The energy/bit metric, while useful, must be evaluated in parallel with receiver sensitivity as well as average and peak power consumption, as generally lower energy/bit values are achieved at higher data rates and worse sensitivity. Table 1 and Fig. 3 present key performance metrics of recently published low power receivers, both narrowband and wideband as well as coherent and noncoherent.

2.3 Design Example: Implementation of a 3-to-5 GHz IR-UWB Receiver

Based on Fig. 3 one can see that a clear trade-off exists between receiver sensitivity and energy per bit. Daly et al. [14] and Mercier et al. [33] achieves a good balance between receiver sensitivity and energy per bit and its implementation is described in detail in this section. The IR-UWB receiver is designed for insect flight control system, with the goal to be able to wirelessly receive commands that control the flight direction of an insect. This system has extremely stringent weight, volume and power consumption requirements, due to the limited carrying capacity of insects. These requirements are similar to distributed sensor network applications.

Figure 4 shows a block diagram of the wireless receiver. The receiver is a noncoherent, energy detection based IR-UWB receiver designed for the 802.15.4a wireless standard. The receiver operates at a peak data rate of 16 Mbps in the

Table 1 Key performance metrics of recently published low power receivers

Author	Data rate (kbps)	Power (mW)	E/bit (nJ/bit)	Sens. at data rate (dBm)	Sens. scaled to 100 kbps (dBm)
Porret [39]	24	1	41.6	−95	−89
Choi [12]	200	21	105	−82	−85
Emira [18]	11,000	114	10.3	−86	−106
Otis [36]	5	0.4	80	−101	−88
Darabi [15]	11,000	360	32.7	−88	−108
Chen [10]	500	2.8	5.6	−80	−87
Lee [25]	16,700	42	2.5	−77	−99
Marholev [28]	3,000	43	14.3	−83	−98
Pletcher [38]	100	0.052	0.5	−72	−72
Zheng [61]	15,600	102	6.51	−75	−97
Weber [50]	2,000	36	17.8	−90	−103
Bohorquez [5]	120	0.4	3.3	−93	−94
Retz [43]	250	30.25	121	−96	−100
Verhelst [49]	20,000	3.1	0.159	−65	−88
Daly [14, 33]	16,000	22.5	1.4	−76	−98
		11	0.7	−50	−72
Brown [7]	30	0.037	1.2	−67	−62

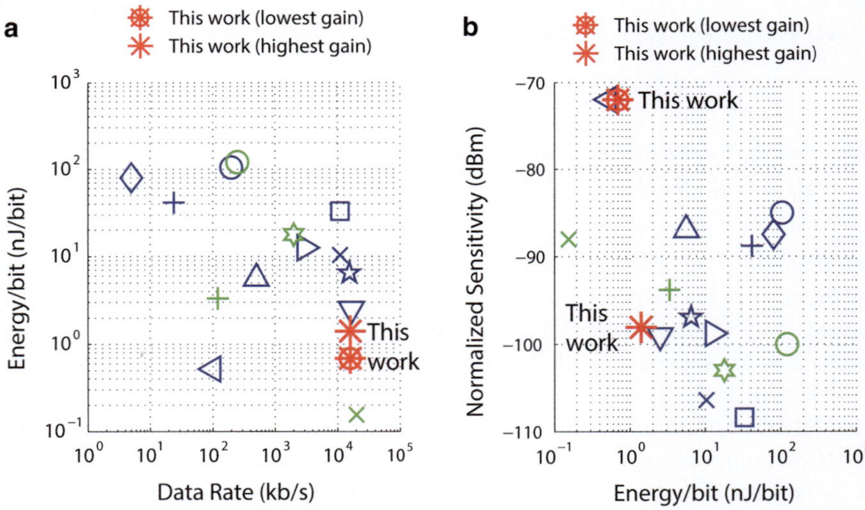

Fig. 3 Two comparison plots of receiver with previously published work: (**a**) energy/bit versus data rate, and (**b**) normalized sensitivity versus energy/bit. In both plots, a point is shown for the receiver at its highest and its lowest gain setting. Data for these plots are found in Table 2

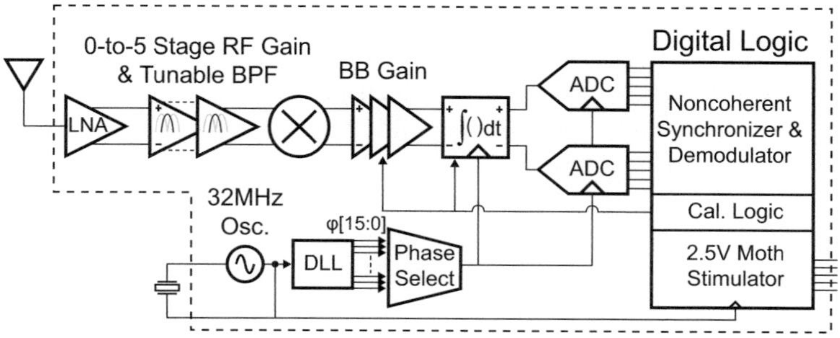

Fig. 4 Detailed block diagram of receiver SoC

3-to-5 GHz UWB band, communicating in one of three 500 MHz channels at 3.5, 4.0, and 4.5 GHz. Through duty cycling, the receiver can operate at lower data rates, thereby reducing average power consumption.

Non-coherent signaling is employed to reduce power consumption on the receiver as it allows for a simple, energy detection architecture without any high frequency clocks. The receiver mixes the received signal with itself at RF, and a windowed integrator and analog-to-digital converter (ADC) at baseband generate a digital signal representing the total energy received in a given time window. This architecture allows for demodulation of both on-off keying (OOK) and PPM signals.

The first stage of the receiver signal chain is an RF front end that amplifies the received signal by up to 40 dB while attenuating out-of-band interferers. This amplified RF signal is then squared, resulting in the RF signal being mixed to baseband. Following the squarer is a baseband amplifier, and then the amplified signal is integrated and quantized by an ADC. The ADC values are passed to a digital backend, which performs packet detection, synchronization and decoding. Also included in the receiver system-on-chip (SoC) is a crystal oscillator and a delay-locked loop (DLL). The entire receiver is clocked by a fixed, 32 MHz clock. After synchronization, the appropriate DLL phase is selected and is used by the windowed integrator and ADC. Each of the specific components of the receiver SoC is described in the following subsections.

2.3.1 RF Front End

For noncoherent receivers, significant gain is required prior to the squarer to obtain a sufficient signal swing such that semiconductor device nonlinearity can be exploited in the squaring element. Passive and active squarers require input voltages on the order of milli-Volts whereas low noise amplifier (LNA) input voltages can be on the order of tens of micro-Volts, thus requiring voltage gain of approximately 40 dB. To achieve such large gain, noncoherent receivers typically employ one of two methods: a super-regenerative architecture [48] or a multi-stage linear amplifier [25]. Although a multi-stage linear amplifier requires more power than a super-regenerative amplifier, it allows for simple support of any arbitrary squaring and integration interval. Moreover, a multi-stage linear amplifier is less subject to RF leakage out of the antenna, which can potentially result in FCC spectrum violations or require the use of an RF isolation amplifier. Based on these advantages, a multi-stage linear amplifier topology is selected, with a per-stage gain of approximately 8 dB.

A key design choice is whether to implement the multi-stage amplifier with single-ended or differential circuits. As the RF front end is integrated on the same chip as digital logic and baseband analog circuits, a differential architecture offers significant advantages in terms of substrate noise and power supply immunity. In addition, reduced decoupling capacitance is required, and a differential structure allows for higher quality factor inductors and virtual ground 'center-tap' nodes. Thus, a differential RF architecture is selected; however, as all commercially available UWB antennas are single ended, and thus the LNA has a single ended input. Single-ended to differential conversion is realized by the LNA and all later stages are differential. Resonant *LC* loads are used instead of non-resonant loads as they offer superior gain in the 3-to-5 GHz frequency band at the same power consumption and also have a second order bandpass characteristic which rejects out-of-band interferers [25].

The schematic of the LNA is shown in Fig. 5. When the LNA is enabled, the switch *en* is closed, connecting the dc output of the differential inverters with the dc input of the inverters. Through negative feedback, the dc voltages at all of the nodes normalize to the same value, V_{CM}. To allow the LNA to turn on rapidly, switches are placed in parallel with R_{S1} and R_{S2} and these switches are briefly enabled while the LNA turns on. In normal operation, R_{S1} and R_{S2} are sufficiently large that the negative feedback does not degrade gain. When the LNA is disabled, the switch *en* is opened, I_{DC} is set to 0 A, and V_{CM} is actively driven to V_{DD}. This allows the output dc voltage to freely float, which is necessary for proper calibration of the receiver.

Following the LNA are five stages of RF gain. Figure 6 presents the schematic of the multi-stage RF amplifier, including the LNA. To dc bias the RF gain stages, the center tap of each stage's inductor is connected to the center taps of adjacent stages' inductors. Due to the differential voltage across each inductor, these center tap

Fig. 5 Schematic of low noise amplifier

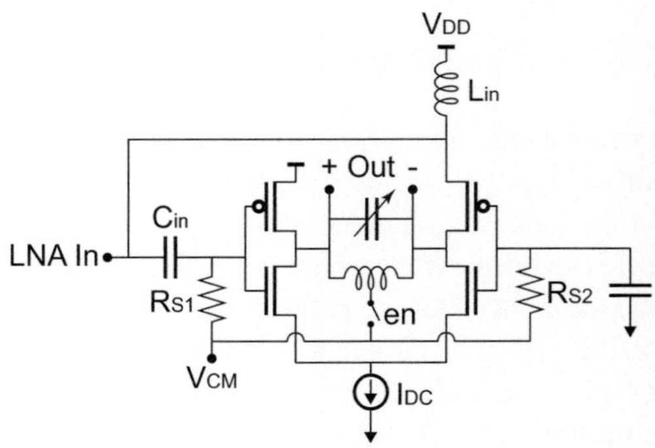

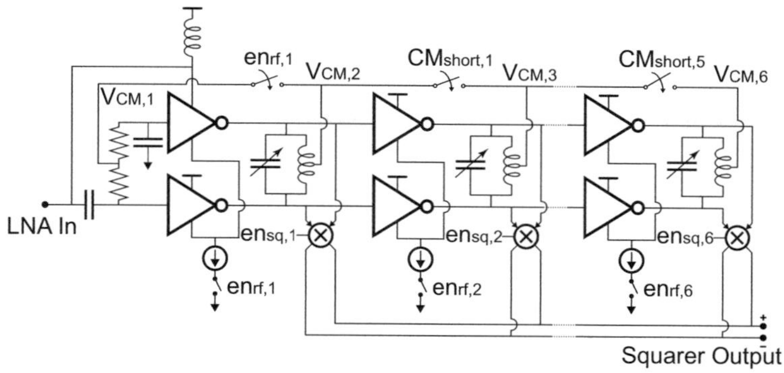

Fig. 6 Schematic of 6-stage RF amplifier, including the LNA. A variable number of stages can be enabled depending on the gain required

nodes are virtual grounds. Moreover, as all RF amplifiers are biased with the same current density, these nodes are nominally at the same dc voltage. By connecting these nodes together with a low impedance connection, the common-mode rejection ratio (CMRR) is superior to what is achieved with more traditional common-mode feedback (CMFB) techniques like resistive feedback. The Monte Carlo simulated common-mode gain of the five stages of RF gain after the LNA has a mean of 7.7 dB and a standard deviation of 7.5 dB at the RF resonant frequency and a mean of 2 dB and a standard deviation of 0.1 dB at low frequencies.

Each gain stage has a squarer at its output, although at any time only one squarer is enabled. Depending on how much RF gain is needed, a variable number of RF gain stages are enabled, as well as the appropriate squarer.

2.3.2 Squarer

A squarer serves two functions in the receiver: to frequency shift (or mix) the received RF signal to baseband and to square its amplitude. It is possible to design an entirely passive squarer that consumes no dc bias current; however, these passive squaring circuits are traditionally single ended [25] or pseudo-differential [37]. In this work, a passive, differential squarer is employed that uses transistors biased in the triode region (Fig. 7). The differential squarer is made possible by the inverter-based RF amplifier, as the output voltage of the RF amplifier is nominally mid-range, thereby allowing both NMOS and PMOS devices to have sufficient gate overdrive. The squarer consumes no static bias currents or active power and has near zero dc output voltage offsets. A key advantage of this structure is that fairly well matched differential outputs are generated. Due to its nonlinear transfer function, the squarer requires RF inputs with amplitudes above approximately 10 mV. At a 10 mV RF input, the single-ended output voltage amplitude is $\sim$0.7 mV.

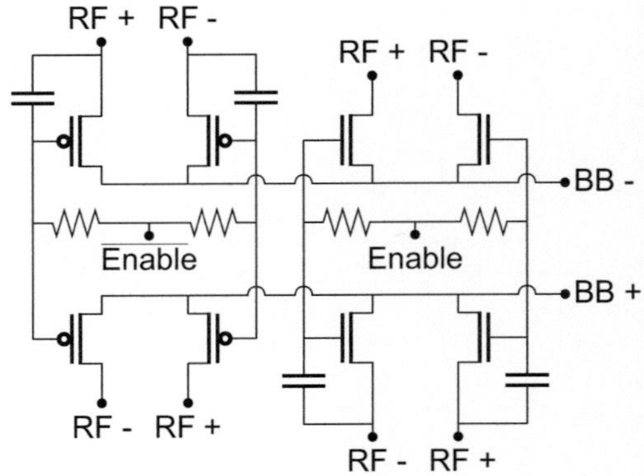

Fig. 7 Schematic of differential, passive squarer

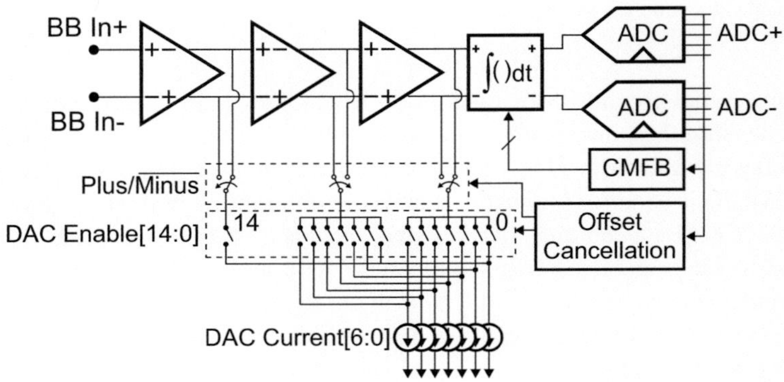

Fig. 8 Baseband signal chain, consisting of a baseband amplifier, an integrator, an ADC, a current-mode DAC that is used to cancel baseband amplifier offsets, and digital calibration logic

2.3.3 Baseband Amplifier

Following the squarer is a baseband signal chain consisting of a three-stage amplifier followed by an integrator and ADC (Fig. 8). The baseband amplifiers are simple differential pairs with resistive loads. The cumulative differential gain of the baseband amplifier chain is simulated to be 83 V/V and the 3 dB bandwidth is 230 MHz. The large baseband gain is required to amplify the squarer output from amplitudes as low as 0.5 mV. Each differential pair operates off a 1 V supply, is supplied 320 μA of current, and has resistive and capacitive loads of 2.5 kΩ and 150 fF, respectively. A multi-stage amplifier is used rather than an op amp due to the wide signal bandwidths and because a high Q filter is not required.

Due to the small input levels and high gain, offset compensation is a critical component of the baseband amplifier. An input referred offset of merely 10 mV would saturate the baseband amplifier. Traditionally, the goal of offset compensation is to establish a 0 V differential output voltage given a 0 V differential input voltage; however, in this system a fixed offset at the output needs to be established to maximize dynamic range. This fixed output offset is required because the baseband signal generated by the squarer is monopolar, meaning that the positive squarer output only increases from its 'zero-input' level and the negative squarer output only decreases. Thus, the positive baseband amplifier output should nominally be biased near the bottom of the amplifier's dynamic range.

Offset compensation is implemented digitally with a current-mode digital-to-analog converter (DAC) in a discrete time process. Rather than a traditional architecture of a binary-weighted DAC connected to the output of the first baseband amplifier stage, the DAC consists of current sources that can connect to any of the three baseband amplifier stages. This allows for fine offset control without requiring very small current sources. To ensure monotonicity as the DAC code increases, the current sources transition from being unconnected, to being connected to the final amplifier stage, to eventually being connected to earlier amplifier stages. Depending on whether a positive or negative offset needs to be cancelled, the current sources can connect to the positive or negative output nodes.

During calibration, the LNA is disabled and the baseband inputs are shorted to the same dc value. Next, the integrator and ADC convert the baseband output to a digital value. The ADC output code is processed by a slope tracking state machine to adjust the DAC until the ADC output code approaches the desired ADC value.

2.3.4 Integrator and ADC

Following the baseband amplifier is an integrator and ADC. Both the integrator and ADC are clocked at 32 MHz, resulting in an integration period of 31.25 ns. The output of the ADC is a digital representation of the total RF energy received within the 31.25 ns integration period. This absolute measurement of energy is preferred to a relative measurement of energy, because it allows for demodulation of both PPM and OOK data.

The ADC consists of two single ended ADCs, operating on the positive and negative integrator outputs and each generating 5 bits of information. The difference between these ADC values generates a 6 bit output code, although if perfect matching is assumed, only 5 bits of useful information is generated. Despite this limitation, the pseudo-differential structure offers improved power supply rejection and common-mode rejection compared to a single ended 5 bit structure, while also allowing for a simpler implementation than a fully differential structure.

Having the integration output quantized to multiple bits is useful for gain control and for accurate timing synchronization. Due to the 5 bits of ADC information combined with coding on the transmitter, the receiver is able to synchronize with an accuracy of ± 1 ns while being clocked with a period of 31.25 ns [32].

The integrator and ADC are jointly designed to not require any high frequency clocks, as well as to allow for a simple integrator that does not need op amps, loads with high output impedance, or positive feedback. A detailed block diagram of the integrator and ADC are shown in Fig. 9. Together, the integrator and ADC are similar to a single-slope integrating ADC, but with some key differences.

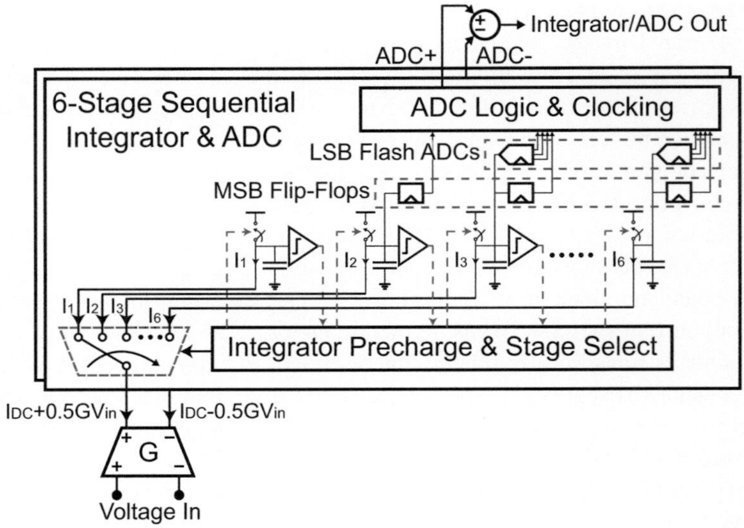

Fig. 9 Six stage sequential integrator and ADC

The differential inputs are first passed through a differential transconductor to convert the input voltage to a current. This current discharges up to six stages from V_{DD} in succession, similar to that of a dynamic inverter. The differential rate of discharge between the positive and negative ADCs is based on the differential input voltage, and thus an integration function is realized. Based on the number of stages that are discharged in the integration period, 2 bits of coarse quantization are generated. Only 2 bits of information are generated from the six stages because the first two stages are not considered in the coarse quantization. The first two stages should ideally always be discharged by the end of an integration period and thus do not contribute information. These first two stages serve to cancel out the static, zero-input dc current of the differential transconductor that is required to appropriately bias the transconductor in a linear region. Additionally, the time while these first two stages are being discharged is leveraged by the final four stages to evaluate the previous integration value.

The ADC generates an additional 3 bits of fine quantization that are combined with the 2 bits of coarse quantization. These 3 bits are generated by quantizing the capacitor voltage of the stage that was being discharged at the end of the integration period with a flash ADC. The capacitor voltages on stages three through six are temporarily held constant while the appropriate flash ADC resolves. During this time period, the next integration period has already begun by discharging stage one. A simple flash ADC with a resistive ladder DAC is used to generate these 3 bits. Thus, 5 bits of data are generated by the integrator and ADC. Both positive and negative outputs of the transconductor are independently processed by this integrator and ADC structure, and thus a pseudo-differential output is generated. The integrator and ADC architecture would only need slight modifications to allow for the use of a differential ADC.

2.3.5 Clocking

The SoC is designed to be clocked off a fixed 32 MHz oscillator that is always enabled. Due to the noncoherent signaling, clock frequency and timing synchronization accuracy requirements between transmitter and receiver are dramatically reduced. Through the use of a Pierce oscillator stabilized with a quartz crystal, it is possible to achieve frequency accuracies on the order of ± 20 ppm [2], allowing the transmitter and receiver to require only one synchronization per packet, without any phase tracking during the packet payload of up to 1,600 bits.

For the receiver to successfully decode data, the integrator and ADC must be phase aligned with the received data. This phase alignment is achieved with a digital synchronization algorithm and a DLL. Based on the result of the digital synchronization, an appropriate phase from the DLL is used to clock the integrator and ADC. During synchronization, the DLL is bypassed and the integrator and ADC are provided the same clock phase as the rest of the digital logic. As the DLL is not being used, the DLL can be calibrated during this time by a successive approximation register (SAR) state machine.

The digital baseband achieves synchronization accuracy of ± 1 ns in an integration window of 31.25 ns, and the DLL is designed to match these specifications. The DLL has 16 outputs, each nominally spaced 1.95 ns apart from one another. Due to the noncoherent signaling, the DLL does not need to have good linearity, and thus it is possible to use very simple delay elements and simple calibration logic. The core delay element consists of a current starved inverter, and a simple DAC is used to control the bias current of the inverter. All outputs of the DLL are passed to a digital, synchronous state machine.

As the integrator and ADC operate from a different clock phase than the rest of the digital logic, there is a potential for timing violations or clock offsets at the interface. To address this problem, the ADC outputs are retimed with registers. These retiming registers can be either positive or negative-edge triggered to ensure sufficient setup and hold time.

2.3.6 Digital State Machine and Duty Cycling

Since the receiver peak data rate of 16 Mbps is much larger than the required data rate in the system, the receiver is designed to be duty cycled. Duty cycling is implemented through the use of a programmable digital state machine. Between packets, the radio and modem are disabled and all digital logic is clock gated except for a sleep counter. This low power sleep mode continues until the sleep counter reaches a programmable count value. At this point, the receiver state machine is triggered, and the receiver attempts to receive a packet.

To receive a packet, the digital state machine first enables the RF and analog circuits, which turn on within one clock cycle. Before the receiver modem performs packet detection, the receiver state machine performs calibration of the DLL, baseband amplifier and integrator. This calibration only takes a few microseconds, and is performed before every packet reception to account for any change in temperature or supply voltage since the last packet reception attempt.

2.3.7 Digital Baseband Synchronizer

Since the transmitter and receiver are not normally phase synchronized, there is no guarantee that the integration window of the receiver front end lines up perfectly with the pulses generated by the transmitter. Thus, the primary purpose of the digital baseband is to perform this synchronization in order to maximize the SNR seen at the receiver ADC. In addition, the digital baseband should understand where the start and end of a packet is in order to properly demodulate the received payload.

In conventional RF systems, synchronization between the transmitter and receiver is typically achieved by transmitting a known preamble code, and having the receiver compare all possible time-shifts of this known code to the signal it is receiving; in doing so the receiver will acquire the precise phase of the incoming signal. Comparison between all time shifts of the code under the presence of noise is typically achieved using a correlator structure—specifically, a matched filter. While this is indeed the optimal solution under a linear, AWGN channel assumption, the squaring element employed by the energy detector in this architecture is inherently non-linear. As a result, a matched filter is not the optimal solution. As discussed in [4], the optimal maximum likelihood solution involves the computation of Bessel functions, which are computationally inconvenient in a low-power implementation. To overcome this, the receiver employs a quadratic correlation technique that simplifies the maximum likelihood expression into one that is amenable to a low-complexity implementation while offering improved performance compared to a simple matched filter.

The digital baseband is comprised of 512 parallel quadratic correlators that computes all 2,048 possible preamble code shifts in a minimum of 14 μs. A detailed description of the digital baseband design, including considerations regarding code choice and circuit-level optimizations, is discussed in [33].

2.3.8 Measurement Results

The receiver is implemented in a 90 nm CMOS process and a die photo of the chip is shown in Fig. 10. The die area is 2.6 mm by 2.1 mm, and the area is dominated by digital logic, which occupies the right side of the die. Due to the significant amount of digital logic integrated on the same die as the RF front end, there is significant potential for digital supply and substrate noise to result in degraded analog and RF performance. This motivated the use of a differential receiver architecture. Additionally, substrate contact rings are used to isolate the digital and analog blocks, as well as reduce the potential for feedback coupling in the high gain RF front end. The receiver is packaged in a 40-lead quad flat no-leads (QFN) package and mounted on an FR4 printed circuit board (PCB).

Fig. 10 Die photograph of pulsed UWB receiver SoC

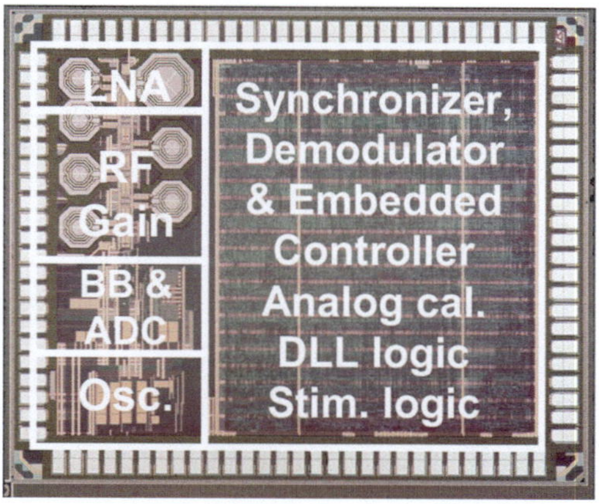

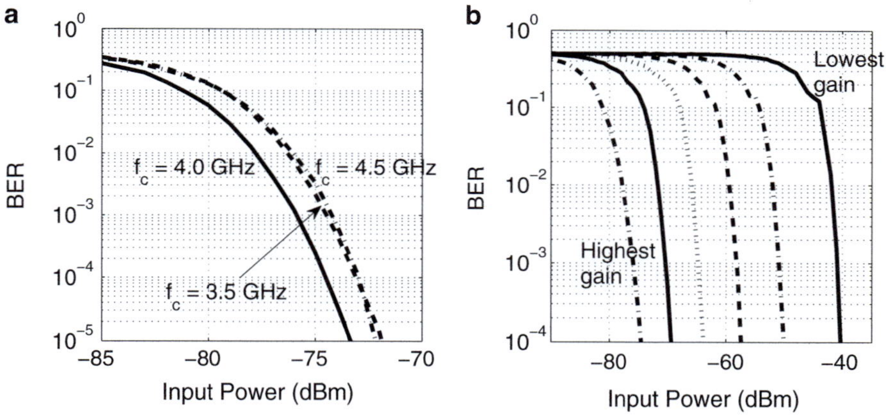

Fig. 11 BER of receiver (**a**) at its highest gain setting at the three center frequencies, and (**b**) at the different gain settings with f_c =4.0 GHz

Figure 11 presents the BER of the receiver in different frequency bands at its highest gain setting and at different gain settings with f_c = 4.0 GHz. The receiver achieves a maximum sensitivity of −76 dBm at a data rate of 16 Mbps and a BER of 10^{-3}. The sensitivity scales by 35 dB from the lowest to highest gain setting, allowing for a trade-off of power consumption for sensitivity.

As the receiver SoC is targeted for low power, highly energy constrained applications, significant effort was spent to minimize overall power consumption and energy/bit. A breakdown of power consumption is shown in Table 2. Due to the extensive digital logic and the absence of power gating switches, the total leakage power is 0.64 mW. The always-on crystal oscillator consumes 0.15 mW. When the receiver is in idle mode, the majority of the clock tree is gated; however, an additional 0.13 mW of power is still consumed. The overall receiver power consumption is dominated by the LNA and the RF amplifiers that follow the LNA. Each individual RF amplifier consumes approximately 2.85 mW of power

Table 2 Receiver instantaneous power consumption breakdown

Receiver component	Power consumption (mW)
Leakage	0.64
Crystal oscillator	0.15
Clock tree (idle)	0.13
Delay locked loop	0.05
Baseband amplifier & ADC	1.51
LNA	5.90
RF Amplifier	0–14.30
Total idle power	0.92
Total active power	8.38–22.69

consumption, and the five-stage RF amplifier consumes a total of 14.30 mW of power when all five stages are enabled. At a data rate of 16 Mbps at the lowest gain setting, the entire receiver consumes 8.38 mW of power and at the highest gain setting, the receiver consumes 22.69 mW of power. When the receiver is duty cycled to low, kb/s data rates, the average power consumption is reduced to the order of a few milli-Watts, ultimately limited by leakage power. By adding power gating switches, the average receiver power consumption could approach the micro-Watt level at kb/s data rates. The receiver power consumption is constant regardless of the RF center frequency and includes the power of the digital backend when decoding data; however, these power measurements do not account for the energy required for synchronization at the start of a packet. As the receiver operates at an instantaneous data rate of 16 Mbps, the energy/bit of the receiver is 0.5-to-1.4 nJ/bit depending on the gain setting.

2.3.9 Receiver System Implementation

For the insect flight control system, some additional electronic components are required alongside the SoC. Figure 12 shows a block diagram of the electronics that are used. The key components include the receiver SoC, a microcontroller, 2.5 V DC-DC converter, 1 V low-dropout regulator (LDO) regulator, miniature coin cell battery, on-off switch, crystal resonator, LED, antenna, and discrete inductors, resistors and capacitors. The electronic components are soldered to a flexible, 4-layer PCB. A flexible PCB allows for a 60–70 % reduction in weight and thickness compared to a rigid PCB. Photos of the PCB are shown in Fig. 13. The entire system consumes an average power of 2.5 mW when the receiver attempts to receive a 68 bit synchronization packet every 1 ms.

The electronics are powered by a 1.4-to-1.6 V Silver-Oxide, size 362 coin cell battery that is capable of sourcing the 2.5 mW consumed by the electronics. The battery has a typical capacity of 27 mAh, weighs 0.32 g, and has a impedance at 40 Hz of 10-to-20 Ω. As the receiver SoC requires 1.0 and 2.5 V supply voltages, dc-dc converters are used to generate the required voltages from the battery.

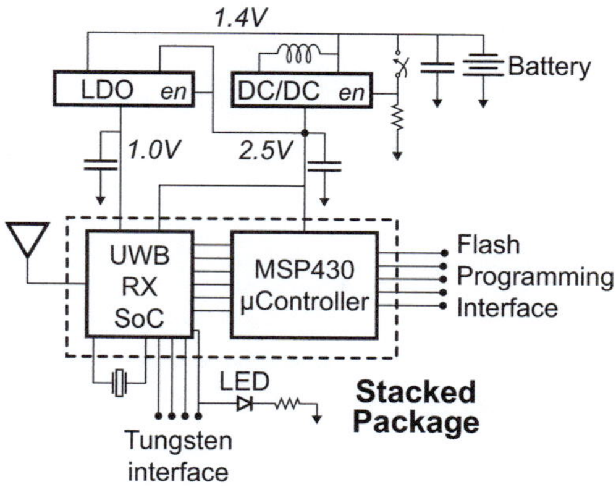

Fig. 12 Block diagram of electronics mounted on a flexible PCB and attached to a moth

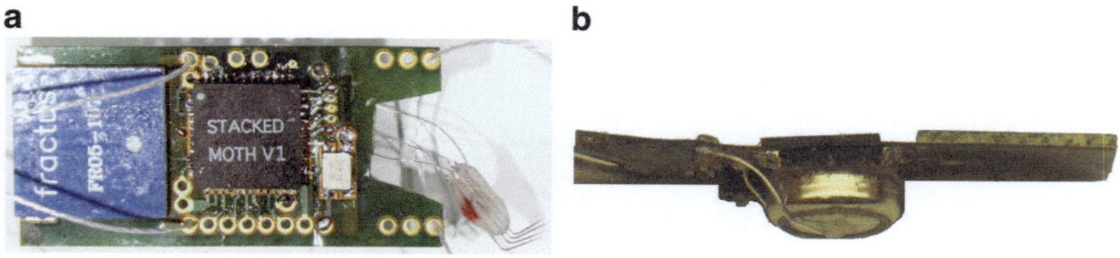

Fig. 13 Flexible PCB (**a**) top and (**b**) side

To further reduce form factor and weight, only a single decoupling capacitor is used for each supply voltage. A miniature on-off power switch is used to enable the dc-dc converters, so that the receiver does not consume any static current when turned off. Additional details on the flight control system implementation and measurement results are presented in [14].

2.4 *Design Example: Implementation of a 9.8 GHz IR-UWB Receiver*

Future biomedical and internet-of-things applications are driving the volume of wireless sensors into the cubic-mm regime, with power and volume requirements significantly more stringent than those demonstrated by the 3-to-5 GHz receiver described in the previous section. At the mm-scale, complete integration is necessary, and operation within the limits of a micro-battery becomes a primary challenge [11]. With CMOS scaling and ultra-low-power circuits reducing battery volume, the antenna and crystal quickly become the largest components in a cubic-mm node. In [7], a UWB transceiver for cubic-mm sensor nodes is demonstrated that achieves average receiver power levels of 37 µW at 30 kbps. Such a low average power consumption is achieved through extensive circuit and system optimizations,

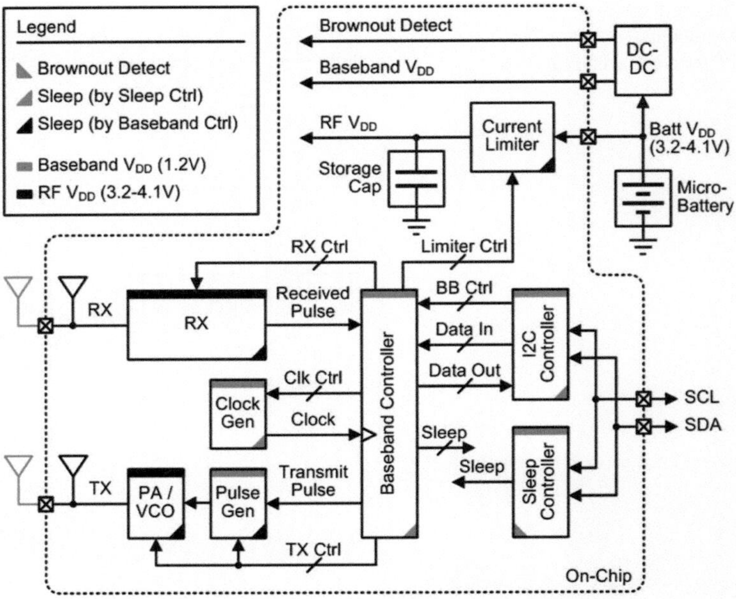

Fig. 14 System block diagram of the entire crystal-less UWB radio

including removing the need for a crystal oscillator, minimizing leakage current, fast duty cycling, operating circuits directly off a battery, and allowing for degraded sensitivity and range. This section briefly summarizes the receiver implementation and measurement results.

A block diagram of the receiver and its associated transmitter is shown in Fig. 14. The transmitter and receiver operate at the battery voltage, through a current limiter (CL) to protect the micro-battery from over-current and under-voltage. An internal storage capacitor allows higher current draws from the transmitter (TX) and receiver (RX) during duty-cycled operation. Digital baseband blocks operate from a 1.2 V VDD to reduce power consumption. To survive on the limited resources of the micro-battery, all blocks on the radio have a low-power sleep state. RF and other analog blocks are duty-cycled at the bit-level by the baseband controller, while baseband blocks are duty-cycled at the packet-level by a separate sleep controller. The sleep controller remains on-continuously unless an under-voltage condition occurs. The sleep controller begins and ends the wake-up procedure for each packet via I2C communication with modified I/Os to eliminate pull-up resistors. The I2C controller provides bidirectional communication with other stacked die in a sensor node.

The receiver uses the non-coherent, energy-detection architecture shown in Fig. 15, similar to the 3-to-5 GHz receiver described in the previous section. Four RF gain stages amplify the 9.8 GHz UWB pulses before down-converting with a squaring mixer. The signal then passes through a baseband gain stage before the

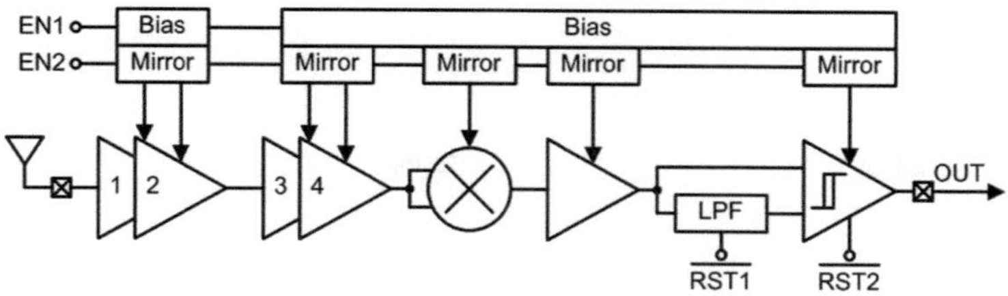

Fig. 15 Block diagram of the crystal-less UWB RX

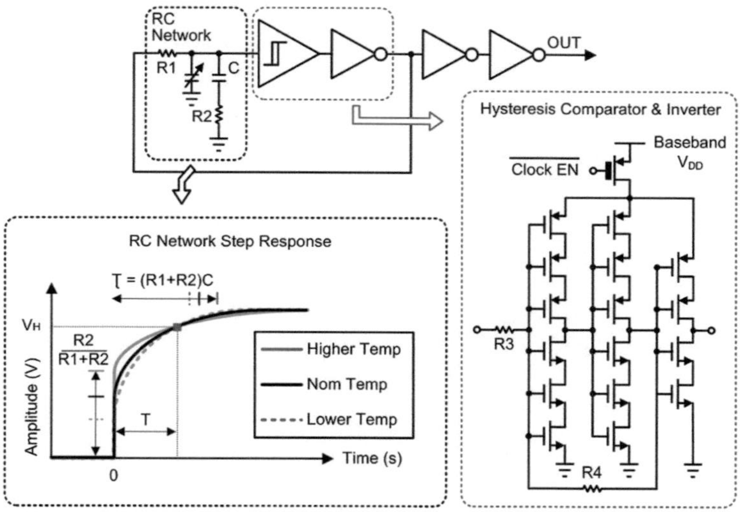

Fig. 16 Schematic of the temperature-compensated relaxation oscillator

signal path is split. Along one path, the pulses are passed directly to a comparator. The other path low-pass filters (LPFs) the signal to provide an auto-zeroed, DC-compensated reference level for comparison. A reset signal enables fast settling of the LPF for fast RX turn-on. Finally, a continuous-time latching comparator with controllable hysteresis digitizes the incoming pulses. BJTs are used for higher RF gain efficiency (gm/I), while the RF gain stages are stacked in order to reuse current and better utilize the supply voltage. The RF center frequency is tunable via 4 binary-weighted control bits. After RF amplification, the signal is self-mixed to dc using a common emitter amplifier with resistive load as a squaring mixer.

To reduce both power and area, the radio includes a relaxation oscillator with a modified RC network and a single-ended hysteretic comparator for on-chip clocking (Fig. 16). The RC network adds an additional zero in the transfer function from R2 over conventional relaxation oscillators, providing an additional degree-of-freedom for temperature compensation. As temperature increases, the initial step at $t = 0$ from the zero increases, but the time constant of the exponential decay also increases, offsetting the step and resulting in a constant time, T, to trigger

General		
	Process	0.18µm BiCMOS
	Modulation	PPM
	Total Area	2.73mm^2
	Data Rate	30kb/s
	RF Voltage	3.2-4.1V
	Baseband Voltage	1.2V
	Sleep Power	1.0nW @3.6V 1.8nW @1.2V

Baseband		
	Active Power	269µW
	Clock	13µW
	I2C & Sleep Ctrl	10µW
	Baseband Ctrl	246µW

Clock		
	Frequency	3MHz
	Supply Sensitivity	±1.5% @±4% V_{DD}
	Temp Sensitivity	±1.0% @0 to 50°C
	Temp Coefficient	-584ppm/°C

RX Frontend		
	Average Power	37µW @3.6V
	Sleep Power	347pW @3.6V
	Center Frequency	9.7-10.2GHz
	Sensitivity	-67dBm @10^{-3} BER

TX Frontend		
	Average Power	22.4µW @3.6V
	Sleep Power	170pW @3.6V
	Center Frequency	9-12GHz
	Output Power	0.1dBm
	Pulse Width	1.2-6.0ns

Current Limiter		
	Active Power	223nW @3.6V
	Sleep Power	550pW @3.6V
	Output Current	6-38µA
	Temp Sensitivity	-34nA/°C
	Line Sensitivity	0.21µA/V
	Efficiency	94% @3.6V

Fig. 17 Summary of the Radio Performance

the switching threshold, VH, so that the overall period remains unchanged. The comparator consists of two stacked inverters with hysteresis levels set by R3 and R4. Stacking the FETs reduces leakage power while the oscillator is asleep, and a 5-bit capacitor bank is added to the oscillator for one-time process calibration of frequency. The oscillator has a measured variation of 1 % over a range of 0–50 °C that allows the TX and RX to be heavily duty-cycled between pulses in order to give the on-chip storage capacitor time to fully recharge and also sufficient accuracy to maintain network synchronization.

The radio was fabricated in 0.18 µm BiCMOS with MIM capacitors. At a 10^{-3} BER, the RX has a sensitivity of −67 dBm and a 30 kb/s data rate while consuming an average of 37 µW from a 3.6 V supply with 6 % duty-cycling. The modem uses PPM and includes early/late tracking of pulses for each PPM window to maintain synchronization. At a 3 MHz oscillation frequency, the entire baseband system consumes 269 µW, of which the clock consumes 12.7 µW. The CL has a 6–38 µA tuning range, which is sufficient for sustained operation of the TX and RX. The CL consumes only 223 nW, yielding a 94 % efficiency. Each block consumes <1 nW while asleep by carefully including thick-oxide headers on all blocks, making this system ideal for heavily duty-cycled cubic-mm sensor nodes. A complete performance summary is provided in Fig. 17. The die occupies approximately 2.73 mm^2, dominated by the modem (Fig. 18). The entire radio is designed to operate from just the seven pads on the left edge to enable die stacking; the remaining pads are for debugging and may be left open.

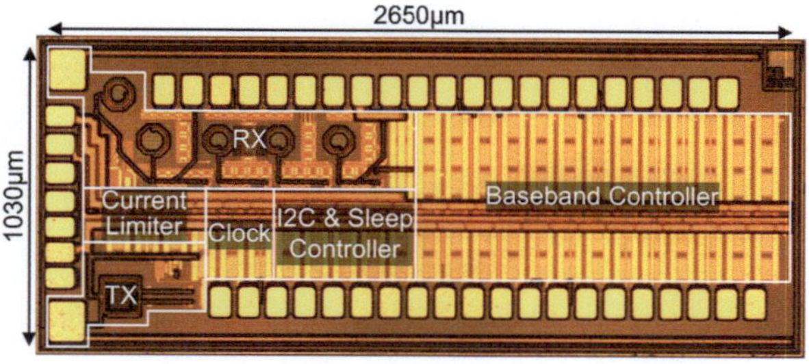

Fig. 18 Die photo of the radio

3 IR-UWB Transmitter Design

This section will first review general classes of IR-UWB transmitters, followed by a detailed description of an example architecture [29, 31].

3.1 IR-UWB Transmitter Architectures

At the simplest level, an IR-UWB transmitter must generate narrow RF pulses and interface these pulses with an antenna, often through a power amplifier. Generation and radiation can typically be distinguished using two sets of criteria:

1. *RF generation*: There are two primary techniques used to synthesize UWB pulses at RF frequencies. The first technique involves mixing a baseband pulse with a local oscillator (LO) running at the desired RF center frequency. The second technique involves generating UWB pulses to lie directly at the desired RF center frequency. In other words, the second technique does not use a local oscillator.
2. *Power amplification*: There are two different techniques used to amplify and interface pulsed signals with an antenna. The first technique involves using analog circuits biased in their active regions for small-signal amplification and balanced conversions. The second technique uses digital circuits to buffer pulses at the interface to the antenna.

This criteria will be used to classify various pulse-generation techniques into four different categories. As a forewarning, it should be mentioned that it is sometimes difficult to make clear classifications, as some pulse generators use a combination of different techniques.

3.1.1 Traditional Small-Signal, Mixer-Based Transmitters

In these types of architectures, baseband data is typically converted from the digital to analog domain and subsequently mixed with an LO. The output of the LO is then amplified by an analog power amplifier (PA), often biased as class A or class AB in

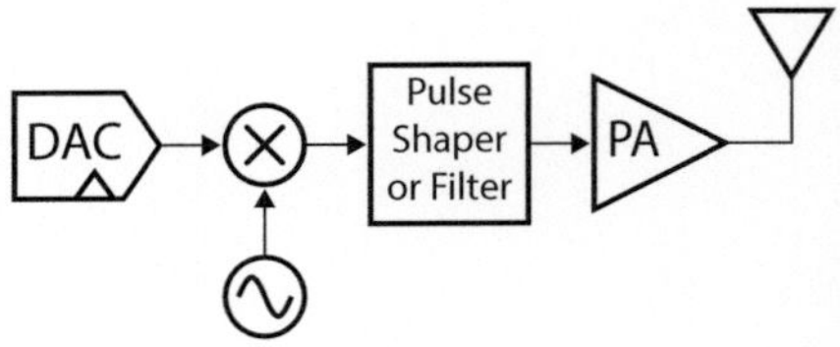

Fig. 19 A traditional small-signal, mixer-based pulse generator architecture

order to meet linearity requirements. A simplified example architecture can be seen in Fig. 19. The initial popularity of this technique stemmed mainly from the fact that similar techniques are well established in traditional narrowband radio design.

From a signaling point of view, this type of architecture is the most robust, as both phase and amplitude modulation are possible.[2] Pulse shaping, used to attenuate RF sidelobes in order to meet FCC spectral masks, is also easily achieved in these types of architectures by either shaping the baseband data, or the RF data before the PA. For instance, the transmitter considered in [53] employed approximate Gaussian pulse shaping in the mixer by utilizing the exponential response of bipolar transistors.

The transmitter considered in [60] operates in dual bands by simultaneously up-converting two data streams onto two separate RF carriers. This is made possible by a wide bandwidth power amplifier employing shunt peaking and inductive feedback. A similar design is shown in [61], however only one band is operated in at a time. The transmitter supports all of the 802.15.4a specifications and reduces the power consumption over [60] by aggressively duty-cycling the class A power amplifier.

3.1.2 LC-Based Transmitters

LC-based transmitters use an LO to generate RF content, yet an explicit mixer is not necessarily required. For instance, a simple switch can either pass or block the LO output, thus effectively mixing the RF signal with a rectangular baseband pulse. As an example, the transmitter considered in [37] operates in a similar fashion to a superregenerative receiver; that is, the output of an LC oscillator is the transmitter output itself. A schematic is shown in Fig. 20. The oscillator can directly connect to the antenna (as a "power oscillator"), or an explicit PA can be employed.

In this example, a rectangular quenching pulse train acts as the baseband mixing signal. Like most oscillators, this circuit can be modeled as a second order system with poles in the right half portion of the s-plane. It is well known that the oscillatory output of such systems grow exponentially in time until circuit non-linearities limit the output swing. This oscillation growth can be leveraged to employ simple, low-overhead pulse shaping.

[2]Note that in-phase and quadrature paths are often used to enable quadrature amplitude modulation (QAM) for high data-rate communication.

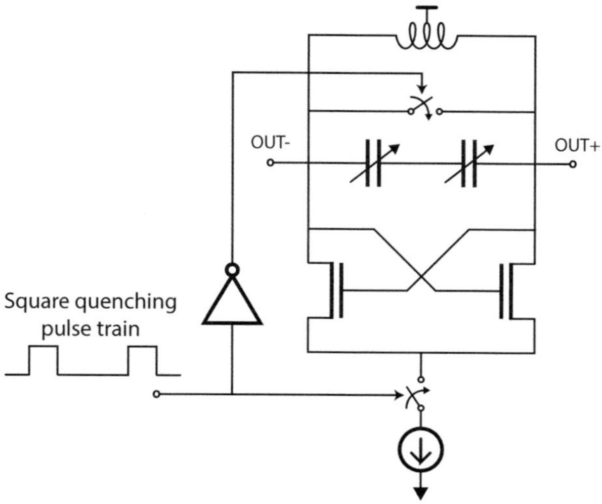

Fig. 20 An LC-based transmitter

3.1.3 Carrier-Less Transmitters

Carrier-less transmitters do not have an explicit local oscillator to mix baseband data up to RF. Instead, baseband data typically triggers a pulse generator to synthesize a pulse directly in the RF band of interest. One advantage over traditional mixer-based architectures is that the carrier frequency generation is inherently duty cycled; that is, RF energy is only generated when it is required. A disadvantage of this approach is that an integrated downconverting receiver typically cannot share the RF generation circuits and therefore must have a separate LO.

Implementation strategies typically involve generating pulses by combining edges of various delay elements, then amplifying the result using a power amplifier [24, 59]. Architectures that use delay lines to synthesize RF frequencies typically have half-RF cycles available at the output of each delay cell. By exploiting the fact that each half-RF cycle can be manipulated, simple pulse shaping schemes and differential-to-single-ended conversions are possible [17, 34, 62]. A popular architecture involves feeding half-RF cycles to alternating sides of a wideband balun, as shown in Fig. 21. This architecture ensures there is close to zero DC content at the transmitter output, thus enabling clean BPSK modulation.

3.1.4 All-Digital Transmitters

All-digital pulse generators attempt to reduce the power consumption over their analog counterparts by eliminating large static currents required to bias transistors in their active regions. Instead, digital static CMOS gates are used to generate high frequency rail-to-rail voltage swings. These digital architecture dissipate only $CV^2 f$

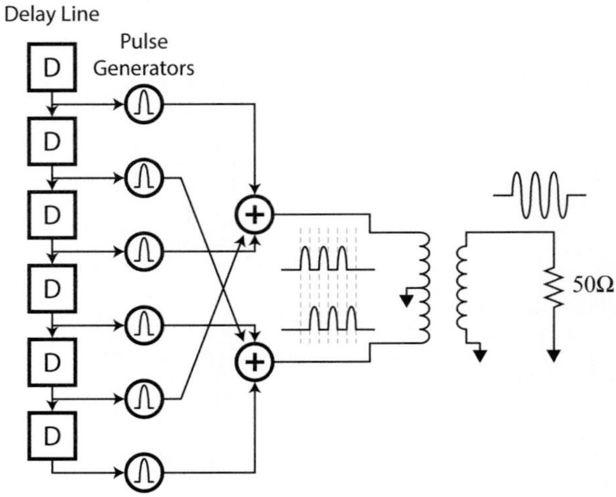

Fig. 21 A carrier-less architecture employing a balun for zero-DC voltage pulse generation

switching power and subthreshold leakage power. Since digital edges have harmonic content, pulse shaping and filtering may become necessary to reduce RF sidelobes.

A similar all-digital technique is popular in narrowband radio design, where linear power amplifiers are replaced by switched-mode power amplifiers. A major drawback of this approach is that constant-envelope modulation schemes must often be used, as switched-mode PAs have poor linearity and thus cannot support amplitude modulation techniques. Similarly, all-digital UWB transmitters are often restricted to phase and position modulations schemes only, unless clever pulse-shaping techniques are introduced.

Pulses in all-digital transmitters can either be synthesized using carrier-less techniques, or by modulating the output of a digitally-controlled oscillator (DCO). Examples of carrier-less techniques include the transmitters presented in [46, 47]. In these examples, UWB pulses are generated directly in the band of interest by NOR-ing two delayed edges together, converting from single-to-differential, and applying the differential signal to a dipole antenna. Similarly, the transmitter considered in [23] generates pulses by combining inverter gate delays using NOR and NAND structures. Relying on uncalibrated gate delays, however, leads to significant deviations in frequency and bandwidth targets over process voltage and temperature (PVT) variation.

The transmitter considered in [45] generates pulses in a carrier-less fashion by combining output edges from a delay line. BPSK modulation is achieved by applying full-swing pulses to either input of a balun, as illustrated in Fig. 22. However, a bandpass filter is required to reduce the low-frequency content typically associated with digital pulse generation driving a single-ended antenna.

The other technique to generate UWB pulses digitally is to modulate the output of a DCO. For instance, the transmitter considered in [44] generates pulses by

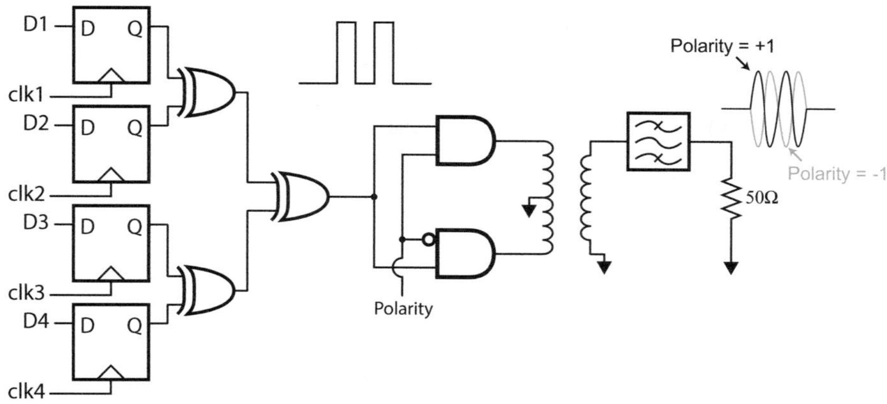

Fig. 22 An all-digital architecture employing a balun for BPSK modulation

directly modulating the output of a three-stage inverter-based ring oscillator. By utilizing the phases of an on-chip frequency divider, discrete two-level pulse shaping is employed. Since digital circuits are used in this architecture, reconfigurability and calibration are easily implemented.

3.2 Design Example: An All-Digital Non-coherent IR-UWB Transmitter Meeting FCC Spectral Masks Without Off-Chip Filters

3.2.1 Motivation for Non-coherent Transmitter Architecture

Coherent modulation schemes (e.g., BPSK or QAM) are generally more spectrally efficient than non-coherent modulation schemes (e.g., OOK or PPM), and thus in theory should be preferred. However, coherent modulation requires phase synchronization between the transmitter and receiver, resulting in more complex implementations that may not feature superior energy per bit. In addition, coherent IR-UWB systems suffer from significant multi-path fading, requiring high-complexity and power-hungry RAKE-based techniques for path consolidation [3].

While less spectrally efficient, non-coherent architectures often feature lower-complexity architectures, resulting in circuits that may consume lower power. Importantly, since precise phase information is not required and the RF bandwidth is large, a precise oscillator derived from a PLL is not required; instead, low-complexity, low-power RF generation techniques are available for use. In addition, non-coherent architectures can have inherent robustness to multi-path effects [51]. In an energy-detecting architecture, for example, the incoming signal is squared, then integrated over a set window of time. If the integration time window is set to be larger than the width of the pulse, the energy of several propagation paths will be collected. Furthermore, the shape of the received pulse is no longer of concern to the receiver.[3] Thus, non-coherent IR-UWB architectures have considerable promise in terms of energy efficiency, and thus a non-coherent architecture is employed in this design example.

3.2.2 Digital Pulse Generation

In principal, it is actually quite simple to design an all-digital IR-UWB pulse generator. For example, the transmitter shown in Fig. 23 uses an LO, a switch, and an inverter-based PA to generate and radiated pulsed-RF waveforms. Data in this transmitter can be modulated using OOK, or PPM, as illustrated in Fig. 24.

Naturally, this overly simple architecture suffers from several drawbacks including lack of programmability and calibration. Additionally, it is difficult to control the spectra, and as a result it is nearly impossible to meet the FCC spectra mask without dramatically reducing the average output power. As shown in Fig. 25, the resulting power spectral density of a square pulse train with non-zero DC content centered at 4 GHz clearly surpasses the FCC indoor mask.

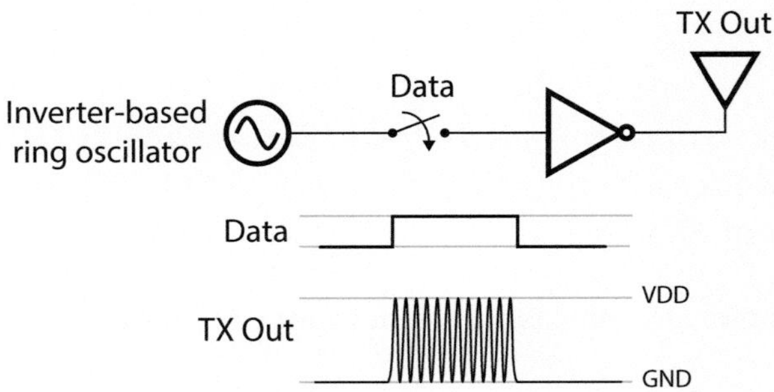

Fig. 23 A simple way to generate UWB pulses using all-digital circuits

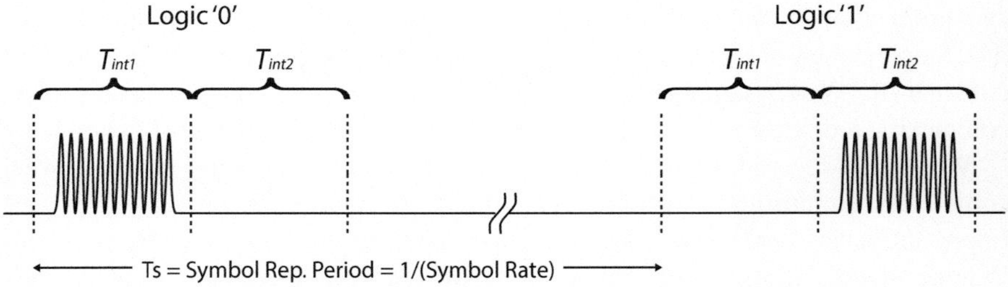

Fig. 24 Pulse position modulation represents data by the presence of a pulse in a particular window in time

[3]The transmitter must still adhere to any pulse shape regulations to be standards-compliant.

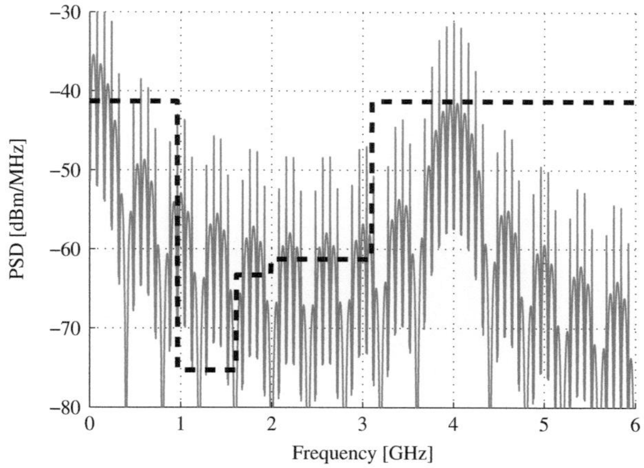

Fig. 25 Power spectral density of a train of PPM-modulated square UWB pulses

3.2.3 Achieving Spectral Compliance

There are three main problems with the spectrum shown in Fig. 25, all three of which must be addressed in order to meet the FCC spectral mask:

1. Large spectral lines spaced at integer multiples of the pulse repetition frequency.
2. Sidelobes centered at the carrier frequency.
3. Sidelobes centered at DC.

Spectral Lines

The problem of spectral lines is conceptually easy to fix. If the UWB pulses were phase modulated with random (or pseudo-random) data during transmissions, the tones would be scrambled out. This effect is most easily achieved by implementing a BPSK scrambler or modulator. The resulting power spectral density is illustrated in Fig. 26.

RF Sidelobes

This spectrum of Fig. 26 still suffers from drawbacks two and three: namely, it contains undesired sidelobes centered at both RF and DC. Although these sidelobes can easily be eliminated by bandpass or highpass filters, the area penalty of this approach is significant. For instance, this particular example would require at least a fourth order passive filter to ensure the necessary roll-off of roughly 20 dB in 0.29 decades (from the 1.61 to 3.1 GHz mask boundaries). A fourth order passive

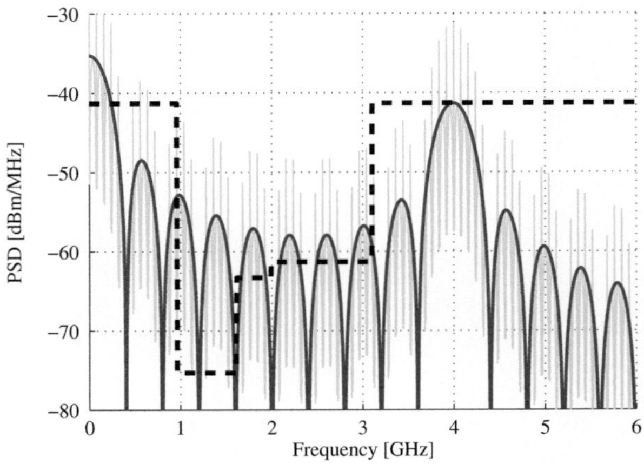

Fig. 26 Power spectral density of a train of PPM-modulated, BPSK-scrambled, square UWB pulses overlaid on top of the non-BPSK-scrambled case

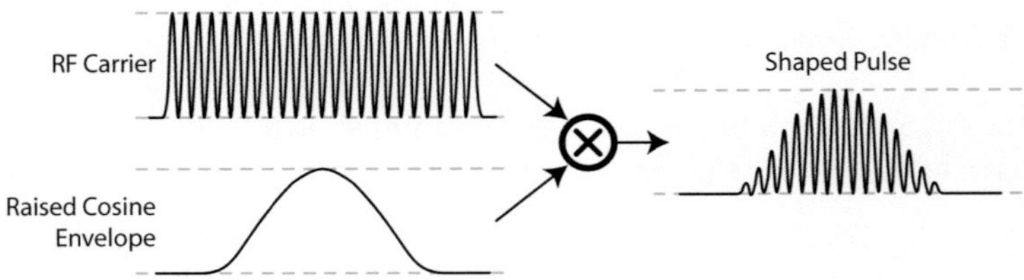

Fig. 27 Time domain view illustrating how to generate a raised-cosine pulse

filter requires several inductors and capacitors, which are not only lossy in modern semiconductor technologies, but also consume significant area.

An alternative to filtering is to employ pulse shaping to reduce the sidelobes centered at RF. As demonstrated in [9, 16, 34, 35, 62], and many other designs, pulse shaping is a very good method for obtaining high-order roll-off without the use of large passive filters. An excellent overview of several popular pulse shapes can be found in [52].

To illustrate the virtues of pulse shaping, consider shaping a pulse with a raised-cosine envelope, as illustrated in Fig. 27. The resulting spectrum achieves up to 17 dB of sidelobe rejection, as shown in Fig. 28.

Low Frequency Sidelobes

The raised cosine envelope greatly suppresses the sidelobes centered around the carrier frequency. However, the sidelobes centered at DC remain. This problem does not depend on pulse shape, but is rather fundamentally related to the method in which the digital pulses are synthesized. The issue stems from the fact that single

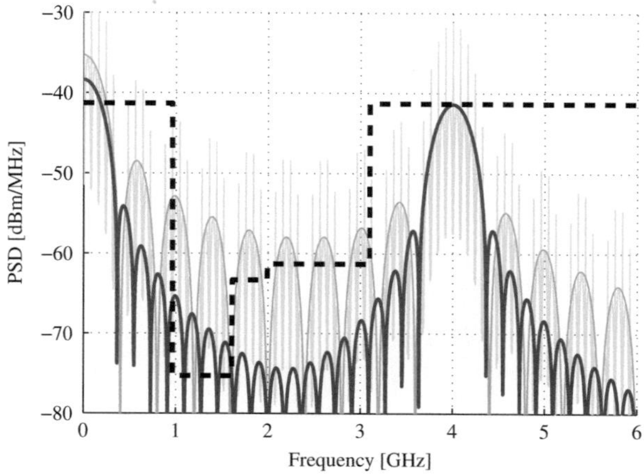

Fig. 28 Power spectral density of a train of PPM-modulated, BPSK-scrambled, raised-cosine UWB pulses overlaid on top of the spectrum in Fig. 26

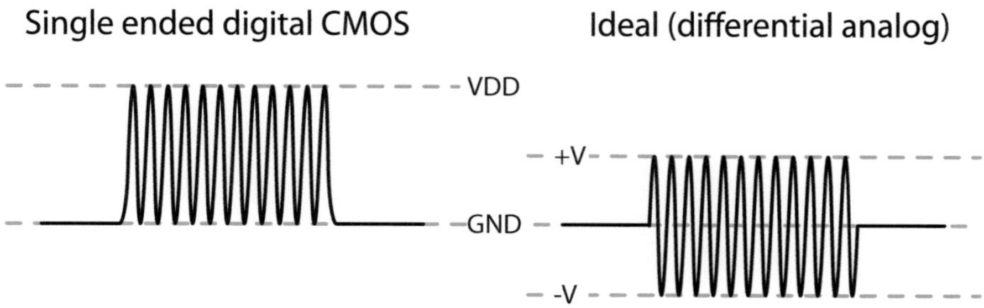

Fig. 29 Digital CMOS circuits can only generate one of two different reference levels. On the other hand, differential analog circuits can generate multiple reference levels at different bias voltages

ended digital circuits have only two stable operating points: the lowest and highest potentials in the circuits (typically GND and VDD). To eliminate the DC content and its associated sidelobes, the generated pulses must have *three* effective levels: GND, +V, and −V, as illustrated in Fig. 29.

For continuous wave systems, this is a relatively easy problem: simply insert an AC-coupling capacitor before the antenna, as illustrated in Fig. 30. This solution is unfortunately not ideal for pulses of short duration, the reason for which will become clear momentarily. Digitally generated pulses with two reference voltage levels (e.g. GND and VDD), can be decomposed into an RF carrier and a baseband pulse, as illustrated in Fig. 27. The baseband pulse will require a finite amount of time to charge and discharge the voltage across the capacitor, as shown in Fig. 31. The time required to charge and discharge, given by t_{charge} and $t_{discharge}$ respectively, is proportional to the RC time constant of the circuit.

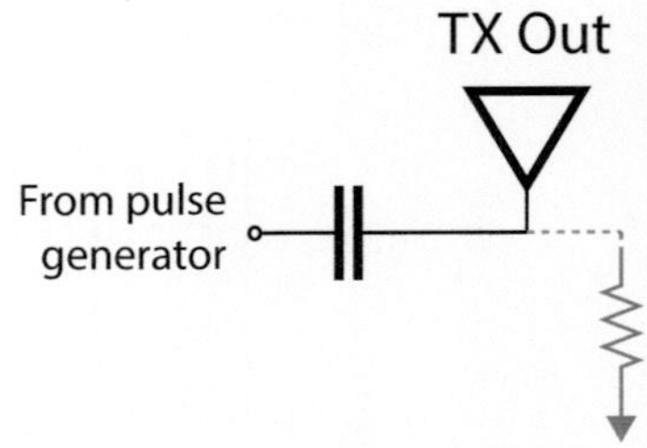

Fig. 30 A coupling capacitor providing a DC block to a wideband antenna. A large resistor may be connected from the output to ground in order to provide a stable DC voltage to the antenna if necessary

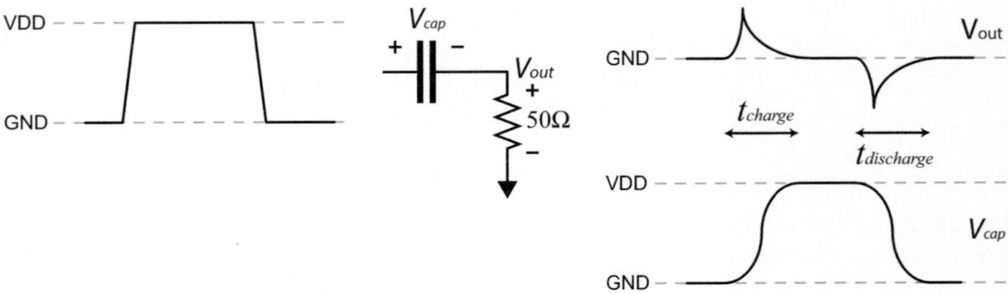

Fig. 31 A baseband pulse requires a finite amount of time to charge and discharge a coupling capacitor

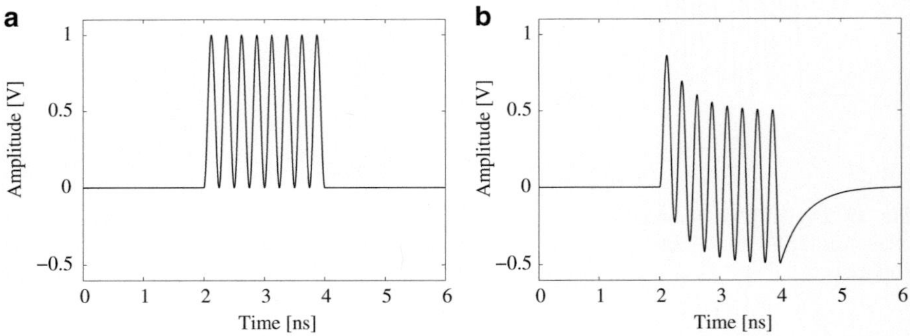

Fig. 32 The effect of passing a UWB pulse through an AC-coupling capacitor. (**a**) Digitally generated pulse. (**b**) After AC-coupling filter

The effect of finite low frequency capacitor charging and discharging times when AC-coupling UWB pulses is illustrated through a time-domain simulation of a square pulse in Fig. 32. It can be noted here that the AC-coupled pulse has a non-zero DC value, as well as some low-frequency turn-on and turn-off transients. The power spectral density of a train of BPSK-scrambled raised-cosine UWB pulses before and after the AC-coupling filter is shown in Fig. 33. The AC-coupled spectrum does not comply with the FCC mask, since the first order roll off is not sufficient to eliminate all of the low frequency sidelobes.

There are several techniques to reduce or even eliminate the low frequency-content of digitally generated UWB pulses. The most common technique relies on

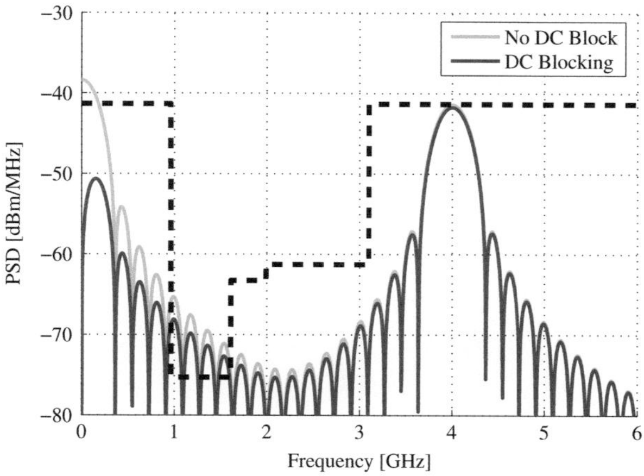

Fig. 33 Power spectral densities of raised-cosine pulses before and after an AC-coupling filter

generating individual half-RF cycles and applying them differentially to a wideband balun [34]. This technique produces excellent spectral results, however it requires the use of inductors which consume more-than-desired chip area. Another potential drawback of this type of architecture is that the half-RF cycles are generated from a delay line instead of a free-running ring oscillator. This can be seen as a benefit if the designed system is only a transmitter which generates a single pulse, then immediately turns off for a period of time. If, instead the designed system is a transceiver that transmits multiple pulses back-to-back, it is beneficial to design a single oscillator which is shared between the receiver and transmitter.

The proposed solution to attenuating the low frequency content using scalable digital structures involves capacitively coupling two paths which have differential baseband signals, yet contain in-phase RF tones. To elaborate, consider the network shown in Fig. 34a, where the two capacitors nominally have opposite DC voltages across them (GND and VDD, generated from digital logic). If they are driven with a differential baseband pulse, the upper capacitor will ideally charge at the same rate that the lower capacitor is discharging, thereby inducing zero voltage at the output.

If the low frequency baseband pulses are multiplied with in-phase RF tones as illustrated in Fig. 34b, then the low frequency common-modes will cancel, and the in-phase RF components will propagate to the output.

Since the two inputs into the capacitive combination network start off with opposite common modes, there is an inherent half RF cycle delay between the start of the effective baseband pulses shown in Fig. 34b. This, combined with circuits mismatches, will create non-idealities including turn-on and turn-off transients leading to spectral impurities. Ideally, the output spectrum will contain zero low frequency content, as illustrated by the spectrum of ideal raised cosine pulses in Fig. 35. In practice, the output spectrum will have a small amount of low-frequency content.

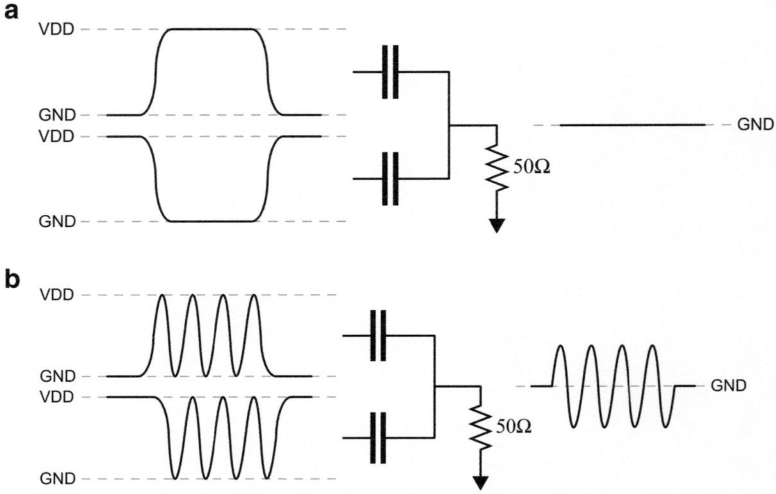

Fig. 34 Differential baseband pulses cancel as shown in (**a**), while in-phase RF signals propagate relatively undisturbed to the output, as shown in (**b**)

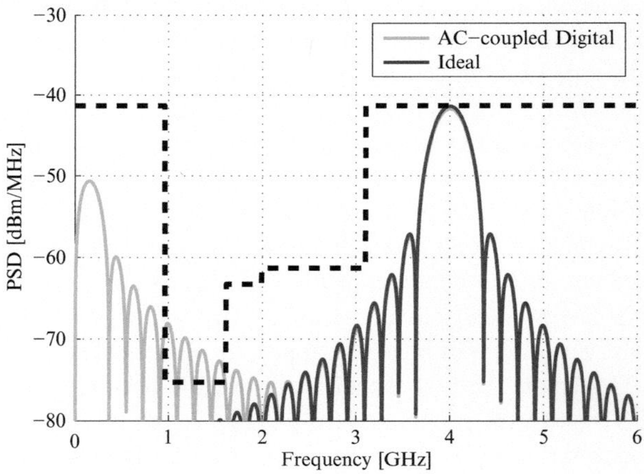

Fig. 35 Power spectral densities of ideal raised-cosine pulses and AC-coupled digitally generated raised cosine pulses

3.2.4 Transmitter Architecture

Given the dual capacitively-coupled paths, a block diagram of the presented transmitter is shown in Fig. 36 [29, 31]. The transmitter is designed to operate in all three channels of the low-band group of the 802.15.4a standard. As per the 802.15.4a specifications, payload data is modulated using time-hopped (TH)-PPM, where a PPM symbol is represented by a burst of several back-to-back pulses contained in a fixed window of time [27]. In idle mode between bursts, all transmitter circuits are off and the transmitter consumes only leakage power.

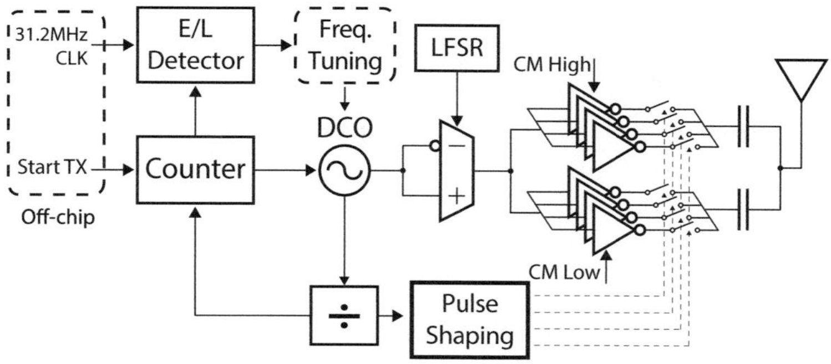

Fig. 36 Transmitter block diagram

Pulse bursts are generated on the rising edge of the off-chip *Start-TX* signal. This edge enables a DCO, whose output is BPSK-scrambled via an linear feedback shift register (LFSR) and subsequently buffered through dual single-ended digital PAs employing capacitive combination.

The DCO output frequency is calibrated and dynamically adjusted using an early-late detector in a digital frequency locked loop (FLL). The DCO output is also synchronously divided to a 499.2 MHz clock as specified by the 802.15.4a standard [22]. Several phases of the divided clock are used by pulse shaping circuitry to dynamically shape the PA envelope to one of four discrete levels. The 499.2 MHz clock sets the pulse repetition frequency (PRF) within a burst, and is also used in conjunction with a counter to program the number of pulses transmitted per burst.

3.2.5 Dual Digital Power Amplifiers

The circuit shown in Fig. 37 implements the dual capacitor technique in order to generate low-DC content RF pulses by driving two 2 pF coupling capacitors with two separate digital PAs.

Each PA consists of 30 tri-state inverters. A single oscillator signal is fed as an input to all 60 tri-state inverters, thus ensuring both paths receive in-phase RF signals. Each tri-state inverter is sized such that all 60 inverters operating in parallel can drive the antenna and associated parasitics up to 800 mV when switching at 4 GHz. Output power control can be configured by programming the number of tri-state inverters enabled at a given time. Furthermore, by dynamically adjusting the number of enabled tri-state inverters during pulse transmission, pulse shaping can be realized. Section 3.2.7 discusses the implementation details of the pulse shaping logic.

The differential baseband pulses (i.e. opposite common mode low-frequency pulses) are generated by ensuring that the outputs of the two PAs are at opposite supply rails immediately before and after pulse generation. Thus, during pulse

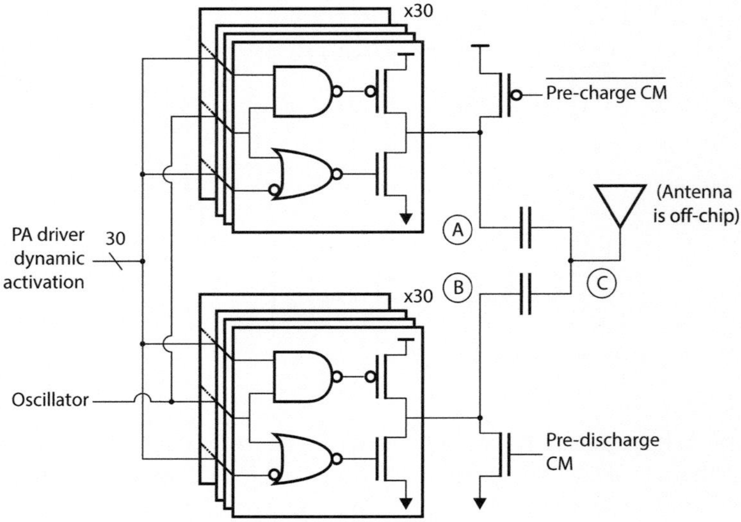

Fig. 37 Dual digital power amplifiers

generation the two capacitively coupled paths begin to charge and discharge low-frequency content at the same rate, resulting in close to zero low-frequency content on the output.

The opposite common modes for the two PA outputs are set by pre-charge and pre-discharge transistors during the idle mode between pulses (i.e. when the PA outputs are tri-stated). The dynamic PA control logic should ensure that the pre-charge and pre-discharge transistors are never turned on during pulse generation in order to avoid static power dissipation. It should be mentioned that since the PA outputs can be tri-stated, the transmitter can easily share the antenna with an integrated receiver without requiring an explicit transmit/receive switch. If required, the DC voltage of node C can be set with a large resistance or inductor to GND in order to eliminate any potential build up of charge.

Figure 38 shows a representative timing diagram of the dual digital power amplifiers with pulse shaping applied. Since the coupling capacitors are charging and discharging at roughly the same rate, the average voltage of nodes A and B approach the same value (ideally VDD/2) during pulse generation. For this reason, a very visible low-frequency transient is seen on nodes A and B at the end of pulse generation when the pre-charge/discharge devices are turned on. If the two paths are matched and the pre-charging and pre-discharging begin at the same voltage on nodes A and B, this low-frequency transient will not be seen at the output (node C).

However, if the two paths are not matched, nodes A and B will discharge with different initial conditions, thus leading to some low-frequency content on output node C. For example, Fig. 39 shows the simulated output spectrum of the dual PAs with ideally-matched paths overlaid on top of Monte Carlo process variation, showing up to 4 dB degradation at both DC and 1.2 GHz. Since there are relatively

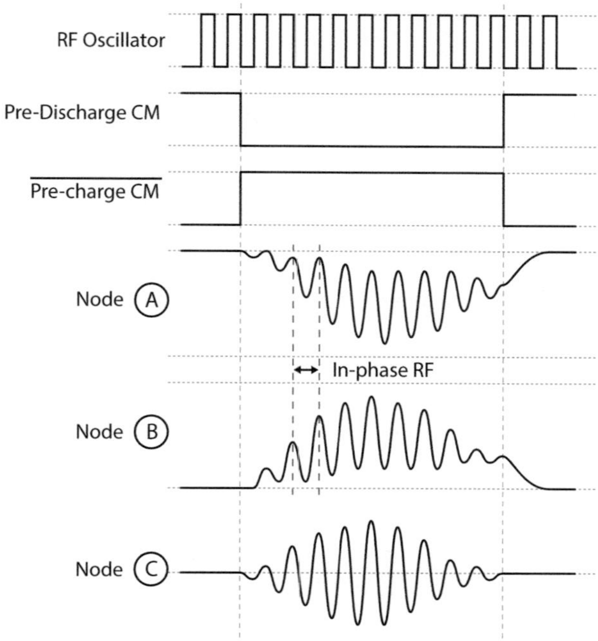

Fig. 38 Timing diagram for the dual digital power amplifiers

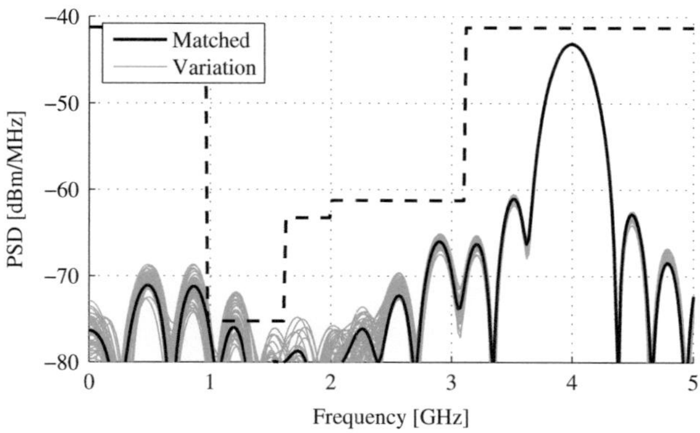

Fig. 39 Simulated output spectra with Monte Carlo process variation

small spectral differences across process variation and device mismatches, a large number of pulse shape configurations should be able to guarantee, with a reasonable degree of confidence, that the FCC mask will be met. This idea of implementing *redundancy* in order to guarantee desired operation is almost necessary in high density memory design, and is becoming more popular for other types of circuits such as ADCs [13, 21].

3.2.6 Clocking

Since non-coherent pulsed-UWB receivers have large input bandwidths *and* discard phase information, precise transmitter frequency tolerances are not necessarily required. To quantify this claim, consider a UWB transmitter with a 6,000 ppm accuracy. At a 4 GHz center frequency, this corresponds to a maximum frequency error of ± 12 MHz. If an ideal receiver with a 500 MHz brick wall input filter received a 500 MHz input signal offset by 12 MHz, a loss of only 0.1 dB would be incurred. The situation typically improves when dealing with non-ideal signal bandwidths and filters. As an example, the system presented in [55] has a transmitter center frequency accuracy of 6,000 ppm. While unacceptably large for coherent and/or narrowband systems, the receiver still achieves a sensitivity of -99 dBm at a BER of 10^{-3} and a data rate of 100 kbps. As a result, the transmitter oscillator design requirements can be relaxed considerably with the ultimate goal of improving energy efficiency.

A DCO that meets the needs of this design is shown in Fig. 40. The current-starved inverter-based three-stage ring structure is designed to have a fast turn-on time on the order of 2 ns to reduce energy consumption in duty-cycled operation. Furthermore, the delay elements are all single-ended to further reduce energy consumption over a differential structure. Although single-ended structures are more susceptible to power supply noise compared to their differential counterparts, the resulting increase in phase noise is of negligible concern to a non-coherent energy-detecting receiver.

Coarse frequency tuning is provided by switchable load capacitors, while fine frequency tuning is provided with NMOS and PMOS current starving DACs. To simplify the frequency locking algorithm, all three current starving DACs are set to the same digital value, except that the second and third stage DACs can be individually incremented by one for increased resolution. This technique results in a resolution of 7.5 bits from the DACs and 2 bits from the three thermometer encoded capacitors, totaling 9.5 bits.

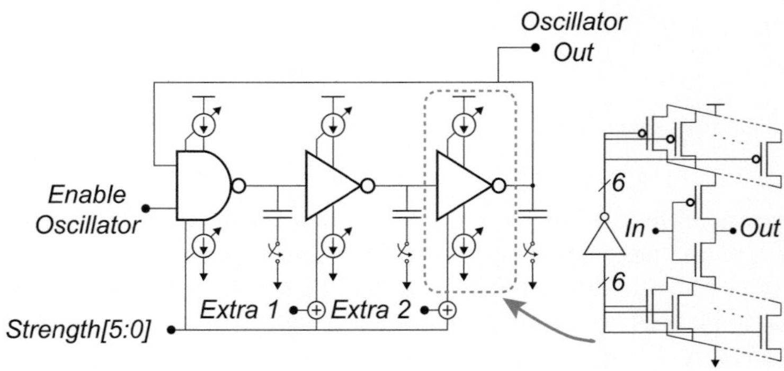

Fig. 40 Digitally controlled oscillator

The output of the DCO is fed to a programmable synchronous frequency divider. The divider is realized using true single-phase clock (TSPC) logic [57] in order to accommodate inputs up to 6 GHz. The divider consists of fourteen half-transparent latchs (HTLs) which can be individually bypassed, thus allowing a programmable divide ratio of up to fourteen. The design is based on the work presented in [44] and [58].

The output of the divider drives the pulse shaping circuitry, which in turn determines the effective transmitted pulse width (and thus the bandwidth). To comply with the 802.15.4a standard, the transmitted pulse width is maximally set to one over the 499.2 MHz PRF within a burst (i.e. 2 ns). If the DCO frequency is set to one of the 802.15.4a channels, an integer division will always yield the required 499.2 MHz clock.

The transmitter contains an early/late detector which can be used for periodic frequency calibration in a frequency locked loop. In addition, the transmitter employs BPSK scrambling in order to smooth out spectral lines associated with non-phase modulated PPM signaling while maximizing peak power. More details of both of these blocks and their design considerations can be found in [30, 31].

3.2.7 Pulse Shaping Logic

The output phases of the frequency divider do not necessarily have 50 % duty cycles. In fact, the duty cycles of the internal phases vary approximately linearly from 10 % to 90 %, depending on the divide ratio. Since the period of each phase is set to be equal to the pulse width of 2 ns, it is possible to combine several of these phases in order to generate the timing required for pulse shaping. This is illustrated in Fig. 41,

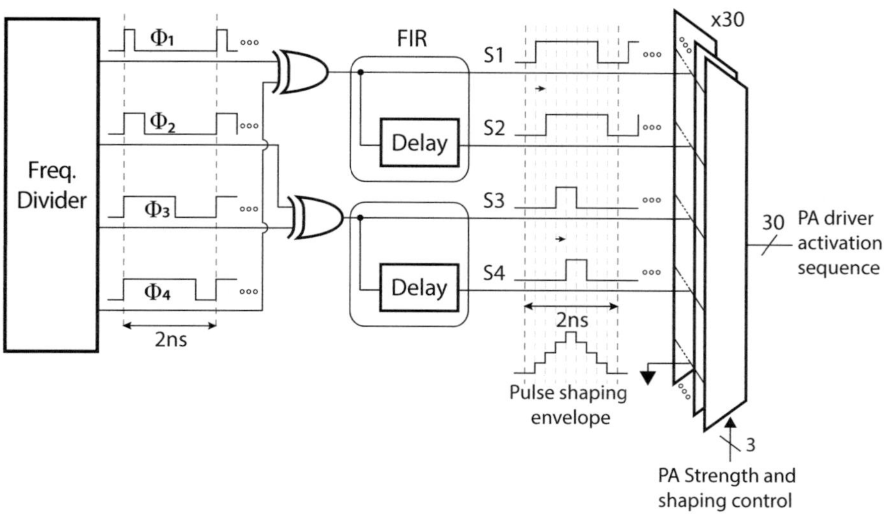

Fig. 41 Pulse shaping logic

where signals Φ_{1-4} are appropriately chosen to have duty cycles of approximately 20 %, 40 %, 60 %, and 80 %, based on which HTLs are enabled or disabled.

By XOR-ing Φ_1 with Φ_4 and Φ_2 with Φ_3, two pulse shaping signals are generated. These two pulse shaping signals are each passed through simple one-tap finite impulse response (FIR) filters to increase the number of pulse shaping signals to four (signal S1-S4). The delay elements of the FIR filters are simply comprised of a programmable number of inverter-based buffers.

Each tri-state inverter of the dual PAs is individually programmed through a five-input multiplexer network to receive one of the four pulse shaping signals as a dynamic activation input. The fifth multiplexer input is grounded in order to allow statically disabled tri-state inverters, as the inverters are typically disabled to perform gain control. The four pulse shaping signals can be thought of as the output of FIR filter taps which are added together at the input of the coupling capacitors via the parallel combination of PA tri-state inverters. Maximum PA output swing is achieved when all four pulse shaping signals are high, i.e. the maximum number of PA inverters are enabled in parallel simultaneously. This pulse shaping configuration also ensures that the output signal amplitude is zero during BPSK phase transitions in order to avoid common-mode glitching and inter-pulse interference.

The actual pulse shape can be modified through two different methods. One method involves changing the relative times at which the shaping signals arrive. This can be accomplished by selecting different frequency divider phases or changing the FIR filter delay element. This does not typically produce a desired pulse shape; an example is shown in Fig. 42a. The FIR filter delays have $2^3 = 8$ possible permutations. The second method involves changing how many tri-state inverters receive a particular pulse shaping signal. This is similar to choosing the weights of the FIR tap coefficients, and can be used to more closely approximate a raised-cosine shape [35]. An example pulse shape is shown in Fig. 42b. There are approximately $2^{30} \approx 10^9$ total pulse shape strength permutations. Many of these permutations

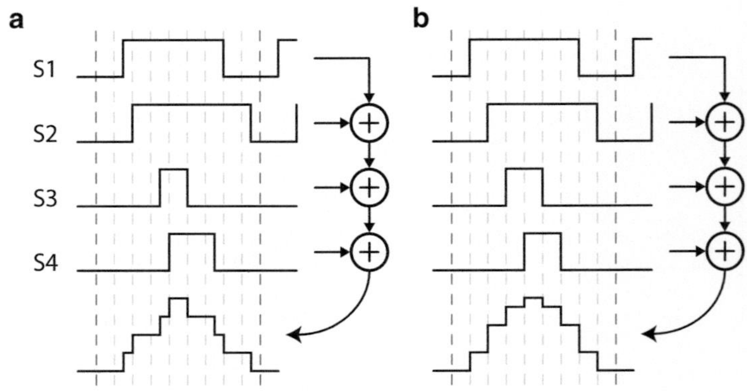

Fig. 42 Modifying pulse shapes by changing (**a**) delays and (**b**) weights

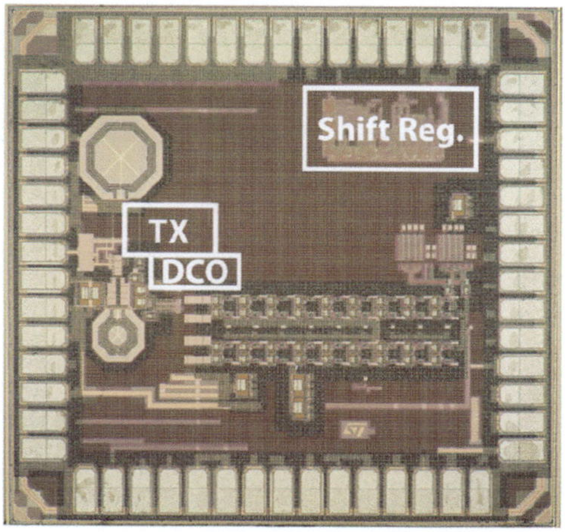

Fig. 43 Die photo of fabricated transmitter

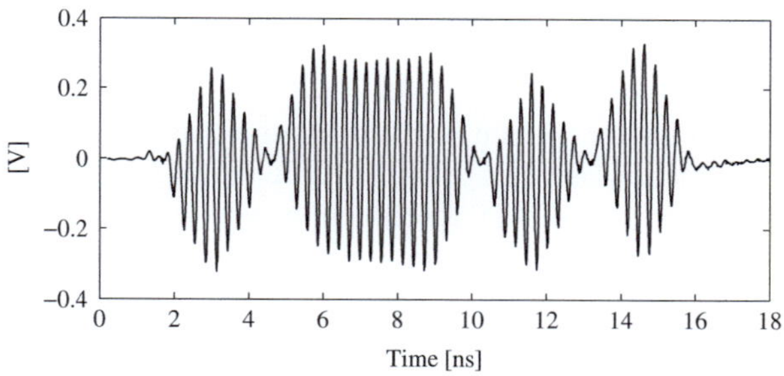

Fig. 44 Measured transient waveform of a burst of five individually BPSK-modulated pulses

are not practically useful, however as discussed in Sect. 3.2.5, there are more than enough possible configurations to guarantee FCC spectral compliance with a reasonable degree of confidence.

3.2.8 Measurement Results

The transmitter was fabricated in a 90 nm bulk CMOS process and packaged in a 40-lead, wirebonded QFN package; all measurement results were taken from the packaged chip. A die photograph is shown in Fig. 43b. The transmitter core and DCO consume an area of 0.07 mm^2.

The transmitter operates at data rates from 0-to-15.6 Mbps on a 1 V power supply. It has a turn-on time of 7.2 ns, measured as the time it takes pulses to appear at the output of the dual PAs after the rising edge of *Start-TX* has arrived.

Figure 44 shows the output when the transmitter is configured to generate a burst of five individually BPSK-modulated pulses at a time as measured by a Tektronix TDS 8000 Sampling Oscilloscope with an 80E04 sampling module. The dual PAs were measured to have an output voltage swing range from 160-to-710 mV, resulting in an output power control range of 13 dB.

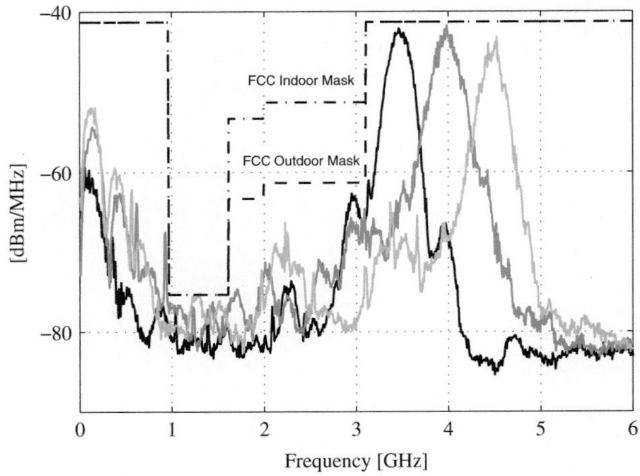

Fig. 45 Overlaid power spectral densities of the three channels low 802.15.4a bands

Frequency domain measurements were taken at the output of the capacitive combination network using an Agilent MXA N9020A Spectrum Analyzer. Unless otherwise noted, the resolution bandwidth was set to 1 MHz. During normal operation, the proposed transmitter achieves both indoor and outdoor FCC spectral compliance in all three low-band 802.15.4a channels. The resulting spectra for sixteen-pulse bursts are superimposed together over the FCC spectral mask in Fig. 45. Note that no off-chip filters were used to make this measurement.

To determine how effective capacitive combining and pulse shaping are at reducing both low frequency content as well as RF sidelobes, Fig. 46 shows the output power spectral density of two-pulse bursts with: capacitive combining and no pulse shaping, pulse shaping and no capacitive combining, and capacitive combining and pulse shaping. Here it can be seen that capacitive combination achieves up to 12 dB of low-frequency attenuation. Furthermore, pulse shaping achieves 15-to-20 dB of sidelobe rejection.

Operating on a 1 V supply, the transmitter draws 4.36 mW when generating 16-pulse bursts at an symbol repetition frequency (SRF) of 15.6 MHz. The total output in this configuration is −16.4 dBm. This results in an energy efficiency of 280 pJ/burst, or 17.5 pJ/pulse. Since all transmitter circuits are inherently off between pulse transmissions, the power consumption scales with data rate. However, the impact of leakage power becomes significant at symbol rates below 1 MHz. The standby (i.e. idle-mode) leakage power is 123 μW. Table 3 summarizes the transmitter's performance.

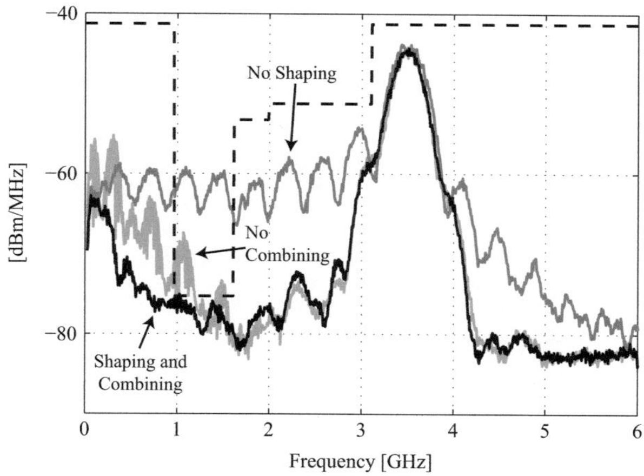

Fig. 46 Overlaid power spectral densities with shaping disabled, combining disabled, and normal operation

Table 3 Transmitter performance summary

Specification	Value	
Process	90 nm CMOS	
Active die area	0.07 mm^2	
Modulation	PPM+BPSK	
SRF range	0-to-15.6 MHz	
Supply	1 V	
	Standby power	Active E/pulse
Power amplifier	22 μW	2 pJ
DCO/clock/control	83 μW	15 pJ
Shift register	11 μW	–
I/O and ESD	7 μW	<1 pJ
Total	123 μW	17 pJ
Output voltage swing	165 mV-to-710 mV	
DCO frequency range	2.1 GHz-to-5.7 GHz	
Turn-on time	7.2 ns	
Symbol rate (SRF)	100 kHz	15.6 MHz
Energy/16-pulse-burst	1.6 nJ	280 pJ
Energy/pulse	103 pJ	17.5 pJ
Total power	164 μW	4.36 mW
Output power	−25 dBm	−16.4 dBm

4 Summary

In this chapter, we presented two IR-UWB receivers and one IR-UWB transmitter that consume low power for low data rate and short range radio applications. We have shown that IR-UWB is a viable technology choice that provides competitively low energy consumption per bit (17.5 pJ/bit TX, 700 pJ/bit RX) when compared to narrowband radio counterparts with similar operating requirements. Furthermore, we have shown that IR-UWB systems require fewer off-chip components for a full transceiver implementation. IR-UWB transmitters do not require much analog design effort since they can be built with all-digital logic and digitally synthesized. They do not require high Q filters; instead, simple digitally-configurable pulse-shaping conforms the transmit power to the spectral mask requirements. Though the average power density for UWB signaling is restricted to be below the noise floor (−41.3 dBm/MHz), it is more energy efficient to transmit the largest allowable instantaneous power in each pulse (less than 0 dBm in a 50 MHz resolution bandwidth) rather than a series of lower power pulses. Finally, since the transmitters are built with all-digital logic, they consume no static power and are inherently duty-cycled between pulse transmissions.

IR-UWB receivers can leverage non-coherent signaling and large signal bandwidth to duty cycle the receiver between pulses, much like the transmitter. However, since RF/analog circuits must be used in the LNA, RF, and baseband gain stages, it requires some effort to ensure these blocks can power up/down as fast as possible for maximum energy savings. Because the signal bandwidth is >500 MHz wide, the LNA and RF bandpass filters are much more robust to any changes in absolute frequency than their narrowband radio counterparts, thereby easing the design effort. The pulses can be converted from RF to baseband without an LO by utilizing a self-mixer. However, since the self-mixer requires at least 10 mV of input amplitude to produce a reasonable output (0.7 mV), RF gain stages are required between the LNA and the self-mixer. Both radios that were presented leverage these aspects of IR-UWB. In addition, we have shown two methods of demodulating data after the self-mixer. One method uses a simple DC-tracking analog threshold voltage and hysteretic comparator. These two blocks operate together to function as a single-bit data converter. This method is the simplest if sensitivity and flexibility are not as important as power consumption. The other method involves an integrator and a 5-bit digitizer, both operating at a 31.5 ns period long enough to capture a typical UWB channel impulse response. The advantage of this more complex demodulator is the increased level of visibility into timing and voltage amplitudes of the RF channel. With this level of detail, the digital baseband is able to synchronize with an accuracy of ± 1 ns, and have the flexibility to demodulate both PPM and OOK IR-UWB signals. The main outstanding challenge for UWB receivers lies in the fact that since the UWB signal bandwidth overlays with many other narrowband standards with much larger allocated power levels, immunity to these narrowband interferers must be developed at the signaling, circuit, and even network levels.

At the system level, we have shown that IR-UWB transceivers do not require accurate RF synthesizers nor external crystal references, because the large signal bandwidth along with non-coherent signaling allows the system to tolerate larger RF frequency differences (>1,000 ppm) with a negligible (approximately 0.1 dB @ 6,000 ppm) impact on sensitivity; however, it should be noted that if the bit-period clock accuracy between TX and RX is improved, the system could exhibit improved performance and further power savings. At the network layer, the order of magnitude difference in energy/bit between TX and RX may also spur development of new networks of wireless sensors where the overall system can be energy-optimized by leveraging the cheaper cost of data transmission using IR-UWB signaling.

From the work presented here, it is clear that IR-UWB radio systems are a viable technology choice in applications that require energy efficient, low rate wireless communication. Additionally, we have shown IR-UWB systems can be reliably implemented in silicon, and can operate with fewer off-chip components than narrowband radio counterparts because they do not need a transmit filter nor an external crystal oscillator. Finally, if the application requires more transmitters than receivers, wireless communication systems built on IR-UWB technology would be a natural fit to allow an overall lower network energy consumption. If these features are important in the application that is under consideration, IR-UWB radios could be the right technology choice.

References

1. M. Baghaei-Nejad, D.S. Mendoza, Z. Zou, S. Radiom, G. Gielen, L.-R. Zheng, H. Tenhunen, A remote-powered RFID tag with 10Mb/s UWB uplink and −18.5 dBm sensitivity UHF downlink in 0.18 μm CMOS, in *Proceeding of the IEEE ISSCC Digest Technical Papers* (February, 2009), pp. 198–199
2. ABM10: Ceramic SMD Ultra Miniature Quartz Crystal, Abracon, Technical Report (2008)
3. M.-G.D. Benedetto, T. Kaiser, A.F. Molisch, I. Oppermann, C. Politano, D. Porcino, *UWB Communication Systems: A Comprehensive Overview* (Hindawi Publishing Corporation, New York, 2006)
4. M. Bhardwaj, Communications in the observation limited regime, Ph.D. thesis, Massachusetts Institute of Technology, 2009
5. J.L. Bohorquez, A.P. Chandrakasan, J.L. Dawson, A 350 W CMOS MSK transmitter and 400 W OOK super-regenerative receiver for medical implant communications. IEEE J. Solid State Circuits **44**(4), 1248–1259.
6. A. Bozkurt, R. Gilmour, D. Stern, A. Lal, MEMS based bioelectronic neuromuscular interfaces for insect cyborg flight control, in *Proceedings of the IEEE International Conference on MEMS Systems* (January, 2008), pp. 160–163
7. J. Brown, K.-K. Huang, E. Ansari, R. Rogel, Y. Lee, D. Wentzloff, An ultra-low-power 9.8ghz crystal-less uwb transceiver with digital baseband integrated in 0.18 μm bicmos, in *IEEE ISSCC Digest of Technical Papers* (February, 2013), pp. 442–443
8. B.H. Calhoun, D.C. Daly, N. Verma, D. Finchelstein, D.D. Wentzloff, A. Wang, S.-H. Cho, A.P. Chandrakasan, Design considerations for ultra-low energy wireless microsensor nodes. IEEE Trans. Comput. **54**(6), 727–749 (2005)
9. X. Chen, S. Kiaei, Pulse generation scheme for low-power low-complexity impulse ultra-wideband. IEE Electron. Lett. **43**(1), 44–45 (2007)
10. J.-Y. Chen, M. Flynn, J. Hayes, A fully integrated auto-calibrated super-regenerative receiver, in *Proceeding of the IEEE ISSCC Digest Technical Papers* (2006), pp. 1490–1499
11. G. Chen, S. Hanson, D. Blaauw, D. Sylvester, Circuit design advances for wireless sensing applications. Proc. IEEE **98**(11), 1808–1827 (November, 2010)
12. P. Choi, H. Park, I. Nam, K. Kang, Y. Ku, S. Shin, S. Park, T. Kim, H. Choi, S. Kim, S.M. Park, M. Kim, S. Park, K. Lee, An experimental coin-sized radio for extremely low power WPAN (IEEE802.15.4) application at 2.4GHz, in *Proceeding of the IEEE ISSCC Digest Technical Papers* (February, 2003), pp. 92–480

13. D.C. Daly, A.P. Chandrakasan, A 6b 0.2-to-0.9V highly digital flash ADC with comparator redundancy, in *IEEE ISSCC Digest Technical Papers* (February, 2008), pp. 554–555
14. D. Daly, P. Mercier, M. Bhardwaj, A. Stone, Z. Aldworth, T. Daniel, J. Voldman, J. Hildebrand, A. Chandrakasan, A pulsed UWB receiver SoC for insect motion control. IEEE J. Solid-State Circuits **45**(1), 153–166 (2010)
15. H. Darabi, S. Khorram, Z. Zhou, T. Li, B. Marholev, J. Chiu, J. Castaneda, E. Chien, S. Anand, S. Wu, M. Pan, R. Roufoogaran, H. Kim, P. Lettieri, B. Ibrahim, J. Rael, L. Tran, E. Geronaga, H. Yeh, T. Frost, J. Trachewsky, A. Rotougaran, A fully integrated SoC for 802.11b in 0.18 μm CMOS, in *Proceeding of the IEEE ISSCC Digest Technical Papers* (February, 2005), pp. 96–586
16. M. Demirkan, R.R. Spencer, Antenna characterization method for front-end design of pulse-based ultrawideband transceivers. IEEE Trans. Antennas Propag. **55**, 2888–2899 (2007)
17. M. Demirkan, R.R. Spencer, A 1.8Gpulse/s UWB transmitter in 90nm CMOS, in *IEEE ISSCC Digest Technical Papers* (February, 2008), pp. 116–117
18. A. Emira, A. Valdes-Garcia, B. Xia, A. Mohieldin, A. Valero-Lopez, S. Moon, C. Xin, E. Sanchez-Sinencio, A dual-mode 802.11b/Bluetooth receiver in 0.25μm BiCMOS, in *Proceeding of the IEEE ISSCC Digest Technical Papers* (February, 2004), pp. 270–527
19. FCC, First report and order, FCC 02-48 (2002)
20. FCC, Second report and order and second memorandum opinion and order, FCC 04-285 (2004)
21. B.P. Ginsburg, A.P. Chandrakasan, Highly interleaved 5b 250MS/s ADC with redundant channels in 65nm CMOS, in *IEEE ISSCC Digest Technical Papers* (February, 2008), pp. 240–241
22. IEEE 802.15.4a Wireless MAC and PHY specifications for LR-WPANs (2007) [Online]. Available http://www.ieee802.org/15/pub/TG4a.html
23. H. Kim, D. Park, Y. Joo, All-digital low-power CMOS pulse generator for UWB system. IEE Electron. Lett. **40**(24), 1534–1535 (2004)
24. V. Kulkarni, M. Muqsith, H. Ishikuro, T. Kuroda, A 750Mb/s 12pJ/b 6-to-10GHz digital UWB transmitter, in *Proceeding of IEEE Custom Integrated Circuits Conference* (September, 2007), pp. 647–650
25. F.S. Lee, A.P. Chandrakasan, A 2.5nJ/b 0.65V 3-to-5GHz subbanded UWB receiver in 90nm CMOS, in *IEEE ISSCC Digest of Technical Papers* (February, 2007), pp. 116–117
26. J.-Y. Lee, R. Scholtz, Ranging in a dense multipath environment using an UWB radio link. IEEE J. Sel. Areas Commun. **20**, 1677–1683 (2007)
27. D. Marchaland, F. Badets, M. Villegas, D. Belot, 65nm CMOS burst generator for ultra-wideband low data rate systems, in *Proceedings of IEEE Radio Frequency Integrated Circuits Symposium* (June, 2007), pp. 43–46
28. B. Marholev, M. Pan, E. Chien, L. Zhang, R. Roufoogaran, S. Wu, I. Bhatti, T.-H. Lin, M. Kappes, S. Khorram, S. Anand, A. Zolfaghari, J. Castaneda, C. Chien, B. Ibrahim, H. Jensen, H. Kim, P. Lettieri, S. Mak, J. Lin, Y. Wong, R. Lee, M. Syed, M. Rofougaran, A. Rofougaran, A single-chip Bluetooth EDR device in 0.13μm CMOS, in *Proceeding of the IEEE ISSCC Digest Technical Papers* (February, 2007), pp. 558–759
29. P.P. Mercier, An all-digital transmitter for pulsed ultra-wideband communication, Master's thesis, Massachusetts Institute of Technology, Cambridge, MA, 2008
30. P.P. Mercier, D.C. Daly, M. Bhardwaj, D.D. Wentzloff, F.S. Lee, A.P. Chandrakasan, Ultra-low-power UWB for sensor network applications, in *Proceedings of IEEE ISCAS* (2008), pp. 2562–2565
31. P.P. Mercier, D.C. Daly, A.P. Chandrakasan, An energy-efficient all-digital UWB transmitter employing dual capacitively-coupled pulse-shaping drivers. IEEE J. Solid-State Circuits **44**, 1679–1688 (2009)
32. P.P. Mercier, M. Bhardwaj, D.C. Daly, A.P. Chandrakasan, A 0.55V 16Mb/s 1.6mW non-coherent IR-UWB digital baseband with $\pm$1ns synchronization accuracy, in *IEEE ISSCC Digest of Technical Papers* (February, 2009), pp. 252–253
33. P.P. Mercier, M. Bhardwaj, D.C. Daly, A.P. Chandrakasan, A low-voltage energy-sampling IR-UWB digital baseband employing quadratic correlation. IEEE J. Solid-State Circuits **45**, 1209–1219 (2010)

34. T. Norimatsu, R. Fujiwara, M. Kokubo, M. Miyazaki, A. Maeki, Y. Ogata, S. Kobayashi, N. Koshizuka, K. Sakamura, A UWB-IR transmitter with digitally controlled pulse generator. IEEE J. Solid-State Circuits **42**, 1300–1309 (2007)
35. A. Oncu, B.B.M.W. Badalawa, M. Fujishima, 22–9 GHz ultra-wideband CMOS pulse generator for short-range radar applications. IEEE J. Solid-State Circuits **42**, 1464–1471 (2007)
36. B. Otis, Y. Chee, J. Rabaey, A 400 μW-RX, 1.6mW-TX super-regenerative transceiver for wireless sensor networks, in *Proceeding of the IEEE ISSCC Digest Technical Papers* (February, 2005), pp. 396–606
37. T.-A. Phan, V. Krizhanovskii, S.-G. Lee, Low-power CMOS energy detection transceiver for UWB impulse radio system, in *Proceedings of IEEE Custom Integrated Circuits Conference* (September, 2007), pp. 675–678
38. N. Pletcher, S. Gambini, J. Rabaey, A 2GHz 52μW wake-up receiver with -72dBm sensitivity using uncertain-IF architecture, in *Proceeding of the IEEE ISSCC Digest Technical Papers* (February, 2008), pp. 524–633
39. A.-S. Porret, T. Melly, D. Python, C. Enz, E. Vittoz, An ultralow-power UHF transceiver integrated in a standard digital CMOS process: architecture and receiver. IEEE J. Solid-State Circuits **36**(3), 452–466 (2001)
40. J.G. Proakis, *Digital Communications*, 4th edn. (McGraw-Hill, New York, 2001)
41. J.M. Rabaey, M.J. Ammer, J.L. da Silva Jr., D. Patel, S. Roundy, PicoRadio supports ad hoc ultra-low power wireless networking. Computer **33**, 42–48 (2000)
42. E. Ragonese, A. Scuderi, V. Giammello, E. Messina, G. Palmisano, A fully integrated 24GHz UWB radar sensor for automotive applications, in *IEEE ISSCC Digest of Technical Papers* (February, 2009), pp. 306–307,307a
43. G. Retz, H. Shanan, K. Mulvaney, S. O'Mahony, M. Chanca, P. Corowley, C. Billon, K. Khan, P. Quinlan, A highly integrated low-power 2.4GHz transceiver using a direct-conversion diversity receiver in 0.18μm CMOS for IEEE802.15.4 WPAN, Feb 2009, pp. 414–415
44. J. Ryckaert, G. Van der Plas, V. De Heyn, C. Desset, B. Van Poucke, J. Craninckx, A 0.65-to-1.4 nJ/burst 3-to-10 GHz UWB all-digital TX in 90 nm CMOS for IEEE 802.15.4a. IEEE J. Solid-State Circuits **42**, 2860–2869 (2007)
45. L. Smaini, C. Tinella, D. Helal, C. Stoecklin, L. Chabert, C. Devaucelle, R. Cattenoz, N. Rinaldi, D. Belot, Single-chip CMOS pulse generator for UWB systems. IEEE J. Solid-State Circuits **41**, 1551–1561 (2006)
46. T. Terada, S. Yoshizumi, Y. Sanada, T. Kuroda, Transceiver circuits for pulse-based ultra-wideband, in *Proceeding of IEEE ISCAS* (May, 2004), pp. 349–352
47. T. Terada, S. Yoshizumi, M. Muqsith, Y. Sanada, T. Kuroda, A CMOS ultra-wideband impulse radio transceiver for 1Mb/s data communications and $\pm$2.5cm range findings. IEEE J. Solid-State Circuits **41**, 891–898 (2006)
48. P. Thoppay, C. Dehollain, M. Declercq, A 7.5mA 500 MHz UWB receiver based on super-regenerative principle, in *Proceeding of the IEEE European Solid-State Circuits Conference* (September, 2008), pp. 382–385
49. M. Verhelst, N. Van Helleputte, G. Gielen, W. Dehaene, A reconfigurable, 0.13μm CMOS 110pJ/pulse, fully integrated IR-UWB receiver for communication and sub-cm ranging, in *IEEE ISSCC Digest of Technical Papers* (February, 2009), pp. 250–251
50. D. Weber, W. Si, S. Abdollahi-Alibeik, M. Lee, R. Chang, H. Dogan, S. Luschas, P. Husted, A single-chip CMOS radio SoC for v2.1 Bluetooth applications, in *Proceeding of the IEEE ISSCC Digest Technical Papers* (February, 2008), pp. 364–620
51. M. Weisenhom, W. Hirt, Robust noncoherent receiver exploiting UWB channel properties, in *Proceedings of IEEE Joint UWBST & IWUWBS* (May, 2004), pp. 156–160
52. D.D. Wentzloff, Pulse-based ultra-wideband transmitters for digital communication, Ph.D. thesis, Massachusetts Institute of Technology, 2007
53. D.D. Wentzloff, A.P. Chandrakasan, Gaussian pulse generators for subbanded ultra-wideband transmitters. IEEE Trans. Microw. Theory Tech. **54**, 1647–1655 (2006)
54. D.D. Wentzloff, A.P. Chandrakasan, A 47pJ/pulse 3.1-to-5GHz all-digital UWB transmitter in 90nm CMOS, in *IEEE ISSCC Digest of Technical Papers* (February, 2007), pp. 118–119

55. D.D. Wentzloff, F.S. Lee, D.C. Daly, M. Bhardwaj, P. Mercier, A.P. Chandrakasan, Energy efficient pulsed-UWB CMOS circuits and systems, in *Proceedings of IEEE International Conference on Ultra-Wideband* (September, 2007), pp. 282–287
56. M.Z. Win, R.A. Scholtz, Impulse radio: how it works. IEEE Commun. Lett. **2**, 36–38 (1998)
57. J. Yuan, C. Svensson, High-speed CMOS circuit techniques. IEEE J. Solid-State Circuits **24**, 62–70 (1989)
58. J. Yuan, C. Svensson, Fast CMOS nonbinary divider and counter. IEE Electron. Lett. **29**, 1222–1223 (1993)
59. Y. Zheng, Y. Tong, C.W. Ang, Y.-P. Xu, W.G. Yeoh, F. Lin, R. Singh, A CMOS carrier-less UWB transceiver for WPAN applications, in *IEEE ISSCC Digest of Technical Papers* (February, 2006), pp. 378–387
60. Y. Zheng, K.-W. Wong, M. Annamalai Asaru, D. Shen, W.H. Zhao, Y.J. The, P. Andrew, F. Lin, W.G. Yeoh, R. Singh, A 0.18μm CMOS dual-band UWB transceiver, in *IEEE ISSCC Digest of Technical Papers* (February, 2007), pp. 114–115
61. Y. Zheng, M.A. Arasu, K.-W. Wong, Y.J. The, A.P.H. Suan, D.D. Tran, W.G. Yeoh, D.-L. Kwong, A 0.18μm CMOS 802.15.4a UWB transceiver for communication and localization, in *IEEE ISSCC Digest of Technical Papers* (February, 2008), pp. 118–119
62. Y. Zhu, J.D. Zuegel, J.R. Marciante, H. Wu, A 10 GS/s distributed waveform generator for sub-nanosecond pulse generation and modulation in 0.18μm standard digital CMOS, in *Proceedings of IEEE Radio Frequency Integrated Circuits Symposium* (June, 2007), pp. 35–38
63. D. Zito, D. Pepe, M. Mincica, F. Zito, A 90nm cmos soc uwb pulse radar for respiratory rate monitoring, in *IEEE ISSCC Digest of Technical Papers* (February, 2011), pp. 40–41

Human Body Communication Transceiver for Energy Efficient BAN

Hyungwoo Lee, Seong-Jun Song, Namjun Cho, Joonsung Bae, and Hoi-Jun Yoo

Abstract Interest on the low energy wireless connections among humans, sensors and mobile devices is getting grown with the start of the phrase, "Internet of Things". The establishment of the Wireless Body Area Network (WBAN) standard by IEEE Standard Association in February 2012 is representing those demands. In the standard, a Human body communication or HBC is newly added and it is regarded as a new promising solution with its low energy consumption and high reliability. The research history of the HBC from the proof of the concept to the high-end standard compatible system with the network consideration will be introduced in this chapter.

Keywords Wireless Body Area Network • Human body communication • Low energy consumption • Transceiver • Standard compatible

H. Lee (✉)
Samsung Advanced Institute of Technology 130, Samsung-ro, Yeongtong-gu, Suwon-si, Gyeonggi-do, 443-803, Korea
e-mail: hwlee010@gmail.com

S.-J. Song • N. Cho
Samsung Electronics Co., LTD., 416, Metan-3dong, Yeongtong-gu, Suwon-Si, Gyeonggi-do, 443-742, Korea
e-mail: sj33.song@samsung.com; cho.namjun@gmail.com

J. Bae
IMEC, Kapeldreef 75 B-3001, Lauven, Belgium
e-mail: Joonsung.Bae@imec.be

H.-J. Yoo
KAIST, 291, Daehak-ro, Yuseong-gu, Daejeon, 305-701, Korea
e-mail: hjyoo@kaist.ac.kr

P.P. Mercier, A.P. Chandrakasan (eds.), *Ultra-Low-Power Short-Range Radios*, Integrated Circuits and Systems, DOI 10.1007/978-3-319-14714-7_9

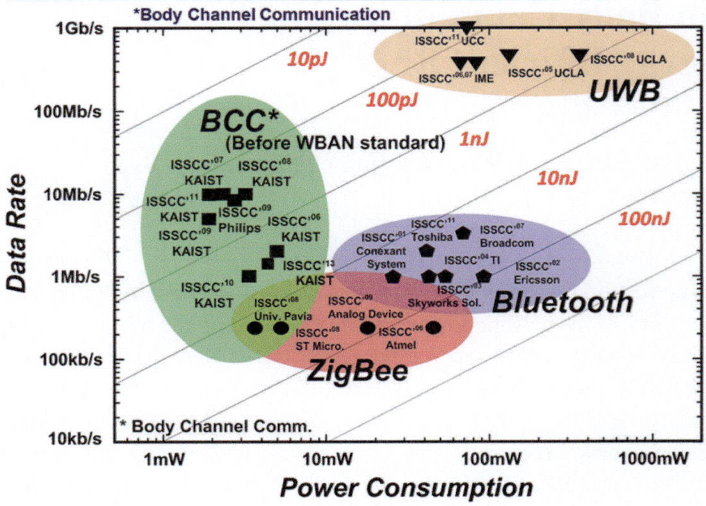

Fig. 1 Energy efficiency of NB, UWB, and HBC TRXs

1 Introduction

1.1 Human Body Communication (HBC)

In February 2012, the IEEE Standards Association established the Wireless Body Area Network (WBAN) standard [1] to accommodate increasing demands on the near body communication technology for many biomedical applications. It standardized 3 PHYs:, human body communication (HBC), narrow band (NB) and ultra-wide band (UWB). Above all, HBC is considered as the most energy efficient PHY with a wide coverage of data-rates compared to NB or UWB as shown in Fig. 1 [2]. HBC uses only electrodes on the body to communicate in place of antennas, without any body shadowing effect [3]. Therefore, it is a new convenient solution for medical services among humans, sensors and mobile devices with low energy wireless connections.

1.2 HBC Transceivers

Since 2006 several types of integrated HBC transceivers (TRXs) have been reported, as shown in Fig. 2. There TRXs can be grouped into five different phases: (1) the technology introduction, (2) the practical application, (3) the channel enhancement, (4) the WBAN standard compatibility and (5) the network consideration phases.

	Phase-1		Phase-2		Phase-3	Phase-4	Phase-5
Parameters	ISSCC 2006 [4]	ISSCC 2007 [5]	ISSCC 2008 [6]	ISSCC 2009 [7]	ISSCC 2011 [8]	ISSCC 2013 [3]	CICC 2014 [9]
Technology	0.25μM CMOS	0.18μM CMOS	0.18μM CMOS	0.18μM CMOS	0.18μM CMOS	0.13μM CMOS	0.13μM CMOS
Power Consumption	0.2mW	2.6mW	4.6mW	2.3mW	2.4mW	5.5mW	2.77mW (@64nodes)
Frequency Band	100kHz – 100MHz	15kHz – 60MHz	30MHz – 120MHz	30MHz – 70MHz	40MHz – 120MHz	21MHz	21MHz
Data Rate	2Mb/s	10Mb/s	60kb/s – 10Mb/s	8.5Mb/s	1kb/s – 10Mb/s	1.3125Mb/s	1.3125Mb/s
TX Mask Restriction	X	X	X	X	X	O	O
RX Sensitivity	-	-30dBm	-65dBm	-60dBm	-66dBm	-97.35dBm	-98dBm
Modulation	WBS	DSSS PPM	AFH FSK	Correlation Direct Digital	Double FSK	FSDT	FSDT
Packet Consideration	X	X	X	X	X	O	O
Spontaneous Scheduling	X	X	X	X	X	X	O

Fig. 2 Summary of previously proposed HBC TRXs

(1) **Phase-1—Proof of concepts**: The first prototype was implemented with digital wideband signaling (WBS) based on a simple phenomenological model of the human body. With the proposed TRX, the feasibility of the communication along the body was demonstrated successfully [4]. Then, a direct sequence spread spectrum (DSSS) with a pulse position modulation (PPM) TRX was used to achieve high data-rate signaling at 10 Mb/s. It supported the N to N communication with power consumption as low as 2.6 mW by adopting an improved empirical channel model of the human body [5].

(2) **Phase-2—Practical applications**: In this phase, an adaptive frequency hopping frequency shift keying (FSK) TRX was proposed to avoid the interference by the outside FM radio or TV broadcasting signals. Its RX sensitivity was improved to −65 dBm [6]. Later, an HBC TRX integrated with the Medical Implant Communication Service (MICS) circuits was proposed to cover the implantable network too. An antenna for the MICS radio was shared with HBC electrodes and was able to be attached directly to the human skin. With the HBC and MICS dual-band TRX, 8.5 Mb/s communications was available with 2.3 mW power consumption [7].

(3) **Phase-3—Channel enhancements**: Hear the focus was shifted to optimize TRXs with a more accurate body channel model. The signal transmission mechanism was examined with detailed analysis of Maxwell's equations when the current dipole excitates electromagnetic waves on the conducting plane. The results show that the quasi-static near field coupling is dominant at low frequency range and the surface wave propagation is dominant at high

frequency range. The model was actively used to design a high performance double FSK TRX [8].

(4) **Phase-4—Standard compatibility**: A fully standard compatible TRX was reported for medical applications in [3]. The WBAN standard requires both stiff TX spectrum mask restriction and high RX sensitivity. To satisfy these requirements, an adaptive digital filter and an RX sensitivity dual-enhanced structure were proposed with 5.5 mW power consumption.

(5) **Phase-5—Network protocol considerations**: Here an intelligent time-domain MAC scheduling TRX was reported for medical body area network (MBAN) applications [9]. The TRX fully satisfies the HBC PHY standard and it also supports a standard compatible MAC controllability with internal MAC scheduler. With the MAC scheduler, the effective power consumption of the TRX system is 2.77 mW with 64 nodes.

In this chapter, each phase of integrated HBC TRXs will be introduced in further detail, including: (1) motivation and design issues, (2) key schemes for the novel operation, and (3) an architecture of the proposed TRX.

2 Phase-1: Proof of Concepts

2.1 Motivation

Several schemes have been suggested for HBC, mostly employing methods to transfer data through the skin of the human body. For example, a near-field electrostatic coupling scheme using a narrowband low frequency signal was firstly introduced by Zimmerman [10], and was expected to significantly reduce the power consumption of BAN nodes. In this case, the proposed coupling scheme is strongly dependent on the conditions of the surrounding environment like the earth ground for the return path and has limited data rate of 2.4 kb/s due to the narrow bandwidth of 400 kHz. Another scheme employing an electromagnetic wave of 10 MHz also suffers from the bandwidth limitation of conventional FM and FSK [11]. Another group reported a transceiver adopting an electro-optic conversion method to achieve a higher data rate of 10 Mb/s by using a special off-chip sensor [12]. However, this leads to high cost, high power consumption and large physical size for human body applications. Moreover, it must have two electrodes, the signal and the ground, which makes it inconvenient to use.

This section presents a novel HBC scheme exploiting wideband signaling (WBS) technique with a direct-coupled interface (DCI) over the optimized HBC channel [4]. The HBC channel is optimized for high speed operation on the human body. The DCI is an interface method connecting the silicon chip with the human body directly. It uses only a single electrode for data transmission without any intentional ground electrode, in contrast to other methods which require an off-chip sensor to detect the feeble electric field and the earth ground path. To validate this approach,

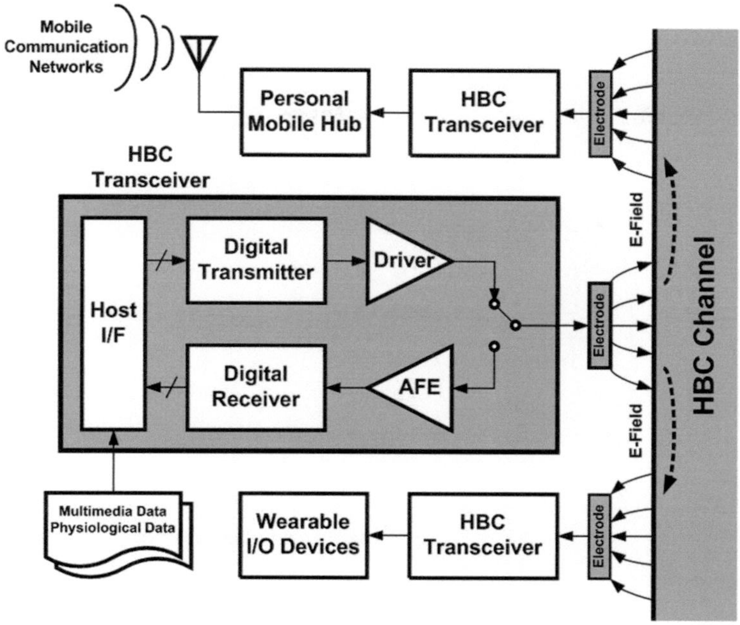

Fig. 3 The HBC using the DCI on the human body

a 2 Mb/s WBS digital transceiver chip is implemented by a 0.25 μm standard CMOS technology. The digital transceiver consumes 0.2 mW from a 1 V supply. In addition to the digital transceiver, a 10 Mb/s WBS receiver analog front-end (AFE) is presented that exploits a wideband symmetric triggering (WST) technique to efficiently recover binary data from the feeble wideband pulse signals. The AFE incorporates with a low-power wideband op amp based on the low-voltage fully complementary folded cascode topology. Realized in a 0.18 μm standard CMOS technology, the AFE chip occupies 0.04 mm^2 and exhibits power consumption less than 5 mW from a 1 V supply. Therefore, the proposed HBC scheme can achieve lower power consumption while boosting data rates compared to other previous HBC schemes in [13, 14], which make it suitable for the application of energy-efficient point-to-point data transmission around the human body. The proposed HBC using the DCI on the human body is illustrated in Fig. 3.

2.2 *Wideband Signaling Communication Link*

In order to improve communication performance, an analytical model for the WBS communication link is necessary for the design and implementation of an efficient transceiver. First, the characteristics of the channel are investigated and, following this, a physically-equivalent lumped RC model is established as shown in Fig. 4.

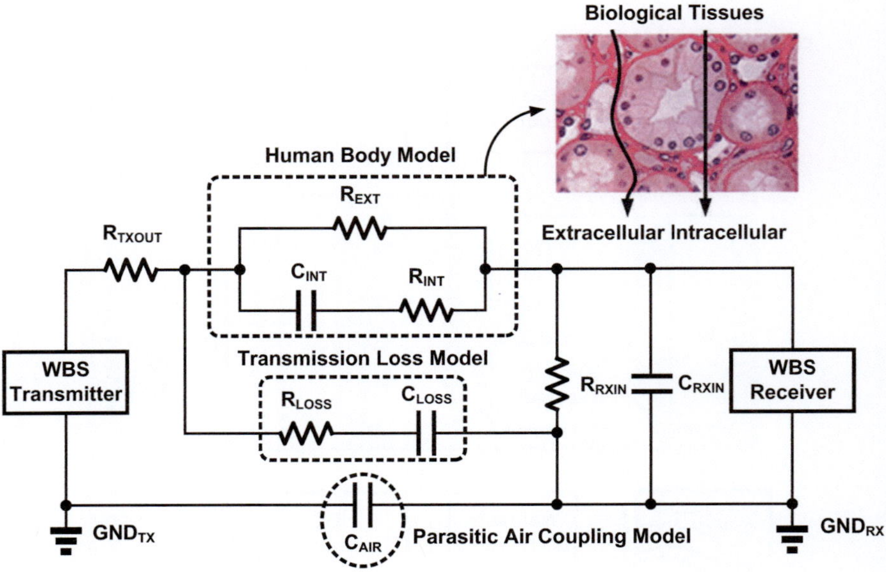

Fig. 4 Design flow diagram of the WBS communication link

Table 1 WBS link parameters

Link parameters	Specification	Link parameters	Specification
Distance	0–100 cm	Transconductance	5 $m\Omega^{-1}$
Receiver input sensitivity	10 mV	Triggering threshold	300 mV
Closed-loop bandwidth	200 MHz	Output capacitance	400 fF
Supply voltage	1 V	Power consumption	6 mW
Closed-loop gain	30 V/V	Bit period	0.5 μs
Bias current	3 mA	Bit energy	3 nJ

R_{EXT}, C_{INT} and R_{INT} are obtained by extending a simplified electrical circuit model for the biological tissues to the human body model. R_{LOSS} and C_{LOSS} are added as the transmission loss model and C_{AIR} is a very small capacitance due to the feeble coupling as the return path between each ground of the TX and RX. R_{TXOUT} and R_{RXIN} are the output and input impedances of the TX and the RX, and C_{RXIN} is the input load capacitance of the RX.

Leveraging this RX model, a simplified link model suitable for the proposed signaling is defined. After that, the link design parameters such as bandwidth, supply voltage, voltage gain, bias current, and threshold are specified [4]. Table 1 summarizes the design specification for link parameters.

According to the specified link parameters, the design specification and the architecture of the transceiver can be defined in regard to energy-efficient data transmission. Finally, the detailed design of each building block is carried out by exploiting circuit techniques to minimize its power consumption.

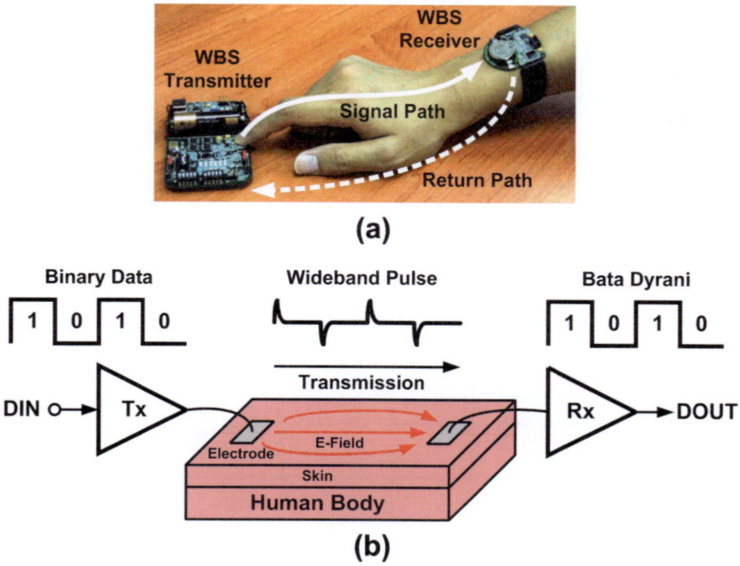

Fig. 5 (**a**) Signal transfer photograph and (**b**) conceptual diagram of the WBS link

Figure 5a, b shows the signal transfer photograph and (b) the conceptual diagram of the WBS communication link, respectively.

2.3 Wideband Signaling Transceiver

The block diagram of the WBS transceiver that comprises a direct digital transmitter and a CDR-based WBS receiver is shown in Fig. 6. The direct digital transmitter consists of a clock synthesizer, a pseudo random binary sequence (PRBS) generator, a 2-to-1 multiplexer (MUX), and a driver. The clock synthesizer has a ring oscillator structure and generates the clock signal with frequency scaling to activate the PRBS generator. The PRBS generates 2^7-1 PRBS data and transmits them through the driver to the human body for on-chip link testing. Also, external binary data such as digitally converted audio data or baseband data can be directly transmitted to the human body by the 2-to-1 MUX. The driver is connected to the electrode and induces the electric field on the skin of the human body. The CDR-based WBS receiver consists of a receiver AFE, a CDR circuit, and a bit error detector. The receiver AFE amplifies, triggers, inverts, and shifts the received wideband pulse signal in order to recover the binary data. The next CDR circuit extracts a clean clock signal from the recovered binary data and latches the data. The bit error detector is integrated for on-chip BER testing and detects error bits of the recovered data from the extracted clock.

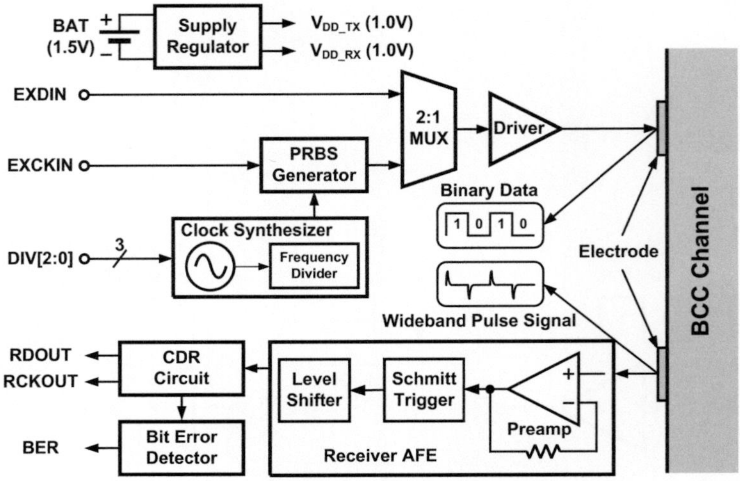

Fig. 6 Proposed architecture of the WBS transceiver

Figure 7 illustrates the timing diagrams of the WBS transceiver operation. As investigated in Sect. 2.1, when binary data is directly inserted into the human body, the channel outputs a narrow small signal that comprises positive and negative pulses with no DC offset. The received pulse signal which may be corrupted by the channel is sufficiently amplified for wide bandwidth, and subsequently, the signal is triggered to positive and negative states by using two symmetric thresholds, V_{TH} and V_{TL}, where the symmetric operation provides the duty cycle of 50 %. Consequently, the binary data can be recovered by inverting and shifting the triggered signal. For CDR at the receiver, the full-rate clock signal is locked at the center point of the bit interval window. The binary data is recovered by latching the level-shifted signal at the rising edge of the clock signal.

2.4 Measurement Results

In this example, all measurements are conducted between the wrist and the fingertip, which corresponds to the distance of about 25 cm as shown in Fig. 8. A WBS transmitter board is attached to the wrist using a single Ag/AgCl electrode on the backside. When the fingertip touches the metal electrode on a WBS receiver board, the stream of the binary data is transferred through the body to the WBS receiver board.

Figure 9 shows the measured eye diagrams of each output of the direct digital transmitter, the HBC channel, the receiver AFE, and the recovered clock and data of the CDR circuit for 2 Mb/s 2^7-1 PRBS data, respectively. The channel output of the binary data exhibits the narrow small pulse signals with the amplitude of about 50 mV and no DC offset.

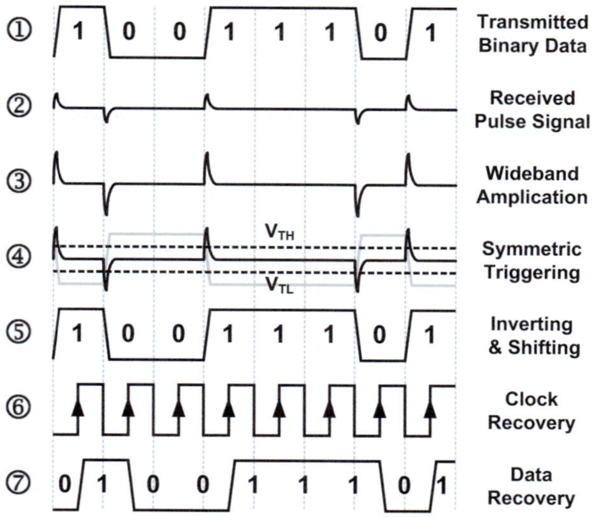

Fig. 7 Timing diagrams for the operation of the WBS transceiver

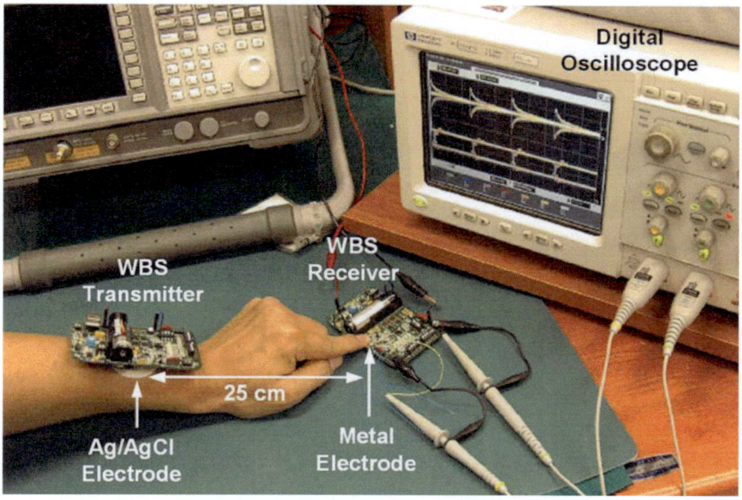

Fig. 8 The test measurement setup

3 Phase-2: Practical Applications

3.1 Motivation

According to the phase-1, the feasibility of the communication through the human body is verified with the simple transceiver system. Since the phase-2, the application is focus to the daily life, especially to the home healthcare service. A number of healthcare devices are willing to be on or inside our body and collaborate

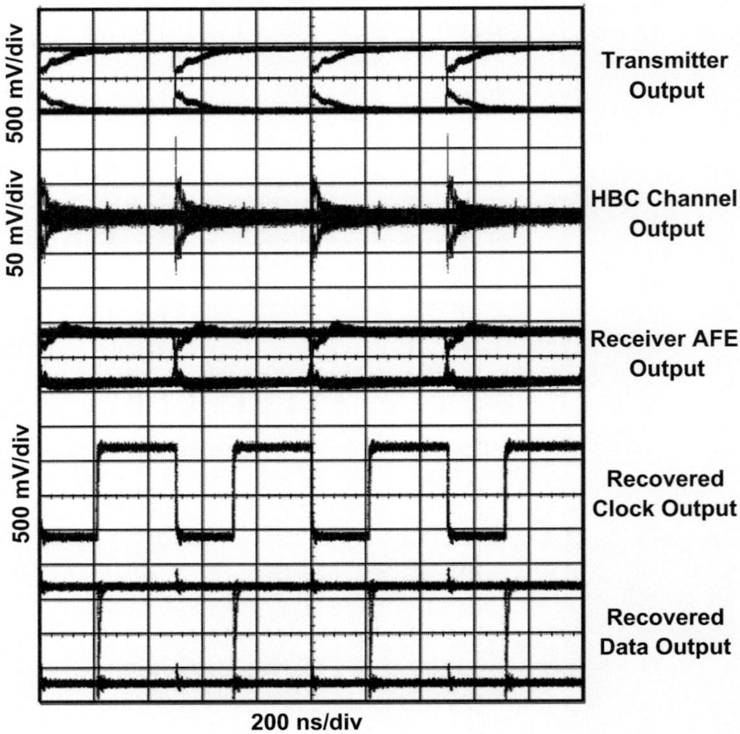

Fig. 9 Measured eye diagrams of the WBS digital transceiver for 2 Mb/s 2^7-1 PRBS

together to give an accurate diagnosis and treatment to us. In order to make this ambulatory health care system feasible, the body sensor network (BSN) which provides wireless connectivity among the wearable and implantable devices needs to be developed. With those needs, the IEEE 802.15.4 organized a new task group to standardize frequency and protocols for BSN at that time, 2009 [15]. They performed extensive channel measurements to decide the optimum frequency range for body-area communication, and found that >1 GHz bands for WPAN are not suitable to BSN due to the severe path loss around the human body.

Another issue discussed in the BSN task group is how to connect the wearable and implantable devices which are under the distinct communication environments—on and inside body, efficiently. Currently, most in-body devices use 402–405 MHz to communicate with an external controller as it is widely accepted that around 400 MHz band shows the smallest path loss through the conductive human tissue [13]. This 3 MHz band is defined as the medical implant communication service (MICS) band by FCC, and implantable radios are regulated to use this band [16]. However, it is not obvious that the 400 MHz is also optimum for communication among on-body devices. In addition, the electromagnetic wave still attenuates significantly with distance inside the human body, and hence the

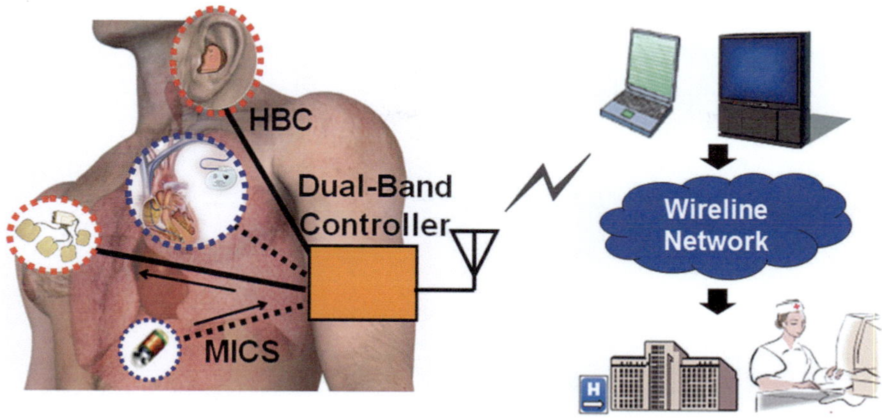

Fig. 10 Unified body sensor network

communication among in-body devices is hardly obtained if the implanted radios do not guarantee sufficient sensitivity.

In this section, a new type of BSN which overcomes the issues related with the body-affected channel environment will be proposed, which achieves the best energy efficiency among the BSN systems published before 2009 [6]. Before proceeding, the current BSN systems and their limitations will be discussed shortly.

3.2 Unified Body Sensor Network

Figure 10 presents the overall structure of the unified BSN which combines wearable and implantable BSNs together. It consists of a network controller attached to human skin and multiple sensors deployed on and inside the human body [17]. The on-body network controller plays significant roles in terms of the patient mobility and the energy consumption. As the network master of the star BSN, it collects vital data sensed by each node and processes them. After the sufficient amount of data is gathered, the controller wirelessly transmits them to a remote access point which is 10–100 m away from the patient. As a result, the coverage of the health care service can be extended without increasing sensitivity of the energy-hungry sensor radios.

The dual-band transceiver for the unified-BSN controller does not just use two frequency bands but supports communications in the distinct mediums. For in-body communication, the transceiver is a normal radio operating in the 400 MHz band. However, for on-body communication, it needs to be configured as an E-field coupler which transmits data in the form of voltage rather than electromagnetic power. Hence, as the first step of the transceiver design, the characteristics of the on-body and in-body channels should be investigated. Then, the channel interfacing

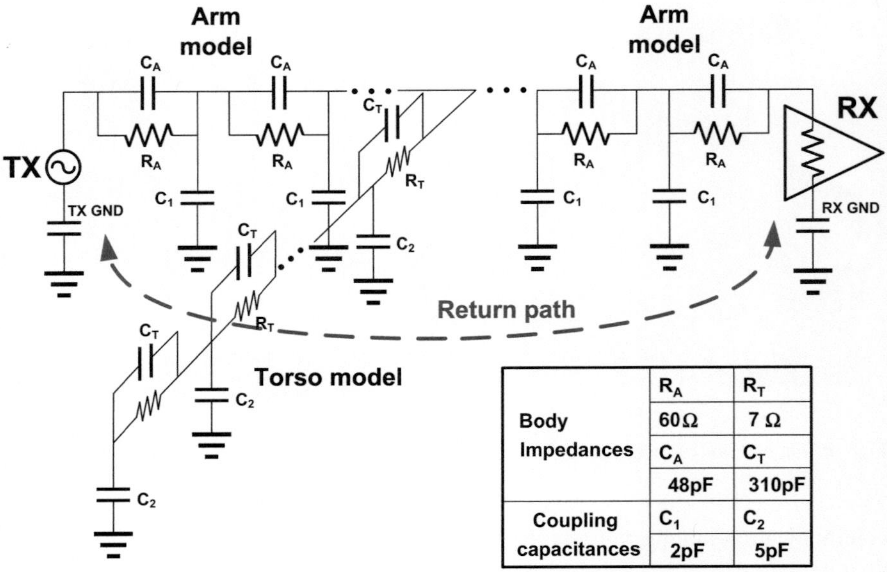

Body Impedances	R_A	R_T
	60Ω	7 Ω
	C_A	C_T
	48pF	310pF
Coupling capacitances	C_1	C_2
	2pF	5pF

Fig. 11 Proposed circuit model of the body channel

unit which functions as both of the antenna and electrode needs to be developed. The design of the building blocks composing the transceiver has to be oriented to low-energy for the longer operating time.

3.3 *Distributed RC Model of Body Channel*

In the previous section, the human body is modeled as a once chunk of biological tissues, converted as a single R and C. However, that model cannot reflect of the effect depending on the distance between the TX and the RX or the frequency of the communication band. In order to enhance the channel model, the segmented T-shaped RX model is suggested. By cascading the multiple RC blocks to construct the model and combining them, a complete circuit model of the human body can be obtained, as shown in the Fig. 11. The R and C values in the distributed network are calculated from the electrical properties in 10–60 MHz and listed together. To consider the channel responses at the various locations on the body, the receiver model can be placed at the corresponding nodes of the circuit model. Because each sub RC circuit is for the 10 cm unit block, N-stage cascaded blocks are equivalent to the N × 10 cm channel length. In the distributed network, the major return path is formed by the electrical coupling between the TX and RX grounds through the external ground. Therefore, using a large ground plane or high carrier frequency is advantageous to enhance the SNR of the received signal.

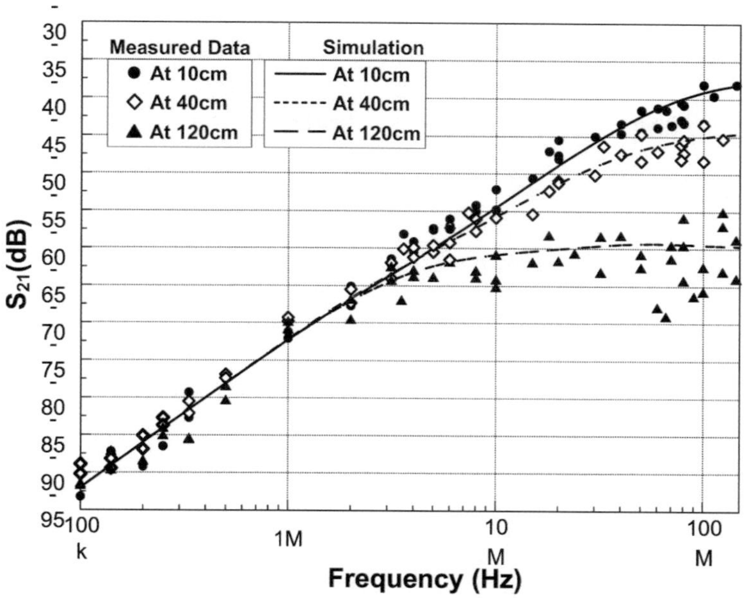

Fig. 12 Measured S_{21} parameters of the body channel

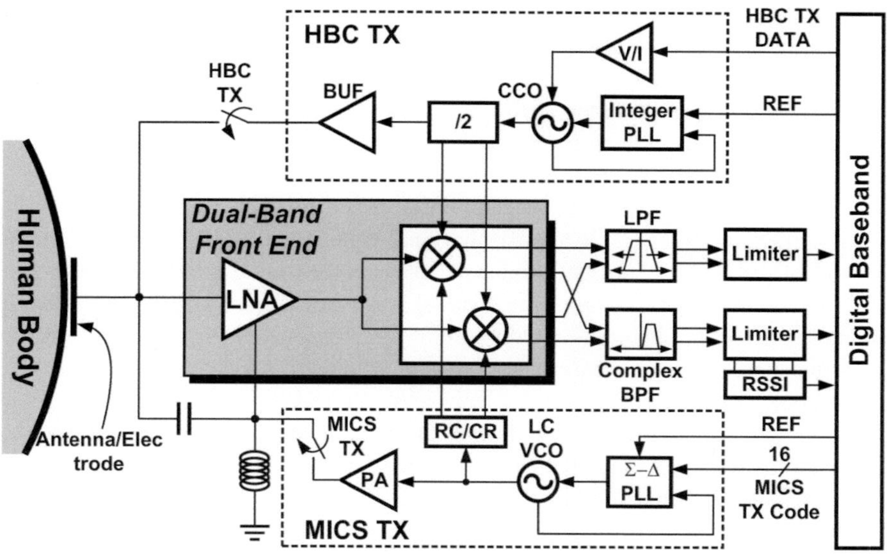

Fig. 13 Proposed unified BSN TRX

The graphs in Fig. 12 show measured S_{21} parameters with respect to the transmitted signal frequency. The effect of the channel distance is considered, and the simulated S_{21} parameters with the distributed RC model are also included. The graph shows that below 4 MHz, the body channel is relatively deterministic, with at most 5 dB of deviation regardless of the distance. Beyond 10 MHz, the channel distance has a great effect on the transferred power. As the channel length increases, the capacitive coupling between the body and the external ground becomes larger, and induces larger signal loss. Especially at 120 cm distance, much of the transmitted power is lost to the external ground through the torso.

3.4 Unified Body Sensor Network Transceiver

Figure 13 shows the transceiver architecture which consists of the distinct two chains for dual-band operation. The LNA and mixer circuits which consume the largest power are shared between HBC and MICS. The MICS band antenna attached to skin also functions as an electrode that couples HBC signal to the human body. Following the dual-band front ends, the HBC and MICS signals are down-converted to zero and low intermediate frequencies (IF) concurrently. The MICS base-band signal with 300 kHz bandwidth is easily corrupted by 1/f noise generated in the base band circuits. Therefore, the 450 kHz IF for MICS RX is selected to position the desired signal band beyond the 1/f corner frequency. The channel bandwidth of the HBC signal changes for the variable AFH scheme. However, the minimum bandwidth is 1.5 MHz which is far higher than the 1/f corner frequency. The direct-conversion receiver is suitable for HBC because 1/f noise and dc offset can be rejected by a simple high pass filter without sacrificing the sensitivity. The base-band filter for MICS should be a complex band pass filter to remove image signals. The 5th order complex Butterworth filter is employed to give 38 and 48 dB image and adjacent channel rejections. The cutoff frequency of the 4th order low pass filter in the HBC base-band chain is programmable for the variable AFH.

The LO signals applied to the dual-band mixer are generated by integer and fractional PLLs. In HBC TX, the BFSK signal is obtained by changing the control current of the CCO according to binary data from the digital circuit. The TX data is DC-free coded not to disturb the channel frequency fixed by negative feedback of the integer PLL. The transconductance of the V/I converter is programmable to make the amplitude of the control current proportional to TX data rate, and thus the modulation index (MI) of the BFSK signal is maintained at 1. In MICS TX, the 16bit input code to the $\Sigma-\Delta$ modulator controls the division ratio of the fractional-N PLL. By changing the code value, the MICS BFSK signal can be generated. The RC/CR network following the LC VCO makes the I/Q LO signals applied to the mixer. As the accurate quadratic relation between the I/Q base-band signals is significant for the image rejection, careful considerations in the layout of the passive elements are necessary to enhance the matching property.

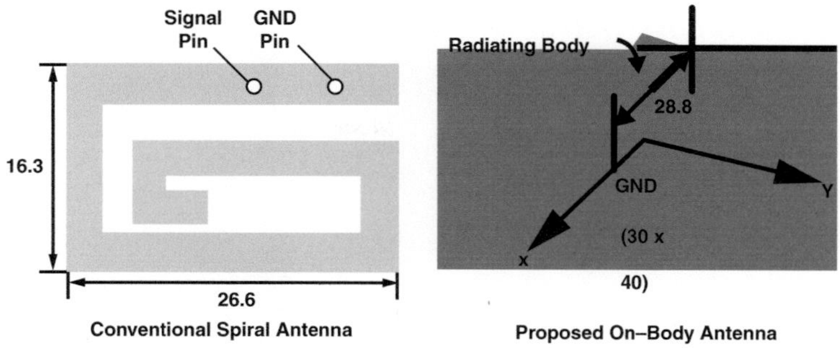

Fig. 14 Comparison with the conventional spiral antenna

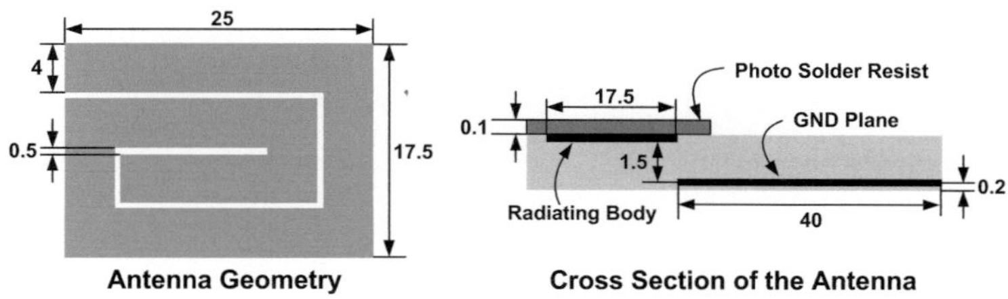

Fig. 15 Detailed antenna structure

3.5 MICS Band Antenna Combined with HBC Electrode

Figure 14 presents the antenna structure for both MICS and HBC. The ground pin on the radiating body is eliminated to make the signal and ground ports open at the low frequencies. Instead, the 3×4 cm^2 ground plane is placed 1.5 mm below the radiating body. The central axes of the two plates are separated by 28.8 mm. The resonant frequency of the antenna is lowered to 650 MHz with the small body area of 2.5×1.8 cm^2 and 1.8 mm thickness because the fringing effect increases the effective antenna size. Arranging the radiating body and the ground plane side-by-side is also advantageous for HBC.

Figure 15 shows the physical dimension of the spiral radiating body tuned to get the best return loss and antenna efficiency. The antenna is intentionally designed to resonate at 650 MHz rather than the desired 400 MHz, considering that the high electrical permittivity of the human body lowers the resonant frequency. The design parameters including the number of spiral turns, line width, line space, and the signal feed point are optimized through EM field simulation.

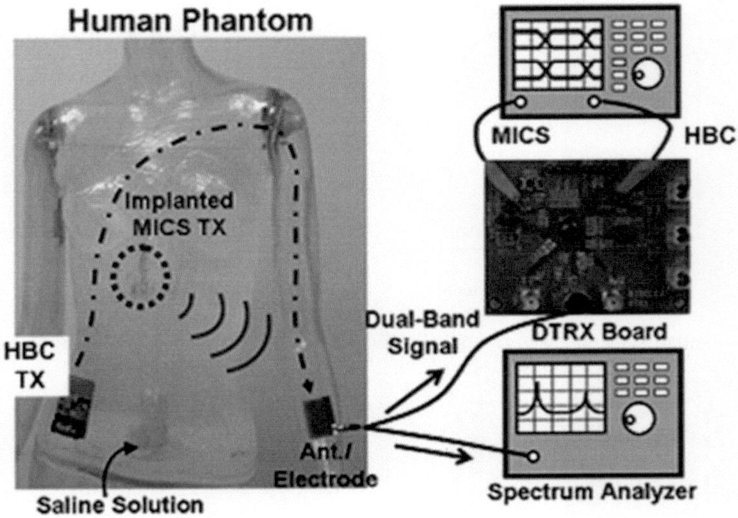

Fig. 16 Experimental setup for sensitivity measurement

3.6 Measurements Results

Figure 16 shows the experimental setup to investigate how much the sensitivities of the HBC and MICS RXs are affected by coexistence of the dual-band signals during concurrent operation. The human phantom is prepared to make the communication channel. Its inner space is filled with saline solution which is often used to simulate the in-body environment, and its outer layer is processed to have the conductivity of the human tissue, which is around 0.5 S/m. The HBC TX on the phantom surface and the implanted MICS TX are battery-powered and emit FSK signals. The HBC signal is electrically coupled to the dual-band transceiver through the conductive layer and the MICS signal in the form of electromagnetic wave propagates the saline solution and goes into the transceiver. The antenna combined with an electrode is an interfacing unit that receives the HBC and MICS signals concurrently and transfers them to the transceiver. The spectrum analyzer measures the power level of the received dual-band signal. The digital oscilloscope displays the HBC and MICS bit streams recovered concurrently. With this experimental setup, the sensitivity of the MICS RX is measured while moving the HBC TX toward the electrode to increase the strength of the HBC RX signal.

As shown in Fig. 17 the HBC sensitivity is −65 dBm when the coexisting MICS signal is low enough. However, above −40 dBm of the MICS signal, the errors in the recovered HBC data start to appear due to the receiver saturation. In summary, the HBC and MICS RXs can withstand the 25 and 30 dB larger MICS and HBC signals without sensitivity degradation.

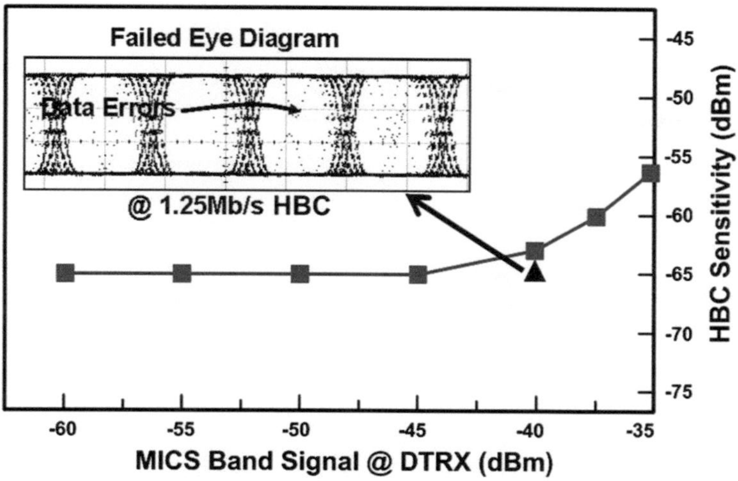

Fig. 17 HBC sensitivity with coexisting MICS signal

Fig. 18 Capacitive coupling mechanism of HBC

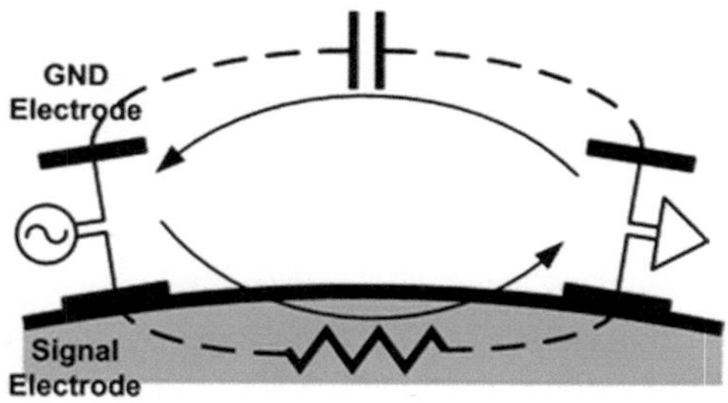

4 Phase-3: The Channel Enhancement

4.1 Motivation

If the previous model was based on the circuit modeling, the channel model for the phase-3 is analyzed based on the principle and the theoretical mechanism. To establish the novel body channel model, two prevailing methods are used. Firstly, Fig. 18 shows the capacitive coupling method. At a frequency lower than tens of MHz whose wavelength is much larger than the size of the human body, the electric field around the human body is almost constant with time, which means its phase is nearly uniform everywhere on the body. In this condition, the time-varying electric field around the human body can be regarded as a quasi-static field. On the other hand, the wave propagation method is used if the frequency is higher than tens of MHz. Figure 19 shows the wave propagation method. In this frequency range, the electrical signal attenuates as the signal propagates through the human body.

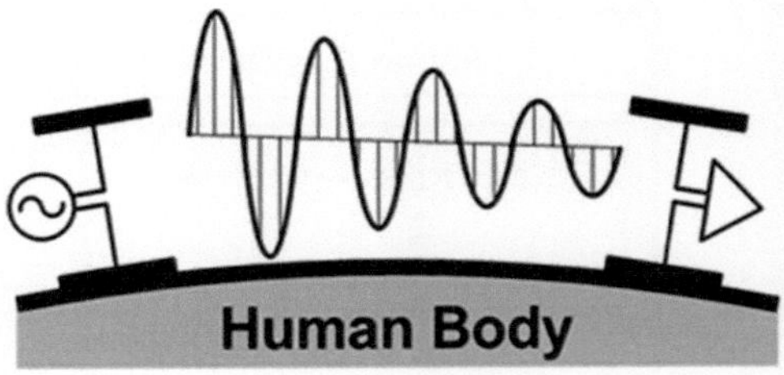

Fig. 19 Wave propagation mechanism of HBC

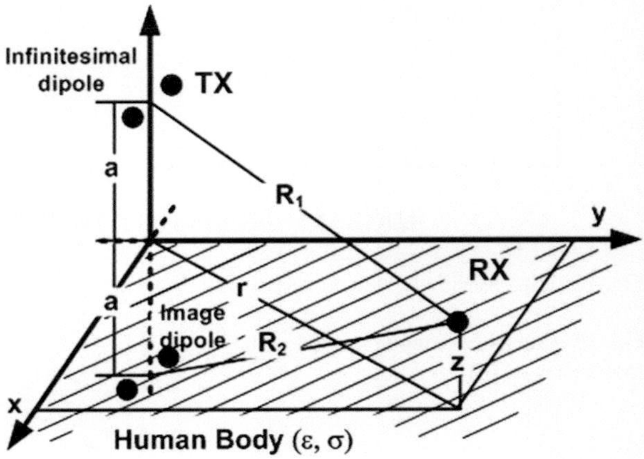

Fig. 20 Electric field from dipole and its geometry on the human body

Previous HBC transceivers were not optimized for WBAN because the only phenomenological circuit/behavior models were used for the body channel analysis, which means there was not a clear understanding of the on-body electric signal transmission mechanism. In summary, previous studies covered only a limited frequency range by limited explanation method.

In this section, the fundamental studies on the HBC will be shown [8]. In addition, an energy-efficient WBAN transceiver system is implemented with dedicated base station node transceiver and sensor node transceiver in 0.18 μm CMOS technology. As a result, optimized WBAN TRX receiver and transmitter consume 2.4 and 2 mW, respectively, at a data rate of 10 Mb/s, corresponding to energy consumption of 0.24 nJ per received bit and 0.2 nJ per transmitted bit.

4.2 Fundamental Channel Model

To investigate physics behind the HBC, Fig. 20 shows the simplified model of HBC and the electric field from infinitesimal dipole for calculating electric field intensity at the point above the surface of the human body, which has finite conductivity (σ) and permittivity (ε).

For the generality and simplicity of the analysis, we assume the surface of the human body as an infinite half-plane with an imperfectly conducting property. To

Fig. 21 Electric field in the condition of HBC derived from the Maxwell's equation

$$\left|E_{Z^V}\right| = 2\left|k \cdot F \cdot \frac{1}{r} + i \cdot \frac{1}{r^2} - \frac{1}{k} \cdot \frac{1}{r^3}\right|$$

obtain a general solution of Maxwell's equation in Fig. 20, the wave potential of a unit vertical dipole and its image dipole is derived by the continuity property of the tangential components from the electric and the normal components from the magnetic field with the boundary surface of the z = 0 plane.

In regard to Maxwell's equation, the electric field intensity in Fig. 21 consists of terms of the first order in 1/r, second order in 1/r, and third order in 1/r. The first term correlates with far-field propagation in combination with attenuation factor of surface wave, kF, which is an inherent property of electric field at the surface of the half-plane with finite conductivity. On the other hand, second and third terms correspond to the induction-field radiation and near-field coupling of the dipole, respectively. Consequently, the mechanism of HBC can be divided into three parts: (1) the surface wave far-field propagation of the first term, (2) the reactive induction-field radiation of the second term, and (3) the quasi-static near-field coupling of the third term. The intensity of the electric field is a function of communication distance and wavenumber. As the wavenumber k or frequency f, and the distance r increase, the surface wave propagation term starts to have significant effect on the overall electric field intensity whereas the quasi-static coupling term is negligible.

4.3 Double-FSK HBC TRX

Figure 22 shows the overall architecture of the double-FSK transceiver. The HBC uses a 40–120 MHz frequency band for the data transmission while the contact impedance sensing (CIS) circuit utilizes a chopper-stabilized AC current-injection source of 1.25 MHz to monitor the differential contact impedance between contact electrode and the human body. CIS also adjusts noise figure, input referred 1 dB compression point, and voltage gain of LNA to support a low power or a high sensitivity mode. The resonance matching (RM) is connected with GND electrode and it provides the high impedance to the input of LNA. With the help of RM, a channel enhancement can be achieved with the power reduction of the driver.

On the transmitter side, from the frequency synthesizer and divider chain, sub-band FSK modulator modulates the TX data, followed by wide-band FSK modulator that drives the electrodes by transmitter driver.

On the receiver side, from the LNA which amplifies the received signal from the electrodes, the wide-band demodulator converts the wide-band carrier signal into a sub-band FSK signal that is demodulated by low-frequency direct-conversion receiver circuits which contain the sub-band demodulator.

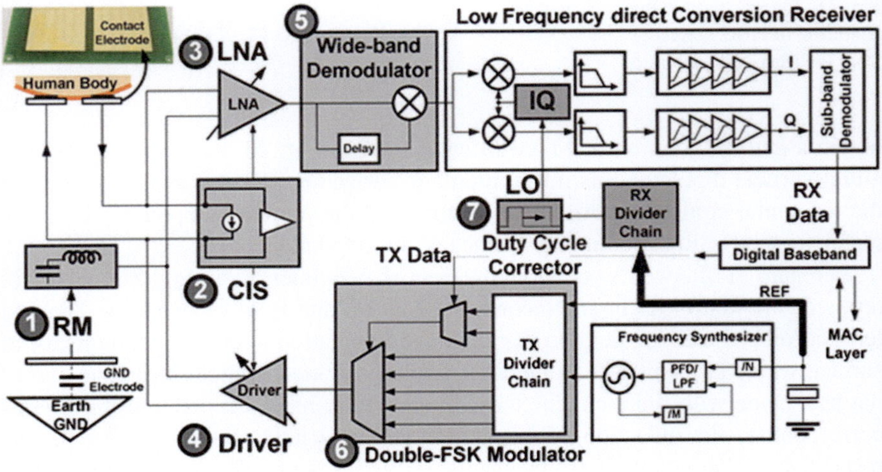

Fig. 22 Overall architecture of double-FSK transceiver

4.4 Measurement Results

Figure 23 shows the measured output spectrum of double-FSK transmitter. The right table shows the operation modes of the BCC transceiver. It supports total 104 operation modes with the data rate from 1 kb/s to 10 Mb/s and the network-coexistence from 1 to 15. Each spectrum represents modulated wide-band FSK signal from 4 sub-bands with various data rates of 10 kb/s, 100 kb/s, 1 Mb/s, and 10 Mb/s. The wide-band modulator spread the sub-band signal into 40–120 MHz with the frequency of 40, 60, 80, 100, and 120 MHz. The output power level is lowered by spreading gain, and currently set to −15 dBm. However, the output level can be raised up to 0 dBm to compensate the channel variation by reconfigurable driver.

Each sub-band is distinguished by a wide-band FSK demodulator with corresponding sub-band carrier frequencies. To demonstrate WBAN coexistence in a low-SNR condition, two users that occupy the same sub-band with different sub-band carrier frequencies of 0.5 and 1 MHz, and different data rates are applied to human body simultaneously with −40 dBm input power as shown in Fig. 23. Figure 24 shows the spectrum of the receiver input. The user 1 and user 2 are in the same sub-band from 0.5 to 2.5 MHz with a data rate of 125 and 250 kb/s as represented in the table.

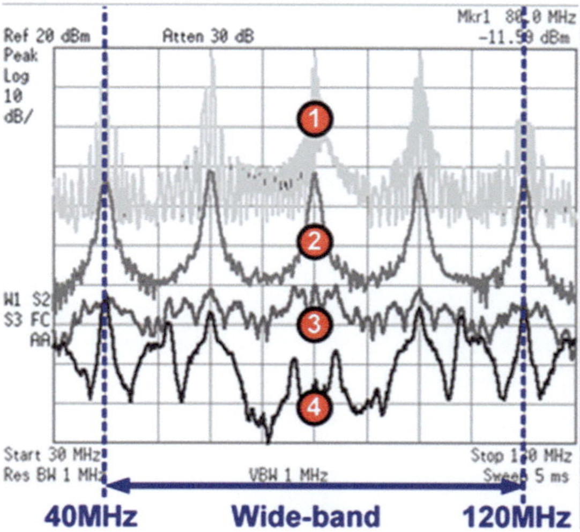

Sub band	# of Users	Data Rate
5kHz – 25kHz	15	1.25kb/s
	7	2.5kb/s
	3	5kb/s
	1 ①	10kb/s
50kHz – 250kHz	15	12.5kb/s
	7	25kb/s
	3	50kb/s
	1 ②	100kb/s
500kHz – 2.5MHz	15	125kb/s
	7	250kb/s
	3	500kb/s
	1 ③	1Mb/s
5MHz – 25MHz	15	1.25Mb/s
	7	2.5Mb/s
	3	5Mb/s
	1 ④	10Mb/s

Fig. 23 Output spectrum of the double-FSK transmitter

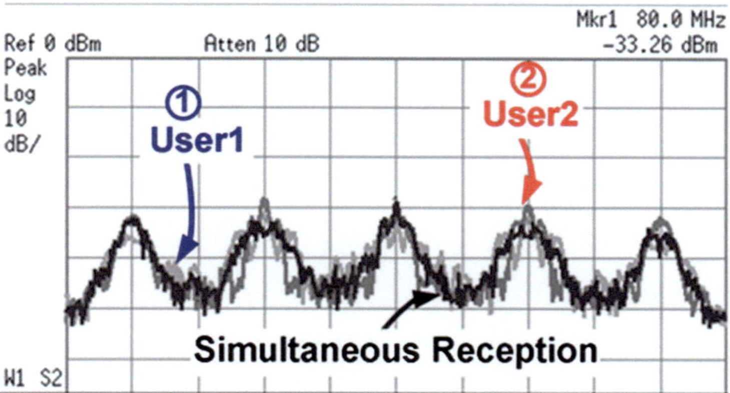

Sub band	# of Users	Data Rate
0.5MHz – 2.5MHz	15 ①	125kb/s
	7 ②	250kb/s
	3	500kb/s
	1	1Mb/s

Fig. 24 Measurement results of coexistence spectrum of the RX input

5 Phase-4: The Standard Compatibility

5.1 Motivation

Before the standardization from phase-1 to phase-3, several HBC TRX SoCs are proposed to verify the prospect of the TRXs to be applied to the network construction without any hard restrictions. Finally, in February 2012, the IEEE Standards Association established a WBAN standard and the HBC is regarded as a new convenient solution for the medical services with its high energy efficiency. A variety of HBC TRXs have been reported before the establishment of the WBAN standard but none of them consider the IEEE 802.15.6 standard, and therefore can satisfy the demanding specifications of the standard, which require (1) a regulated packet structure with the Frequency Selective Digital Transmission (FSDT) modulation, (2) a stiff TX spectrum mask and (3) a high RX sensitivity.

In this section, we achieve first development of the state-of-the-art HBC transceiver satisfying all of the specifications and requirements for the IEEE 802.15.6 standard. First, the FSDT modem with maximum likelihood detection (MLD) is proposed to realize the standard packet structure. Second, the TX with active digital band pass filter (ADF) is adopted to fulfill the tight spectral mask without using external components. Third, the RX with a sensitivity dual-enhanced (SDE) structure is applied to satisfy the tight sensitivity requirement. Lastly, the proposed WBAN transceiver is applied to multichannel electro-acupuncture (EA), one of the most prominent emerging medical applications, which is useful to verify the operation of the TRX.

5.2 WBAN Standard Requirements for HBC PHY

Figure 25 shows three PHY requirements to implement the HBC TRX satisfying the standard. First, a packet, transmitted to the body, should be modulated with the FSDT modulation. The packet consists of a preamble, SFD/RI, header and data. A length of data can increase up-to 255 bytes. The FSDT modulation can be obtained by multiplying the frequency shift code (FSC) and a base-band signal. The FSC is composed of recursive 1s and 0s and its length can be chosen among 8bits, 16bits, 32bits and 64bits due to the data-rate. As a result of the FSDT modulation, the output of TX is spread in the frequency domain centered at 21 MHz with 5.25 MHz bandwidth. Moreover, the TX spectral mask should meet the HBC standard mask. According to the standard, the transmitted power should be less than −75 dB over 400 MHz and −80 dB under 2 MHz compared to the power at 21 MHz, to protect the safety for the human body. In addition, the minimum RX sensitivity should be the level listed in Fig. 25. In case of the lowest data-rate, 164 kbps, the minimum RX sensitivity level shall be −97.35 dBm when it is obtained for a packet error-rate (PER) is less than 1 % with a payload of un-coded 128 octets.

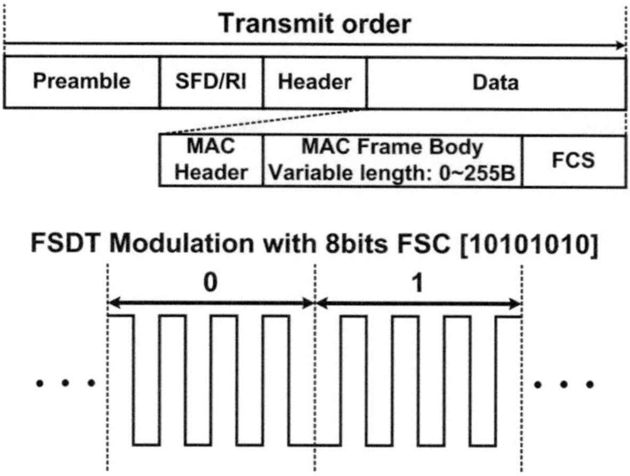

(a)

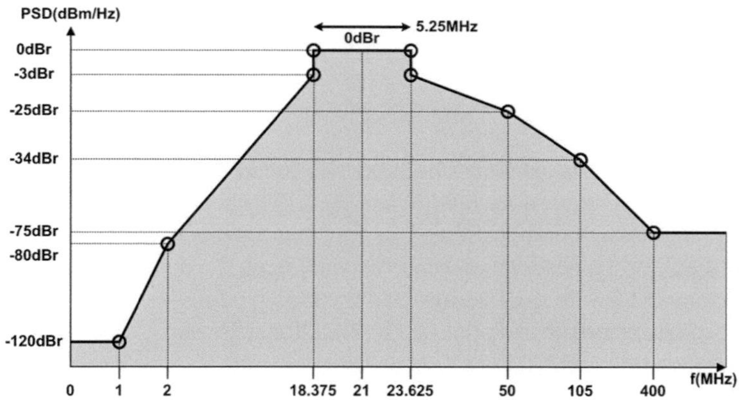

(b)

Frequency band (f_c, BW)(MHz)	Information data rate (kbps)	Minimum Sensitivity (dBm)
(21, 5.25)	164.1	-97.35
	328.1	-94.34
	656.3	-91.33
	1312.5	-88.32

(c)

Fig. 25 WBAN standard requirements of HBC PHY

5.3 *HBC Modem*

In order to construct the regulated packet with the FSDT modulation, a FSDT modem is proposed as shown in Fig. 26. The process of generating packet is start from the preamble and the header and data are followed next. The preamble

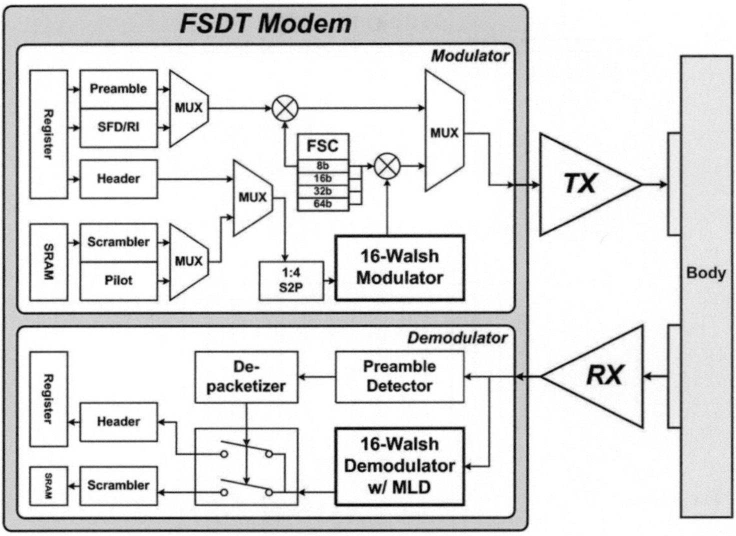

Fig. 26 HBC modem for FSDT modulation scheme

including the SFD/RI is generated from the 64b pseudo-random code, stored in the register. An 8b FSC is used for modulating them. The header and data requires one more step before the multiplication of the FSC, which is a 16-walsh modulation. A length of FSC for the header and data is depending on the data-rate. Especially, the data is scrambled before the 16-walsh and the FSC modulation step for acquiring a randomized characteristic and pilots are inserted for achieving a bit synchronization. For a demodulation part, the preamble detector finds the start of the received packet by comparing it to designated register values. When the packet is detected, the 16-walsh demodulator starts to restore the transmitted header and data. In this step, a MLD scheme helps the 16-walsh demodulator to match 16b orthogonal codes to the expected 4b output with more tolerance to bit errors.

Figure 27 shows the operation of the 16-walsh modulation and the demodulation with the MLD. The 16-walsh modulator converts a 4b input, A[1:4], into a 16b orthogonal code, W[1:16]. When the 16-walsh and the FSC modulated signal is transmitted to the body and restored at the RX, a number of bit errors are added with a noise. The restored signal, W[1:16] with bit errors, enters into the demodulator and it is compared with the 16 candidates codes stored in the 16-walsh code map. Then, only one code is selected when it shows the most of positive comparison results among 16 candidates. The Winner Takes All (WTA) block performs MLD to discard all of the remaining candidates. By the process, the demodulator finds the most highly correlated 4b original code, B[1:4].

A[1:4](B)	W[1:16]
0000	1111 1111 1111 1111
0001	1010 1010 1010 1010
0010	1100 1100 1100 1100
0011	1001 1001 1001 1001
0100	1111 0000 1111 0000
0101	1010 0101 1010 0101
0110	1100 0011 1100 0011
0111	1001 0110 1001 0110
1000	1111 1111 0000 0000
1001	1010 1010 0101 0101
1010	1100 1100 0011 0011
1011	1001 1001 0110 0110
1100	1111 0000 0000 1111
1101	1010 0101 0101 1010
1110	1100 0011 0011 1100
1111	1001 0110 0110 1001

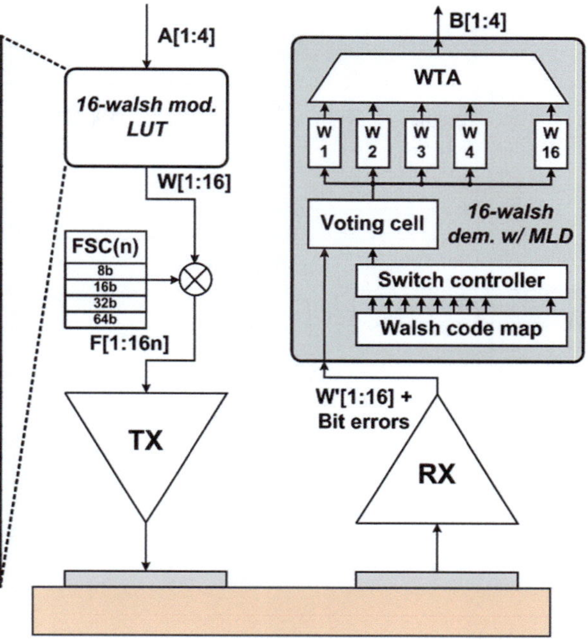

Fig. 27 Walsh modulator and demodulator in HBC modem

5.4 HBC PHY TRX

To satisfy the standard TRX requirements, the structure of TRX is proposed as shown in Fig. 28. First of all, TX satisfies the standard spectrum mask by adopting the ADF, composed of a digital 8th BPF, an 8-to-256 thermo decoder, a DAC and a TX driver. The high and low cut-off frequencies of the filter are set to 23.625 and 18.375 MHz, respectively, to allow 5.25 MHz bandwidth of a HBC communication channel, centered at 21 MHZ. In order to implement the TX filter without off-chip component, the BPF is digitally implemented. The Butterworth IIR type filter is used to design the BPF. The Butterworth type is chosen for the linear characteristic of a pass band and IIR type is adopted to reduce the order of filter. For the RX, interference signals below 2.625 MHz are removed by the down conversion. Both I/Q paths are used for the BPSK synchronization with a Costas loop. In order to satisfy the sensitivity requirement of less than −97.35 dBm, a 4th order LPF is used for each I/Q path. With the proposed HBC TRX structure, the linearity should be maintained during the DAC process for TX because it can distort the frequency characteristic of the transmitted signal so it may not meet the TX mask requirements. Moreover, the noise figure of the implemented blocks on the RX signal path should be considered for satisfying the sensitivity requirement.

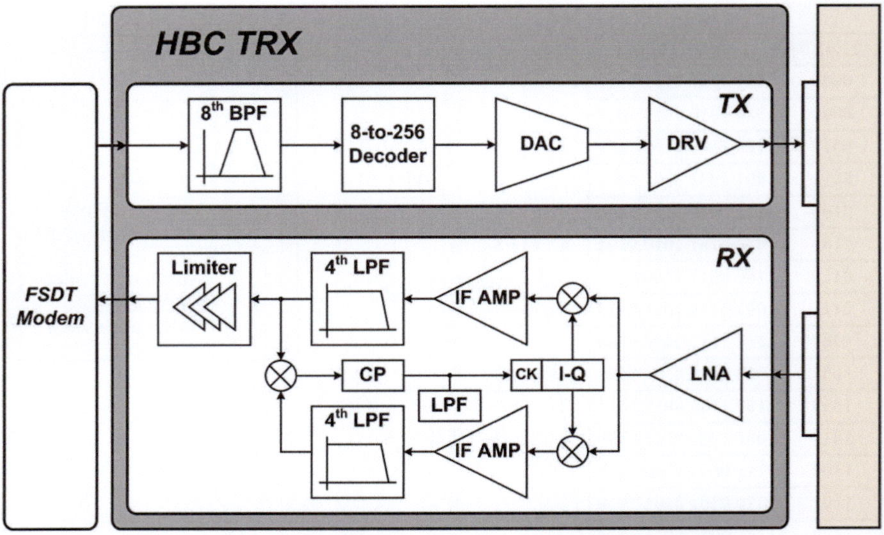

Fig. 28 Architecture of HBC PHY TRX

5.5 Measurement Results

Figure 29 shows measured network control results when the proposed SoC is applied to multi-channel EA application. Two types of control mode, individual and global mode, are demonstrated in Fig. 29. The individual command can manage each node with its own ID. With the single packet by the broadcasting ID, the global command can handle all the node. By applying the WBAN network to the multi-channel EA application, the patterned stimulation, and sequential stimulation can be realized for the effective EA treatment.

6 Phase-5: The Network Consideration

6.1 Motivation

The HBC TRXs in the previous studies achieved high energy efficiency when it is operated independently and separately, but their energy efficiency in the WBAN system applications were not analyzed in detail. Previous TRXs did not take into consideration that WBAN systems are usually composed of multiple nodes, and the transactions among nodes and a hub should comply with network protocol such as the frame protocol. That is, the TRX should achieve low energy consumption in a system with multiple nodes rather than in a single node system. When they designed TRXs, only PHY circuits and its operations were paid attention, and

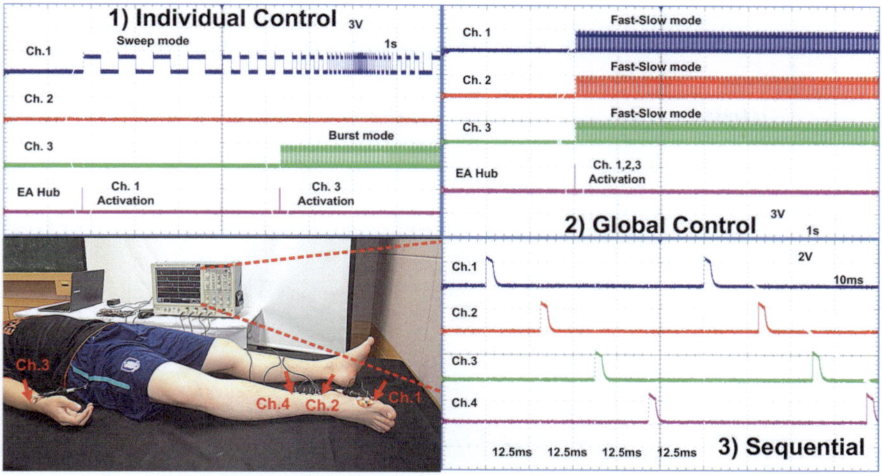

Fig. 29 Measurements for multi-channel EA application with HBC TRXs

LINK and TRANSPORT layers for the seamless data transaction were not. This is especially critical to the HBC WBAN standard because the HBC uses only one communication channel, centered at 21 MHz with 5.25 MHz bandwidth, and it should support up to 64 sensor nodes. If multiple nodes try to access to the hub simultaneously, a probability of the collision increases, causing (1) the failure of a packet transmission, (2) the data loss and (3) the QoS degradation from the network delay. Even though the designed HBC TRX consumes low power, a time domain scheduling is essential to realize the low energy network by optimally arranging the communications among 64 nodes.

In this section, a HBC TRX is proposed to optimize the network system operation rather than just the TRX circuit operation. It adopts (1) Duty Cycle Control (DCC) with a MAC scheduler to save the energy consumption of the network with the smart time domain scheduling policy, and (2) a HBC modem with Maximum Likelihood Detection (MLD) scheme to enhance the RX sensitivity. The RX also adopts a reconfigurable LNA to control its power consumption by monitoring the received signal strength. The RSSI is proposed not only to minimize the power of the front-end, but also to estimate the status of the communication channel whether it is being used or not. Moreover, because the privacy of data is important factor in medical applications, (3) a Zero-delay Cipher (ZDC) is proposed to ensure the security without extra network delay. With the proposed HBC TRX, the hub can coordinate each access of the nodes to support the optimized active period of the network. As a result, the network system with the proposed TRX demonstrates the ultralow energy operation which will be useful in the medical BAN system (Fig. 30).

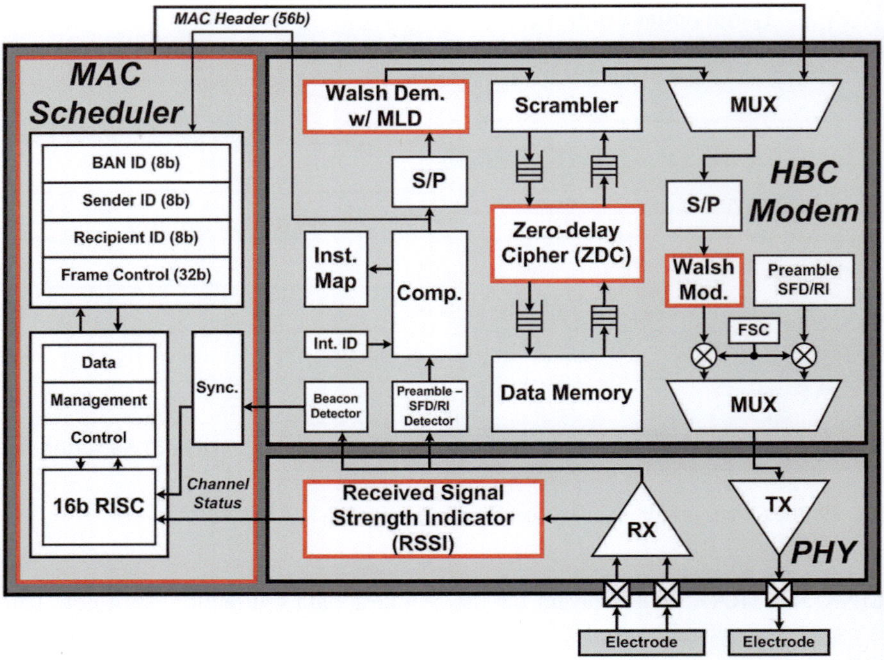

Fig. 30 HBC TRX with the MAC scheduler

6.2 *Duty-Cycle Control with the MAC Scheduler*

Figure 31 shows the operation of DCC when a network system is formed on the body with multiple sensor nodes and a hub. According to the WBAN standard, up to 64 nodes can join a single BAN. The hub broadcasts a beacon at every start of the time reference, the super-frame, to synchronize the sensor nodes and share the accessible time information among nodes. Nodes can transmit data only during a managed access period, T_{MAP}, and their accesses can be granted ahead by the hub during a random access period, T_{RAP}. The TRX of all nodes and the hub are able to stay in a sleep mode during a sleep period, T_{SLP}. The MAC scheduler can reduce $T_{RAP} + T_{MAP}$ to maximize T_{SLP} of each super-frame. It helps the hub to find an optimized active period for the realization of the low energy BAN.

Consequently, an average duty and a QoS are measured for the proposed HBC network system depending on the number of nodes. The QoS is defined as follows.

Success = # of transmissions @ Expected Super-frame
Failure = # of transmissions @ Delayed Super-frame
QoS = (1−(Failure/(Success + Failure)) × 100 %

Measured average duty and QoS shows that 96.6 % QoS with only 38 % average duty is achieved among 64 nodes and a hub.

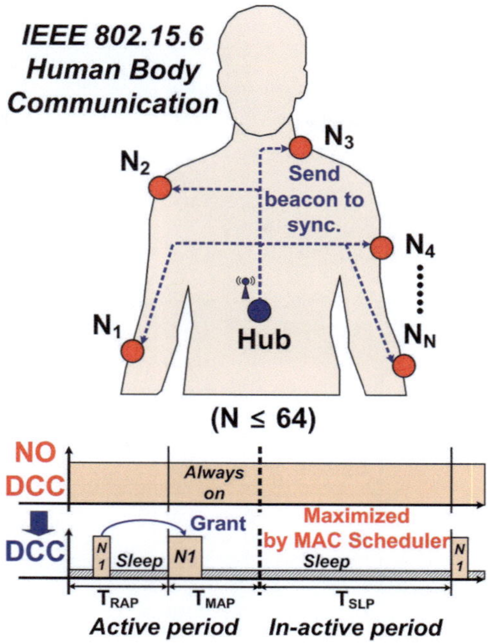

Fig. 31 DCC operation of the MAC scheduler

6.3 *Measurement Results*

In Fig. 32, the proposed HBC TRX chip is applied to form a WBAN system with two sensor nodes and a hub to measure physiological signals such as ECG and APW (Artery Pulse Wave). The hub issues a beacon at every start of super-frame to share the time reference with sensor nodes. Sensor nodes that are willing to transmit data request the hub to grant a data communication time at TRAP. After the request, the hub allows nodes to access in their requesting orders. Due to DCC, sensor nodes can enter into the sleep after transmitting data and the effective power consumption is measured as 61.7 μW with two nodes for the RX mode. In the bottom, the bio-signals are successfully received and reconstructed correctly.

References

1. IEEE Computer Society, *IEEE standard for local and metropolitan area networks: part 15.6 wireless body area networks* (IEEE Standards Association, 29 February 2012)
2. J. Bae et al., The signal transmission mechanism on the surface of human body for body channel communication. IEEE Trans. Microw. Theory Tech. **60**(3), 582–593 (2012)
3. H. Lee et al., A 5.5 mW IEEE 802.15.6 wireless body area network standard transceiver for multi-channel electro-acupuncture application, in *ISSCC Dig. Tech. Papers*, pp. 452–453, February 2013

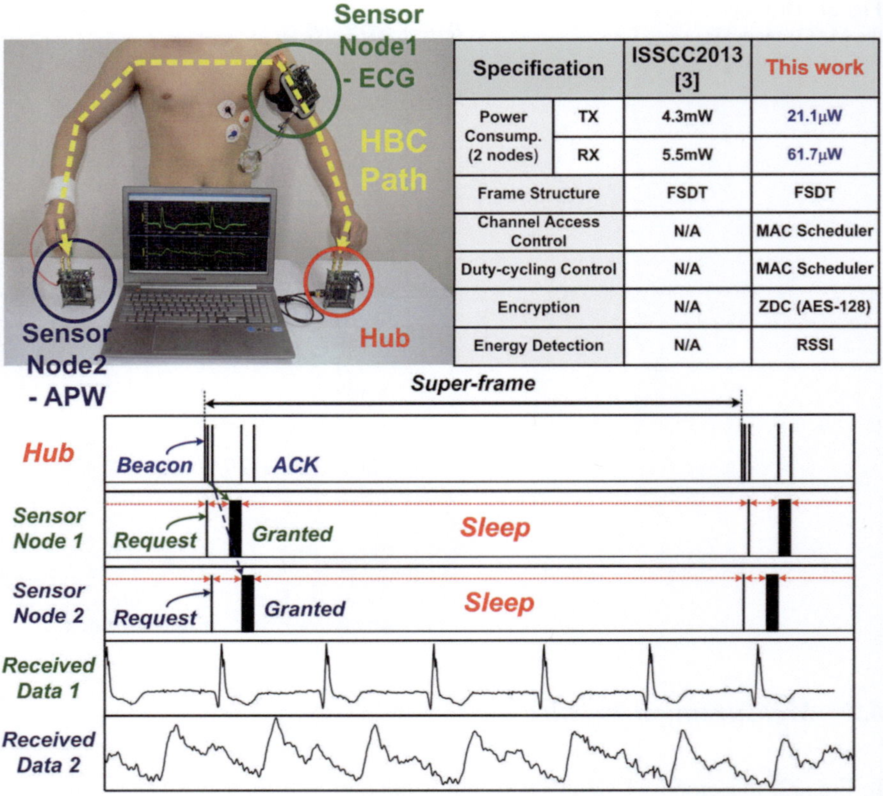

Specification		ISSCC2013 [3]	This work
Power Consump. (2 nodes)	TX	4.3mW	21.1μW
	RX	5.5mW	61.7μW
Frame Structure		FSDT	FSDT
Channel Access Control		N/A	MAC Scheduler
Duty-cycling Control		N/A	MAC Scheduler
Encryption		N/A	ZDC (AES-128)
Energy Detection		N/A	RSSI

Fig. 32 Measurement of HBC TRX with the MAC scheduler

4. S. Song et al., A 2 Mb/s wideband pulse transceiver with direct-coupled interface for human body communications, in *ISSCC Dig. Tech. Papers*, pp. 558–559, February 2006
5. S. Song et al., A 0.9 V 2.6 mW body-coupled scalable PHY transceiver for body sensor applications, in *ISSCC Dig. Tech. Papers*, pp. 336–337, February 2007
6. N. Cho et al., A 60 kb/s-to-10 Mb/s 0.37 nJ/b adaptive-frequency-hopping transceiver for body-area network, in *ISSCC Dig. Tech. Papers*, pp. 132–133, February 2008
7. N. Cho et al., A 10.8 mW body-channel-communication/MICS dual-band transceiver for a unified body-sensor-network controller, in *ISSCC Dig. Tech. Papers*, pp. 424–423, February 2009
8. J. Bae et al., A 0.24 nJ/b wireless body-area-network transceiver with scalable double-FSK modulation, in *ISSCC Dig. Tech. Papers*, pp. 34–35, February 2011
9. H. Lee et al., A 33 μW/node duty cycle controlled HBC transceiver system for medical BAN with 64 sensor nodes, in *IEEE Proceedings of the Custom Integrated Circuits Conference*, 15–17, September 2014
10. T.G. Zimmerman, Personal area networks (PAN): near-field intra-body communication, M.S. thesis, MIT, September 1995
11. K. Hachisuka et al., Development and performance analysis of an intra-body communication device, in *Transducers '03*, pp. 1722–1725, June 2003
12. M. Shinagawa et al., A near-field-sensing transceiver for intrabody communication based on the electrooptic effect. IEEE Trans. Instrum. Meas. **53**, 1533–1538 (2004)

13. S. Song et al., A 0.2-mW 2-Mb/s digital transceiver based on wideband signaling for human body communications. IEEE J. Solid-State Circuits **42**, 2021–2033 (2007)
14. S. Song et al., A 4.8-mW 10-Mb/s wideband signaling receiver analog front-end for human body communications, in *Proc. IEEE Eur. Solid-State Circuits Conf.*, pp. 488–491, September 2006
15. IEEE 802.15 WPANTM Task Group 6: Body Area Networks (BAN), November 2007 [Online]. Available: http://www.ieee802.org/15/pub/TG6
16. S. Kim et al., A fully integrated digital hearing aid chip with human factors considerations. IEEE J. Solid-State Circuits **43**, 266–274 (2008)
17. J. Lee et al., A power management unit with continuous co-locking of clock frequency and supply voltage for dynamic voltage and frequency scaling, in *IEEE International Symposium on Circuit and System*, pp. 2112–2115, May 2007

Centimeter-Range Inductive Radios

Mehdi Kiani and Maysam Ghovanloo

Abstract This chapter describes the fundamental principles of cm-range wireless telemetry through inductive links and provides insight in regards to the methods of analysis, choice of modulation schemes, carrier frequencies, and coil design. After presenting simplified models for the inductance and mutual coupling of conductive loops, the inductive link equivalent network is derived to be used for analysis of inductive data links. Different carrier-based modulation schemes such as amplitude-shift keying (ASK), frequency-shift keying (FSK), and phase-shift keying (PSK) are discussed for near-field simultaneous data and power transmission in different applications such as implantable microelectronic devices (IMDs), radio frequency identification (RFID), and smart cards. Data communication through load-shift keying (LSK) is also discussed followed by presenting the pulse-based schemes for low-power communication. Finally, new pulse-harmonic modulation (PHM) and pulse-delay modulation (PDM) schemes that offer high data rate in IMDs without dissipating much power on the implantable side are presented.

Keywords Inductive coupling • Near field • IMD • Data communication • RFID

1 Introduction

Near-field wideband bidirectional data communication is a viable technique to wirelessly communicate with devices such as sensors and actuators. Moreover, it is possible to use the same short-range wireless link to transfer power to those devices. Wireless implantable microelectronic devices (IMDs) are good examples of where near-field data and power transmission links can be used effectively. IMDs have

M. Kiani (✉)
Assistant Professor, Electrical Engineering Department, The Pennsylvania State University, 111-H Electrical Engineering West, University Park, PA 16802, USA
e-mail: http://sites.psu.edu/icsl/

M. Ghovanloo
Associate Professor, School of Electrical and Computer Engineering, Georgia Institute of Technology, 85 Fifth Street NW, Room TSRB-419, Atlanta, GA 30308-1030, USA
e-mail: mghovan@ece.gatech.edu; www.ece.gatech.edu/ mghovan

P.P. Mercier, A.P. Chandrakasan (eds.), *Ultra-Low-Power Short-Range Radios*, Integrated Circuits and Systems, DOI 10.1007/978-3-319-14714-7_10

been significantly improved by going through many generations since the invention of the first implantable pacemaker in 1958, and their importance in several state-of-the-art medical treatments is on the rise [1]. They have made it possible to treat a wide range of ailments and disabilities from bradycardia [2], to chronic back pain, epilepsy [3, 4], and deafness [5]. IMDs have the potential to alleviate more challenging types of disabilities such as blindness [6–8], paralysis [9], and loss of limbs [10]. These devices need to transmit and receive information wirelessly across the skin barrier since breaching the skin with interconnect wires would be a source of morbidity for the patient and significantly increases the risk of infection.

In sensory prosthetic devices, which interface with the central nervous system (CNS) to restore a sensory function such as hearing or vision, the quality of perception enhances with the number of stimulating sites and electrodes and the rate of stimulation [11, 12]. These devices may stimulate the neural tissue by means of tens to thousands of stimulating channels and they generally require considerably more power and communication bandwidth than autonomous devices, such as pacemakers. State of the art visual prostheses are currently targeting beyond one thousand sites to improve the quality of the visual functions, such as mobility without a cane, face recognition, and reading [12]. Every stimulation command in such prosthesis requires $\sim$10 bits for addressing the stimulating sites, $\sim$6 to 8 bits for stimulation pulse amplitude levels, and $\sim$2 to 4 bits for polarity, parity-checking, and sequencing. This would suggest at least 20 bits per command frame for site selection and stimulus amplitude information. Considering that it might be necessary to stimulate electrodes at rates up to 200 Hz each (for physiological reasons), and the need for up to four commands per biphasic-bipolar stimulation pulse in some microstimulator architectures, raster scanning for all 625 sites of an implantable stimulator at this rate requires a serial data bit stream of 625-sites $\times$ 20-bits $\times$ 4-commands $\times$ 200-frames $=$ 10 Mbps. It is obvious that a high data transmission bandwidth is highly needed for the wireless implantable microstimulator [13].

Radio frequency identification (RFID) takes advantage of inductive links to not only power up the ultralow power RFID tags, which cannot have batteries due to their size, weight, and lifetime limitations, but also read the information, through back telemetry [14]. Yet another new application for inductive data transmission is known as near-field communication (NFC), which is incorporated in state-of-the-art smartphones mainly to perform financial transactions in lieu of credit cards [15]. Because of the significant growth of handheld, wireless, and mobile electronic devices and gadgets, such as smartphones, tablets, mobile laptops, etc. in recent years, cm-range and wideband communication to wirelessly exchange data between such devices would gain considerable attention in near future.

2 Induction Principles

The main physical principle behind the operation of telemetry coils is the Faraday's law, which states that when the total magnetic flux through a conductive loop—defined as the integral over the surface enclosed by the loop of the magnetic flux—

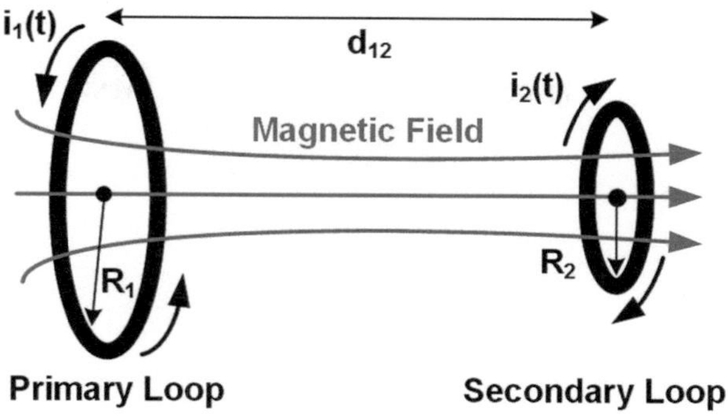

Fig. 1 The principle of inductive coupling, in which a varying magnetic field resulted from a time-variant current in the primary loop with the radius of R_1 can create a current in the secondary loop with the radius of R_2 and separated by d_{12}

varies with time, a current is induced in the loop itself. This, in turn, results in an electromotive force (EMF) induced in the loop [16]. Thus, a primary loop will generate the varying magnetic field, which concatenates with the secondary loop, resulting in an induced current in the secondary loop as shown in Fig. 1.

2.1 *Magnetic Fields*

A magnetic field is associated with any moving charged particles that result in flow of current. The magnitude of the magnetic field regardless of the material properties of the medium can be described by the magnetic field strength, integrating which along a closed curve that encircles the moving charges results in the total current that passes through the curve. For designing a telemetry link, there are a number of parameters that are directly associated with the magnetic field and its strength must be considered, such as self and mutual inductances.

2.2 *Inductance and Inductive Coupling*

Self-inductance is the ratio of the magnetic flux generated in an area enclosed by a conductor loop to the current passing through the loop. Under the condition of $r/R \ll 1$, where r and R are the radii of the wire and the circular loop that it is forming, self-inductance can be approximated by

$$L\left(R,r\right) \approx \mu_0 R\left(\ln\left(\frac{8R}{r}\right) - 2\right) \quad (1)$$

where μ_0 is the permeability of the free space [17].

For the case of circular coils with n turns, if the coil length, d, is much smaller than R, the self-inductance is approximately equal to n^2L, where L is the self-inductance of a 1-turn loop in (1). Whereas, for the case of planar spiral coils having n turns with different radii R_i ($i = 1, 2, \ldots n$) the total self-inductance should be calculated from

$$L = \sum_{i=1}^{n} L(R_i, r) + \sum_{i=1}^{i=n} \sum_{j=1}^{j=n} M_{ij} \left(R_i, R_j, d_r = 0\right)\left(1 - \alpha_{i,j}\right) \qquad (2)$$

where $\alpha_{i,j} = 1$ if $i = j$, and $\alpha_{i,j} = 0$ otherwise [17].

2.3 Mutually Coupled Coils

M_{ij} is the mutual inductance between two conductor loops, which depends on the proportion of the magnetic flux generated by one loop that passes through the other loop (flux coupling). Therefore, it highly depends on their geometries, relative orientation, and magnetic properties of the medium. In a simplified case for two perfectly aligned parallel coaxial circular coils in the air as shown in Fig. 1, separated by relative distance d_{12} and the radius and number of turns of (R_1, n_1) and (R_2, n_2) for the first and second coils, respectively,

$$M_{12}(R_1, R_2, d_{12}) = \frac{\pi \mu_0 n_1 R_1^2 n_2 R_2^2}{2\sqrt{\left(R_1^2 + {d_{12}}^2\right)^3}} \qquad (3)$$

In order to more accurately calculate self and mutual inductances of coils with various geometries, one should either use tabulated parameterized equations [18, 19] or finite element electromagnetic software, such as FastHenry, SONET, or HFSS (Ansoft, Pittsburgh, PA).

In addition to the distance and geometry, alignment of the coils has a significant effect on their mutual inductance. The effects of coils' misalignments have been analyzed in [20]. For example, if one of the coils is tilted by an angle θ, their mutual inductance reduces by a factor of $\cos(\theta)$

$$M_{12}(d_{12}, \theta) = M_{12}(d_{12}, 0)\cos(\theta) \qquad (4)$$

We can normalize the mutual inductance between two coils to get a qualitative sense of how strongly they are coupled and compare the coupling between different pairs of coils. The coupling coefficient, k_{12}, between two coils with self-inductance, L_1 and L_2, is defined as

$$k_{12} = \frac{M_{12}}{\sqrt{L_1 \times L_2}} \quad 0 \le k_{12} \le 1 \qquad (5)$$

Assuming $R_2 < R_1$, the coupling coefficient in this case can be approximated by [14]

$$k_{12}(d_{12}) = \frac{R_1^2 R_2^2}{\sqrt{R_1 R_2}\left(\sqrt{\left(R_1^2 + {d_{12}}^2\right)}\right)^3} \tag{6}$$

As a result of (4), k_{12} also depends on the coils orientation and alignment

$$k_{12}(d_{12}, \theta) = k_{12}(d_{12}) \cos(\theta) \tag{7}$$

According to (7), in order to minimize k_{12} between two planar co-axial coils at a certain distance d_{12}, their plains should be orthogonal. This technique will be discussed later for reducing the undesired coupling between power and data coils. On the other hand, parallel and perfectly aligned co-axial coils provide maximum k_{12}.

2.4 Equivalent Network Models

In an inductive link, shown in Fig. 2a, a time variant current $i_1(t)$ in the primary coil, L_1, generates a time variant magnetic field, part of which passes through the secondary coil, L_2. This part of the time varying magnetic field generates voltage $V_2(t)$ across L_2 and current $i_2(t)$ through the secondary loop due to its mutual inductance M_{12} with L_1. The time domain relationship between these voltages and currents can be found from,

$$\begin{aligned} V_1(t) &= R_1 \cdot i_1(t) + L_1 \cdot \frac{di_1(t)}{dt} + M_{12} \cdot \frac{di_2(t)}{dt} \\ V_2(t) &= M_{12} \cdot \frac{di_1(t)}{dt} + R_2 \cdot i_2(t) + L_2 \cdot \frac{di_2(t)}{dt} \end{aligned} \tag{8}$$

where R_1 and R_2 are the ohmic losses of L_1 and L_2, respectively. In order to find the inductive link equivalent Z-network model, shown in Fig. 2b, all voltages and currents in (8) should be represented in the Laplace domain,

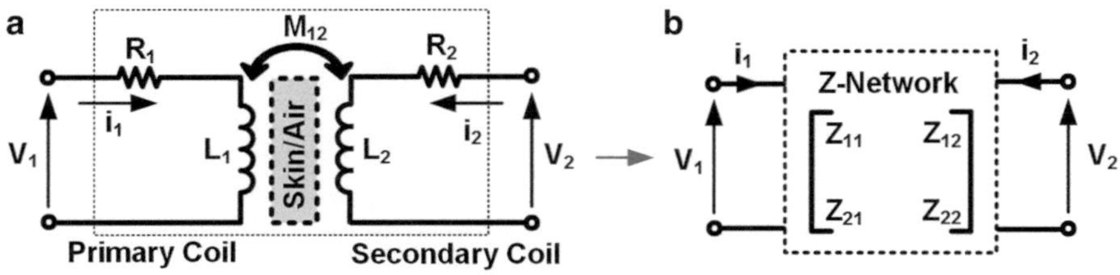

Fig. 2 (**a**) A simplified circuit diagram of an inductive link. R_1 and R_2 represent the ohmic losses of L_1 and L_2, respectively. (**b**) The equivalent Z-network model for an inductive link

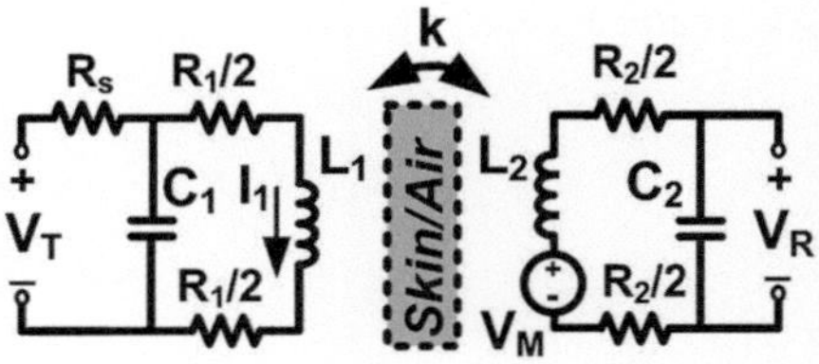

Fig. 3 Lumped model for near-field data telemetry through resonant coupled coils. Since k is small, we can safely neglect the effect of L_2C_2 loading on the Tx circuitry

$$\begin{aligned} V_1(s) &= R_1 \cdot I_1(s) + L_1 \cdot sI_1(s) + M_{12} \cdot sI_2(s) \\ V_2(s) &= M_{12} \cdot sI_1(s) + R_2 \cdot I_2(s) + L_2 \cdot sI_2(s) \end{aligned} \tag{9}$$

Therefore, the equivalent Z-matrix can be written as,

$$\begin{bmatrix} V_1(s) \\ V_2(s) \end{bmatrix} = \begin{bmatrix} R_1 + L_1 s & M_{12}s \\ M_{12}s & R_2 + L_2 s \end{bmatrix} \cdot \begin{bmatrix} I_1(s) \\ I_2(s) \end{bmatrix} = Z \cdot \begin{bmatrix} I_1(s) \\ I_2(s) \end{bmatrix} \tag{10}$$

Figure 3 shows a lumped equivalent circuit model of an inductive link, in which primary (L_1C_1) and secondary (L_2C_2) sides are composed of LC-tank circuits and their associated parasitic components (R_1 and R_2). L_2C_2-tank is tuned at f_r to increase the magnitude of the received voltage across L_2C_2-tank, i.e. V_R, while the L_1C_1-tank, depending on the data rate and transmission range requirements, can be either tuned to f_r or left at its self-resonance frequency (SRF), in which case C_1 in Fig. 3 simply represents the parasitic capacitance of L_1 [14]. Moreover, to maximize V_R, the L_2C_2-tank should achieve a high quality factor (Q). Therefore, the loading on L_2C_2-tank should be negligible that imposes the use of a parallel LC-tank in the secondary side.

One can use the Z-network model in (10) to find the secondary voltage for a given input voltage. However, since L_1 and L_2 are loosely coupled (small k_{12}) and the current in L_2C_2-tank is very small, unlike inductive power transmission links, as shown in Fig. 3 we can safely neglect the effect of L_2C_2 loading on the transmitter (Tx) circuitry to simplify our equations. Hence, the inductive link transfer function in the S-domain can be described as,

$$\begin{aligned} H(s) &= \frac{V_R}{V_T} = \frac{V_R}{V_M} \times \frac{V_M}{I_1} \times \frac{I_1}{V_T} \\ &= \frac{Ms}{(R_sL_1C_1s^2 + (R_sR_1C_1 + L_1)\,s + R_s + R_1)(C_2L_2s^2 + C_2R_2s + 1)}, \end{aligned} \tag{11}$$

where V_T, V_R, I_1, and V_M are the Tx output voltage, receiver (Rx) input voltage, current passing through L_1, and induced voltage across L_2, respectively. Other parameters are lumped circuit elements in Fig. 3.

$H(s)$ is composed of two second-order systems, one originating from the Tx and the other from the Rx LC-tanks, each of which can be expressed as,

$$\frac{1}{s^2+2\zeta\omega_n s+\omega_n{}^2}=\frac{1}{(s-s_1)(s-s_2)},\ s_1=-(\zeta\omega_n+j\omega_d),\ s_2=s_1{}^*,\ 0\leq\zeta<1 \tag{12}$$

where ζ is the damping ratio, ω_n is the natural frequency, and is the natural damping frequency of the system. From (1) and (2), these 2nd-order system parameters can be expressed in terms of the lumped circuit elements in Fig. 3,

$$\zeta_1\omega_{n_1}=\frac{(R_sR_1C_1+L_1)}{2R_sL_1C_1}\cong\frac{1}{2R_sC_1},\ \ \omega_{n_1}{}^2=\frac{R_s+R_1}{R_sL_1C_1}\cong\frac{1}{L_1C_1} \tag{13}$$

$$\zeta_2\omega_{n_2}=\frac{R_2}{2L_2},\ \ \omega_{n_2}{}^2=\frac{1}{L_2C_2} \tag{14}$$

$$\zeta_1=\frac{(R_sR_1C_1+L_1)}{2R_s\sqrt{L_1C_1}}\sqrt{\frac{R_s}{R_s+R_1}}\cong\frac{1}{2R_s}\sqrt{\frac{L_1}{C_1}} \tag{15}$$

$$\zeta_2=\frac{R_2}{2}\sqrt{\frac{C_2}{L_2}} \tag{16}$$

Assuming both 2nd-order systems are under damped, i.e. ζ_1 and $\zeta_2 < 1$, which is often the case for the LC-tanks used in data telemetry links, (11) can be rearranged as,

$$\begin{aligned}H(s)&=\frac{Ms}{R_sL_1C_1(s-s_{11})(s-s_{21})\times L_2C_2(s-s_{12})(s-s_{22})},\\ s_{1j}&=-\left(\zeta_j\omega_{n_j}+j\omega_{d_j}\right),\ s_{2j}=s_{1j}^*,\ j=1,\ 2\end{aligned} \tag{17}$$

Now we can break $H(s)$ up into the sum of its first order components,

$$H(s)=\left[\frac{A_1}{(s-s_{11})}+\frac{A_1{}^*}{(s-s_{21})}+\frac{A_2}{(s-s_{12})}+\frac{A_2{}^*}{(s-s_{22})}\right], \tag{18}$$

where

$$A_1=\frac{Ms_{11}}{R_sL_1C_1(s_{11}-s_{21})\times L_2C_2(s_{11}-s_{12})(s_{11}-s_{22})}=a_1+jb_1, \tag{19}$$

$$A_2 = \frac{M s_{12}}{R_s L_1 C_1 (s_{12} - s_{11})(s_{12} - s_{21}) \times L_2 C_2 (s_{12} - s_{22})} = a_2 + j b_2, \tag{20}$$

and apply the inverse Laplace transform to find the impulse response for the inductive link,

$$\begin{aligned} h(t) = {} & 2e^{-\zeta_1 \omega_{n_1} t} \left(a_1 \cos\left(\omega_{d_1} t\right) + b_1 \sin\left(\omega_{d_1} t\right)\right) \\ & + 2e^{-\zeta_2 \omega_{n_2} t} \left(a_2 \cos\left(\omega_{d_2} t\right) + b_2 \sin\left(\omega_{d_2} t\right)\right) \end{aligned} \tag{21}$$

High-Q is desired for the inductive data transmission links used in IMD applications to improve transmission range and robustness of the link against interference without increasing the transmitted power. Q_1 and Q_2 can be calculated from (15) and (16) by substituting them in $Q = 1/2\zeta$, which indicates that ζ should be small and thus, $\omega_d \approx \omega_n$. As a result, $h(t)$ in (21) can be simplified into two exponentially decaying ringings, one with a long time-constant of $\tau_2 = 1/\zeta_2\omega_{n2} = 2L_2/R_2$ on the Rx and the other with a short time-constant of $\tau_1 = 1/\zeta_1\omega_{n1} \approx 2R_sC_1$ on the Tx. The sum of these two terms will result in a ringing across V_R in Fig. 3, which builds up rapidly but decays slowly.

3 Inductive Data Transmission

Bidirectional wireless data transmission is essential for IMD and RFID systems to establish a short-range wireless communication between the transmitter (Tx) and receiver (Rx) parts of the system. RFID readers use the same inductive link to not only power the passive RFID tags but also interrogate them [14]. The majority of modern IMDs have several adjustable parameters that can be fine-tuned after implantation for every individual patient according to his/her specific needs. In addition to those parameters, research is underway to equip sensory devices with a flow of stimulation commands from the external artificial sensors and signal processing units to build closed-loop neuroprosthetic devices [9]. Sending adjustment and control commands wirelessly from the external unit to the implanted unit is known as the forward telemetry or downlink as shown in Fig. 4a. Moreover, the same devices often need to inform the external processing components about the IMD operating status, possible faults, and in some cases the neuronal response immediately after stimulation for proper adjustment of the stimulation parameters [21, 22]. This direction of dataflow that sends information from inside towards out of the body is often referred to as backward telemetry or uplink (Fig. 4b).

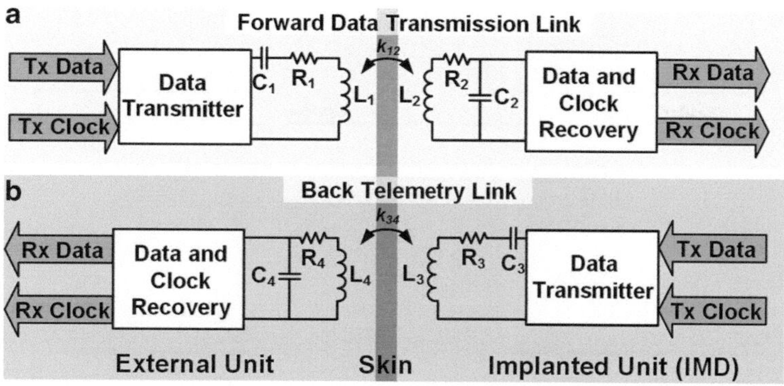

Fig. 4 Bi-directional inductive data transmission. (**a**) Data transmission from the external unit to the implanted unit is called forward telemetry. (**b**) Data transmission from the implanted unit to the external unit is called back telemetry. The same forward/back telemetry link could also be used for power transmission to the implanted unit [14]

3.1 Forward Telemetry

A simple option for forward telemetry, which has been used in most of todays' IMD and RFID systems, is to modulate the same carrier that has been used for power transmission for transmitting data [13, 23–28]. The advantage of this method is less complexity on both external/reader and implant/transponder components of the system, which can result in lower power consumption and smaller size. In the following, three main modulation techniques for forward data telemetry through magnetic coupling are presented: ASK, FSK, and PSK.

3.1.1 Amplitude Shift Keying (ASK)

The majority of the IMD and RFID systems use amplitude shift keying (ASK) for forward telemetry due to its simplicity on both Tx and Rx sides [23–28]. In this method, shown in Fig. 5a, the external data modulates the amplitude of the power carrier, known as the carrier envelope. This can be easily done by changing the supply voltage and consequently the output swing of the power amplifier (PA), which drives the primary coil, based on the modulating signal. The ASK carrier frequency should be several times higher than the data rate to provide the Rx with enough cycles to detect the change in the envelope amplitude. The difference between data rate and carrier frequency also depends on the amplitude modulation index and primary and secondary coils' quality factors. Increasing the modulation index helps with adding to the data rate at the cost of degrading the power transfer efficiency (PTE). The higher the Q-factor of the coils, the PTE improves, however, the longer it takes for the Tx or Rx LC-tank circuits to follow the carrier amplitude, and the smaller the data transmission bandwidth. In general, it can be concluded that

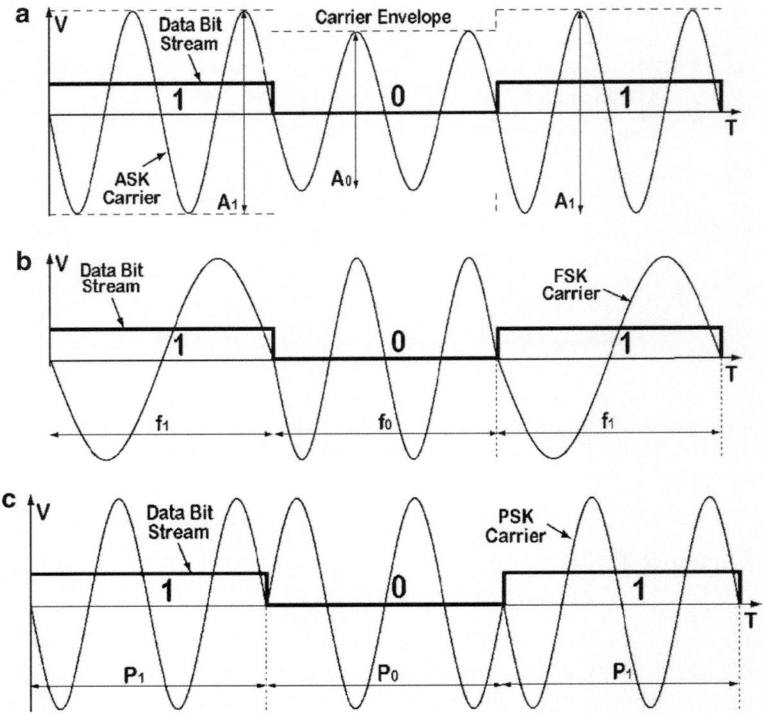

Fig. 5 (**a**) Amplitude shift keying (ASK), (**b**) frequency shift keying (FSK), (**c**) phase shift keying (PSK)

in single-carrier systems there is always a compromise between the inductive link PTE and data bandwidth.

One way to demodulate the ASK signal in the Rx is to use an envelope detector. The ASK signal passes through a low-pass filter (LFP), which cutoff frequency is somewhere between the carrier frequency and data bandwidth. The filter rejects the higher frequency carrier from the carrier envelope, which can then be easily converted back to a serial data bit stream by passing through a high-pass filter (HPF) and a comparator. One factor that affects forward data rate in ASK is the percentage of change in carrier amplitude when transmitting logic '0' and logic '1'. This parameter, which is known as the ASK modulation index, has a direct relationship with the data bandwidth. However, since carrier power is proportional to the second power of its amplitude, a higher modulation index means less average delivered power in the carrier signal. In a particular case, in which the modulation index is 100 % i.e. the power carrier is turned off for logic '1' bits, the ASK method leads to on-off keying (OOK). The OOK mechanism results in higher data rate and easier data detection at the Rx side at the cost of lower PTE.

Another limitation of the ASK method in IMD and RFID applications is its susceptibility to noise, interference, and motion artifacts, all of which mainly affect the carrier amplitude. Equation (3) shows that the mutual coupling between two coils highly depends on their relative distance. Therefore, the voltage across the Rx coil, V_2, is also strongly dependent on coupling distance, d_{12} ($V_2 \propto 1/d_{12}$). When a patient wears an inductively powered IMD and walks, the motion artifacts and vibrations affect d_{12} and consequently the amplitude of the received signal. Even when d_{12} is constant, any instantaneous changes in the IMD current consumption, due to stimulation for example, directly results in V_2 variations, and deteriorates the quality of the ASK signal [29–31]. Hence, the ASK demodulator should be able to distinguish between the amplitude variations that are resulted from noise, interference, and artifacts and those that represent the received data bits. Utilizing subcarriers is one of the methods to get around the aforementioned problems in ASK [14].

3.1.2 Frequency Shift Keying (FSK)

Frequency shift keying (FSK), which is a popular modulation technique in high fidelity (HiFi) audio transmission as well as digital communications, has not been widely utilized in IMDs and RFID systems because of the complexities in implementation of the FSK modulation and demodulation circuits. In this method, shown in Fig. 5b, the external data modulates the frequency of the power transmission carrier and the carrier amplitude remains constant. Therefore, logic '0' and logic '1' are transmitted by sinusoidal signals at frequencies f_0 and f_1, respectively. As a result, the carrier power stays constant regardless of the data contents, which is an advantage of the FSK compared to ASK. In addition to constant power, the superior robustness against various noise sources and interference of FM over AM has been known since the early days of radio engineering. In FSK, it is very unlikely that d_{12} or IMD current variations would affect the frequency of the induced signal.

For FSK modulation, the PA input can be switched between two oscillators operating at f_0 and f_1 depending on the serial data bit stream. One limitation of the FSK technique is that its frequency spectrum occupies a wider bandwidth compared to ASK at the same data rate. Therefore, the Q-factors of both Tx and Rx LC-tank circuits should be lowered to provide enough bandwidth to pass major FSK carrier components at the expense of lowering the PTE. According to Carlson's rule, the bandwidth, BW, required to include 98 % of the total power of an FM signal is,

$$BW \approx 2\,(\delta_{\max} + f_{i\,\max}) \tag{22}$$

where δ_{max} is the maximum frequency shift caused by modulation, depending on the VCO gain and amplitude of the modulating signal, and f_{imax} is the maximum frequency content of the modulating signal. To provide enough bandwidth and improve the inductive link robustness against coupling variations, a stagger tuned circuit is proposed in [32]. Also a variation of the class-E PAs, which can switch the carrier frequency by switching the LC-tank capacitive components, is proposed in [33].

There are several traditional methods for FSK demodulation. One of the basic methods involves a limiter to eliminate noise and interference on the received signal amplitude, a discriminator to convert the FSK signal to an ASK signal, and an envelope detector to demodulate the ASK signal. Phase-locked loops (PLLs) can also be used for FSK demodulation. For high bandwidth forward telemetry, however, these methods may require high-order analog filtering down the signal path, which would consume a large chip area or several off-chip components in the low-end RF application of interest (IMD).

To achieve a high data rate along with synchronization between the Tx and Rx without filtering, a phase coherent FSK (pc-FSK) protocol, shown in Fig. 5b, was proposed in [34]. In the pc-FSK protocol, binary symbols '1' and '0' are transmitted by one and two carrier cycles at f_1 and f_0, respectively. Choosing $f_0 = 2 \times f_1$ provides a constant bit length, which helps extracting a synchronous sampling clock directly from the pc-FSK carrier. The FSK demodulator manages the received carrier as a baseband signal, and directly measures the duration of each received carrier cycle. The measured duration is then compared with a predefined value to indicate its associated binary symbol. Therefore, every single carrier cycle can transfer a data bit, resulting in a data rate to carrier frequency ratio close to one, which is higher than the same ratio in many wideband wireless communication techniques that are currently in use.

3.1.3 Phase Shift Keying (PSK)

In phase shift keying (PSK), shown in Fig. 5c, the serial data bit stream modulates the phase of the power transmission carrier and both carrier amplitude and frequency remain constant. Therefore, PSK has the highest spectral efficiency compared to the other two techniques, which means that using PSK, it is possible to transmit higher data rates per unit available wireless link bandwidth. PSK is also the basis for vector modulation. In binary PSK (BPSK) each phase transition represents one bit, and logic '0' and logic '1' are 180° out of phase. This is equivalent to multiplying the original carrier with a bit stream of '1s' and '−1s' to represent logics '0' and '1', respectively [35].

It is also possible to send more than one bit per phase transition by using smaller phase shifts. For example in quadrature PSK (QPSK) by using four different phases that are 90° apart, it is possible to transmit four symbols, i.e. two bits per phase. As a result of these capabilities, PSK is a popular modulation technique in wideband digital communications and wireless local area networks (WLAN). This is the case especially when PSK is combined with ASK to further increase the number of bits per phase/amplitude transition. This method is yet another type of modulation, which is known as quadrature amplitude modulation or QAM. QAM, however, has not been widely utilized in IMDs since it requires very stable and accurate local oscillators on both Tx and Rx sides, which can add to the volume and power consumption of the IMD.

PLLs can be used for both PSK modulation and demodulation. Traditional PLLs require accurate local oscillators, which in turn require crystals. The intense size constraints in many implantable devices and RFID transponders do not allow inclusion of crystals, which are relatively large and not scalable. Therefore, researchers have tried to either use specific types of PLLs that do not need crystals [35] or extract the phase transitions directly from the incoming carrier signal [36]. The Rx complexity, which results in high power consumption, is one of the disadvantages of the PSK for use in IMD applications [6].

3.2 Backward Telemetry

Back telemetry can be implemented either passively by relying on the mutual coupling between the power coils or actively by adding a Tx and an antenna to the secondary side. Each method has its own advantages and limitations, which are explained in the following.

3.2.1 Passive Telemetry

Load shift keying (LSK) is a common passive back telemetry method in RFID applications [14, 29–31], which has also been used in many IMDs [28, 35, 37–40]. In this method, also known as impedance modulation or load modulation, changing the resistive or capacitive loading of the secondary coil (R_L or C_2 in Fig. 6) based on the modulating back telemetry signal affects the current in the external primary coil due to their mutual coupling. A change in R_L or C_2 in the secondary side can result in variations in the reflected resistance and capacitance, and consequently changes in the primary current especially if k_{12} is sufficiently large. By detecting these primary current variations, which also affect the primary coil voltage, the primary side of the system can demodulate and recover the back telemetry signal. Therefore, the LSK signal on the primary coil should always be picked up and demodulated as an ASK signal, and our earlier discussions apply to the LSK as well.

The main advantage of LSK is its simplicity, especially on the IMD or transponder side where the size matters most. On the other hand, LSK affects the power transmission efficiency by disturbing the resonance circuit and reducing or completely cutting the received power to the main load for finite periods of time. The back telemetry data rate that is achievable through LSK highly depends on k_{12} and power carrier frequency, f_0. It also depends on many other factors including coils quality factors, i.e. Q_1, Q_2, load variations during normal operation, sensitivity of the current or voltage sensing circuit on the external primary side, the amount of noise and interference in the primary coil, and the type of encoding technique that is usually combined with the LSK data [37].

There are three possible configurations for LSK as shown in Fig. 6. Two of these are ohmic and the third one is capacitive. In the series ohmic load modulation, shown

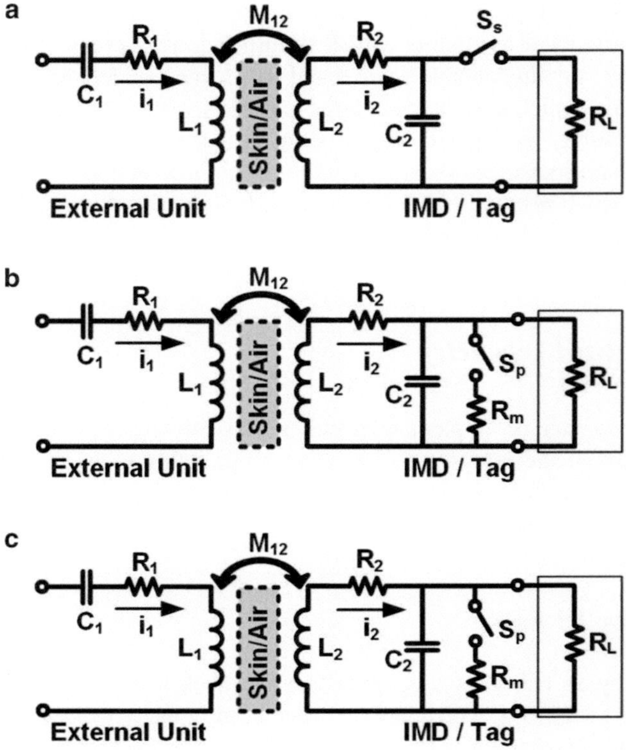

Fig. 6 Passive back telemetry by load shift keying (LSK) using (**a**) series ohmic load modulation, (**b**) parallel ohmic load modulation, and (**c**) parallel capacitive load modulation

in Fig. 6a, the secondary loading is changed between Z_L and open-circuit (infinity) by a series switch based on the back telemetry data. In the parallel configuration, shown in Fig. 6b, the secondary loading is changed between Z_L and short-circuit (close to zero) by a parallel switch. Since a larger change in the secondary loading results in easier detection of the back telemetry signal in the primary coil, the series and parallel configurations are suitable for small and large loads (i.e. value of Z_L), respectively. Most RFID devices have very small power consumption and therefore use the parallel configuration [14]. However, implantable microstimulators may have much higher power consumptions especially when the stimulation is active. Therefore, the series ohmic configuration or a combination of both depending on the loading would be a better choice [41]. Finally, in capacitive load modulation, an additional capacitor, C_m in Fig. 6c, is switched in and out in parallel to C_2 based on back telemetry data bits. This action would result in detuning the Rx LC-tank circuit from its original resonance frequency, which in turn affects the reflected capacitance onto the Tx and consequently i_1.

3.2.2 Active Telemetry

Considering that neural signals have a bandwidth of about 10 kHz, a wideband telemetry link in the order of several MHz is needed to wirelessly record from a large number of sites, simultaneously. Therefore, the small bandwidth provided by the passive back telemetry method is not enough for IMDs that are dedicated to multichannel neural recording. These IMDs are usually equipped with low-power transmitters for active back telemetry and utilize a separate carrier that is significantly higher in frequency than the power carrier. The major challenges in active back telemetry are reducing the power consumption while achieving sufficient range, small size of the IMD, and efficient antenna design for an effective wireless link. Design of the external Rx would be less challenging due to more relaxed size and power constraints outside of the body.

Several research groups have implemented active back telemetry links for neural recording systems using commercial components or custom ASICs [42–46]. In most of these designs the IMD Tx is significantly simplified to reduce the size and power consumption at the expense of more complexity in the external Rx. Use of the high frequency band known as Industrial, Scientific, and Medical (ISM) band along with inductive power has also been proposed [45].

3.3 Single Carrier vs. Multi-Carrier

The main advantage of using a single carrier for both power and data transmission is the relatively robust coupling between power coils, which can lead to more reliable data transfer. Another advantage is the saving in space by reusing power coils for multiple purposes. However, achieving high PTE and high data transmission bandwidth utilizing the same carrier can be challenging because of their conflicting requirements. It was shown earlier that modulating the power carrier in any form or direction complicates the power Tx circuitry and reduces the PTE. Another important issue is the low frequency of the power carrier, which further limits the data transfer bandwidth in either direction to levels that are insufficient for advanced neuroprosthetic devices for sensory substitution and brain computer interfacing (BCI).

As a result, the use of two or three carrier signals for power, downlink, and uplink has been proposed with each carrier having its own pair of coils or antennas in order to decouple the data transfer link bandwidth from the power transmission efficiency [47]. As shown in Fig. 7, three different bands could be used for power transmission, forward and back telemetry. Aside from the size overhead, the use of multiple carrier signals within a space as small as an IMD introduces new challenges, the most important of which is the strong power carrier interference with much weaker data carriers. Several researchers have offered solutions such as using orthogonal symmetrical coils [47, 48], coaxial coils with differential phase shift keying (DPSK) [8, 49], and shifted coplanar coils with offset quadrature phase-shift

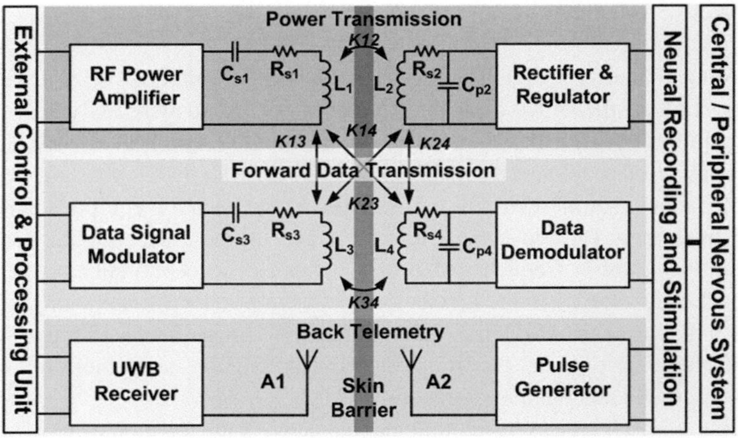

Fig. 7 Block diagram of the multi-band wireless link and its associated blocks in a high-performance implantable neuroprosthetic device [47]

keying (OQPSK) [50]. Nonetheless, the most effective way to reduce interference is to separate out the carrier frequencies and take advantage of the band-pass filtering (BPF) effect of the high-Q LC-tank circuits at resonance.

In the case of the orthogonal coils, a pair of planar spiral coils (PSC), shown in Fig. 8a, which geometries have been optimized based on the power carrier frequency and tissue volume conductor, are used for transcutaneous power transfer [51]. A second pair of vertical coils is wound symmetrically across the PSC pair to establish the data transfer link. Orthogonal orientation and symmetry lowers the undesired cross-coupling between the two pairs without affecting the desired coupling within the pairs [48]. This will minimize the power carrier interference on the data carrier, which can then benefit from any robust data modulation technique. There are also other symmetrical coil geometries, such as figure-8 shown in Fig. 8b, which can attenuate the effects of external common mode magnetic fields and reduce cross-coupling from power coils. In such designs, a pair of planar figure-8 coils is utilized, in which the electromotive force induced from the power carrier in one loop opposes the same in the other loop. Therefore, in a perfectly aligned condition, the power carrier interference becomes negligible [48].

3.4 *Pulse Based Data Transmission*

The main advantage of using a single carrier for both power and data transmission is the relatively robust coupling between power coils, which can lead to higher reliability. The majority of modulation techniques that have been used in near-field inductive links and discussed earlier modify a sinusoidal carrier signal based on the data to be transferred across the link. Even though modulating a carrier

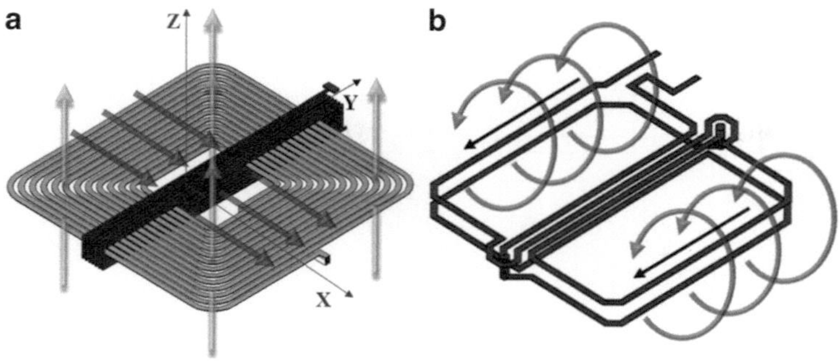

Fig. 8 (**a**) A pair of planar spiral coils (PSC) is used for power transfer. A second pair of coils can be wound symmetrically across the PSC pair for data transfer such that their fluxes are orthogonal and minimize the power carrier interference when the coils are perfectly aligned. (**b**) A symmetrical figure-8 coil geometry to attenuate the effects of strong common mode magnetic fields due to the power carrier interference [48]

signal provides a robust mean to transfer data, generation of the carrier signal at a power level that ensures sufficient signal to noise ratio (SNR) at the Rx involves consuming a considerable amount of power at the Tx, which is scarce on the IMD side. Therefore, carrier based modulation techniques are more suitable for the downlink. Because of the significant electromagnetic field absorption in the tissue, which exponentially increases with the carrier frequency, high bandwidth must be achieved at the lowest possible carrier frequencies. This requirement rules out the majority of commercially available wideband wireless protocols, such as Bluetooth or WiFi, which operate well in the air at 2.4 GHz but not in the tissue. On the other hand, there are specific standards, such as Medical Implant Communication Service (MICS), operating in the 402–405 MHz band, which can only offer a limited bandwidth (300 kHz).

One solution, recently proposed in [52], is to substitute the carrier signal with a series of sharp and narrow pulses, which require much less power to generate them. The timing and amplitude of these pulses have been carefully selected to reduce the inter-symbol interference (ISI) on the Rx side and make it easier to detect and recover the serial data bit stream. This method, which is called Pulse Harmonic Modulation (PHM), takes advantage of the residual ringing in high-Q LC-tanks. To transmit each bit "1", the PHM transmitter generates a sharp pulse at the onset of the bit period to initiate a ringing response in the Rx high-Q LC-tank, as shown in Fig. 9. A second pulse is then generated with specific amplitude, $P < 1$, and delay, t_d, with respect to the initial pulse that suppresses the residual ringing across the Rx LC-tank well before the end of the bit period. No pulses are transmitted in this scheme for bit "0". This method allows for reaching high data rates in excess of 10 Mbps without reducing the inductive link Q-factor, thus significantly improving the transmission range and selectivity of the data link in rejecting out of band interferes, such as the power carrier, without consuming too much power in the IMD.

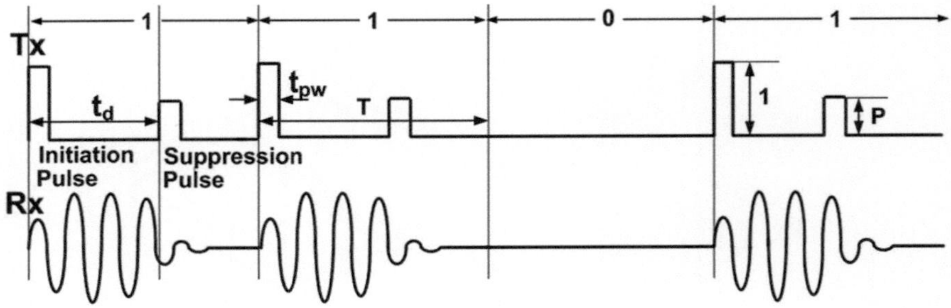

Fig. 9 PHM conceptual waveforms including their key parameters

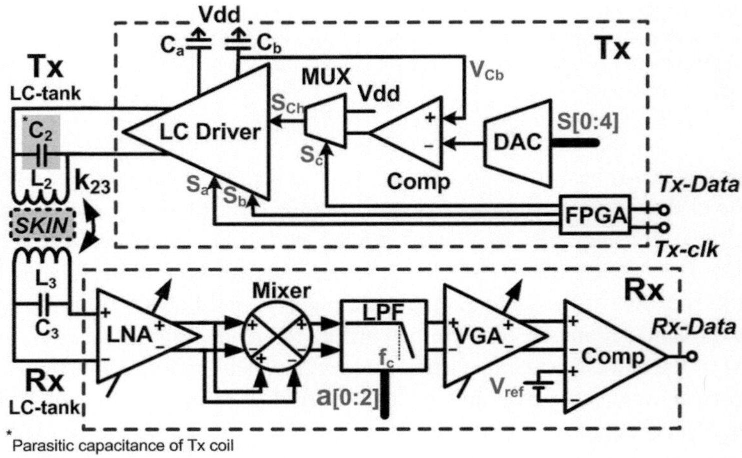

Fig. 10 Block diagram of the PHM-based based transceiver [53]

Figure 10 shows the block diagram of a PHM transceiver. On the Tx side, C_b is charged up to a voltage set by a digital-to-analog converter (DAC) and C_a is charged up to the supply voltage, V_{dd}. C_a is then discharged into the primary data coil followed by C_b via an LC-driver circuit according to a specific timing that is dictated by an FPGA, which accepts the serial data bit stream (Tx-Data). In the Rx block, which operates based on a non-coherent energy detection (ncED) scheme, the received the received ringing signal is amplified, squared, and low-pass filtered. Finally, a comparator recovers the serial data bit stream. This transceiver has achieved a data rate of 10.2 Mbps with bit error rate (BER) of 6.3×10^{-8} at 1 cm coupling distance while consuming 345 and 294 pJ/bit in Tx and Rx, respectively [53].

Utilizing the PHM transceiver in Fig. 10 at higher data rates, i.e. >10 Mbps results in high power consumption, which is not desirable in IMDs. To push the limits on data rate and power consumptions, a fully-integrated low-power PHM-based transceiver has recently been proposed in [54]. The key improvements in this

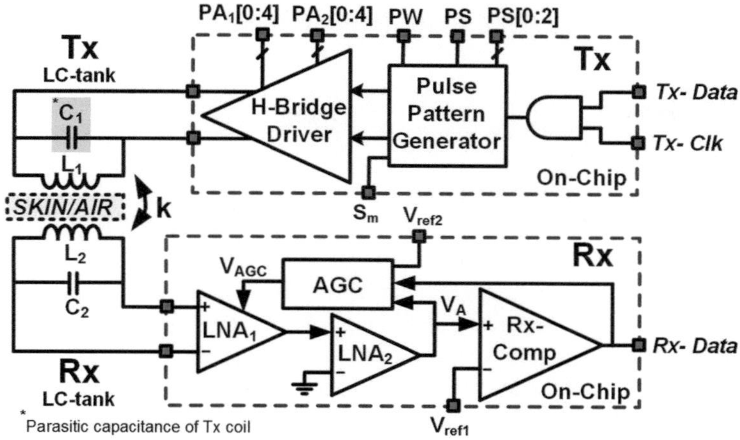

Fig. 11 Block diagram of the fully-integrated low-power 20 Mbps near-field transceiver based on PHM [54]

PHM transceiver, which block diagram is shown in Fig. 11 are: (1) increasing the data rate to 20 Mbps, (2) integrating the entire transceiver on-chip by eliminating the accompanying FPGA used on the Tx side in [53] to generate narrow pulses, (3) a new Rx architecture with higher bandwidth, lower power consumption, and less die area, and (4) an automatic gain control (AGC) mechanism in the Rx to further reduce power and inter-symbol interference (ISI) at short distances, and relax the design requirements and complexity of the Rx building blocks.

Inside the PHM Tx, shown in the upper dashed box in Fig. 11, a pulse pattern generator (PPG) block generates two narrow pulses with specific timing for a bit "1". The H-bridge driver transmits each pulse with specific amplitude across a pair of high-Q LC-tank circuits. Inside the Rx, shown in the lower dashed box in Fig. 11, which operates based on a ncED scheme, the received signal is amplified by two gain stages, followed by a comparator that recovers the serial data bit stream. The gain of the 1st LNA is automatically adjusted by the AGC to ensure that for different coupling distances, the ringing amplitude at the comparator input, V_A in Fig. 11, is at a desired level, V_{ref2}, which should be chosen higher than V_{ref1}. This AGC mechanism ensures that small ringing resulted from previous bits due to non-idealities have amplitudes less than V_{ref1} at the V_A node and remove false data detection, i.e. improving ISI robustness. This transceiver has achieved a data rate of 20 Mbps with BER of 8.7×10^{-8} at 1 cm coupling distance while consuming 180 and 12.5 pJ/bit in Tx and Rx, respectively, due to its simplicity [54].

It should be noted that there are also other pulse-based near-field data transmission methods, developed for chip-to-chip communication and body area networks [55–57]. However, they require an inductive link with a low-Q to achieve wide bandwidth, which is not suitable for the IMD applications, where higher transmission distance and better selectivity for noise and interference rejection are necessary [53].

3.5 Simultaneous Power and Forward/Backward Data Transmission

The majority of modulation techniques that have been used for simultaneous power and forward/backward data transmission are ASK, FSK, LSK, and BPSK/QPSK as discussed earlier. The use of a power carrier signal along with these methods was attractive in early IMDs because the same inductive link could be used for both power and data transmission to/from IMD. In high-performance IMDs that require wider bandwidth, however, a separate power carrier from the data carrier is preferred because increasing the frequency of the high amplitude power carrier can lead to unsafe temperature elevation due to excessive power loss in the tissue. To achieve high power transfer efficiency (PTE) and high data rate, a high frequency carrier (>50 MHz) is required for the data link while the power carrier frequency should be kept below 20 MHz. This has led designers to the use of dual-carrier power/data links with each carrier linking a separate pair of coils [47, 49, 50, 58].

A major challenge in dual-carrier designs is the cross-coupling between the two pairs of coils, which need to be miniaturized and co-located inside the IMD. In particular, the strong power carrier interference can dwarf the weak data signal on the Rx side and make data recovery quite difficult, if not impossible. In other words, to achieve a low BER, a large signal-to-interference ratio (SIR) is needed. While innovative coil designs, such as orthogonal and figure-8 coils in Fig. 8a, b, can help with reducing the coils' cross-coupling [48, 59, 60], it is still necessary to filter out the power carrier interference at the Rx input electronically at the cost of adding to the power consumption and complexity of the IMD [61]. Moreover, achieving high data rates via traditional modulation schemes requires power consuming frequency-stabilization RF circuits, such as PLLs, which are not desired on the IMD side.

A pulse-based data transmission method, called PHM, was described earlier to further push the limits of power consumption and data rate in near-field telemetry links using sharp pulses. Unfortunately none of pulse-based methods in [52–57] including PHM are robust enough against strong power carrier interference, i.e. they operate only when the SIR at Rx input is high.

Recently, a new carrier-less data transmission scheme, called Pulse Delay Modulation (PDM), has been developed for near-field simultaneous data and power transmission [62, 63]. The novel aspect in the PDM is utilization of undesired power carrier interference across the Rx input to deliver the data bits. The proposed method saves the power and space needed for filtering out the power carrier interference on the Rx side, and at the same time enjoys power saving properties of the near-field impulse radio ultra wideband (IR-UWB), particularly on the Tx side, by eliminating the data carrier.

Figure 12 shows a dual-band inductive data and power transmission link using PDM, in which two separate links are used for power (L_1-L_2) and data (L_3-L_4) transfer to keep both PTE and bandwidth high as possible. On the other hand, the use of narrow pulses for data transmission across L_3-L_4 link significantly reduces power consumption on the Tx side.

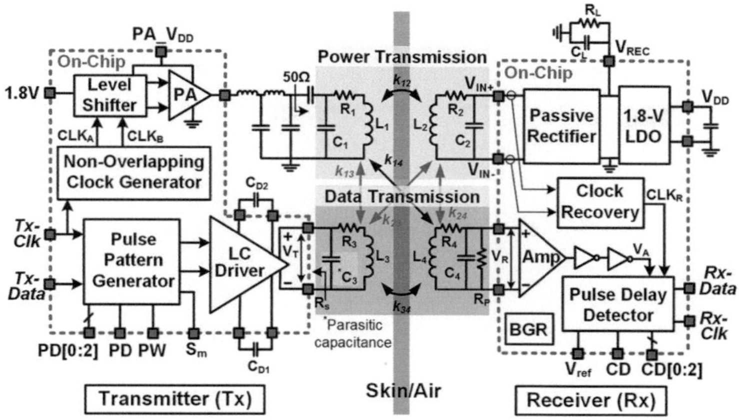

Fig. 12 Block diagram of the wireless power and data transmission circuit across a dual-band inductive link using the PDM scheme. This prototype PDM transceiver ASIC includes all the blocks inside the Tx and Rx dashed boxes. Direct (k_{12} and k_{34}) and cross (k_{13}, k_{14}, k_{23}, and k_{24}) couplings across two pairs of coils are also presented [62, 63]

In the presence of narrow pulses across L_3C_3-tank in Fig. 12, a ringing appears across L_4C_4-tank due to k_{34} at the resonance frequency of the tank, f_r. The inductive data link acts as a band-pass filter (BPF), only letting through the narrow pulse harmonics around f_r with a small bandwidth around it, which results in a decaying sinusoidal waveform at f_r, following each data pulse, which represents a "1". The rate of damping depends on the quality factor (Q) of the L_4C_4-tank. To achieve a high data transmission rate, it is necessary for the ringing to dampen very quickly, such that the presence or absence of the next pulse, which represent the following data bit "1" or "0", respectively, can be detected. One way of achieving this is to reduce the Q_4 of the L_4C_4-tank by increasing R_4 or add a parallel resistance R_p in Fig. 12 [62].

Even though the L_4C_4 BPF attenuates the power carrier at f_p ($\ll f_r$) to some extent, the sizeable amplitude of the power carrier, lowered Q_4, and the power-data coils cross couplings, k_{13}, k_{14}, k_{23}, and k_{24}, due to their proximity result in a considerable power carrier interference being added to the decaying ringing across the L_4C_4-tank. As such, the presence or absence of the data pulse ringing changes the shape of the power carrier interference across the L_4C_4-tank. The novel solution in [62] to detect these changes is to select the timing of the data pulses across L_3C_3-tank in a way that they alter the zero-crossing times of the interfering sinusoidal power carrier across L_4C_4-tank, which are indeed the most sensitive points of a sinusoidal signal to an external disturbance (zero induced fields from L_1 and L_2). Using this method, the SIR in dB can even be negative. Because despite the large interference amplitude, the zero-crossings are still amenable to manipulation.

Figure 13 shows the PDM concept and key waveforms. In order to send a bit "1", two narrow pulses spaced by half a power carrier cycle i.e. $T_p/2 = 1/2f_p$ are

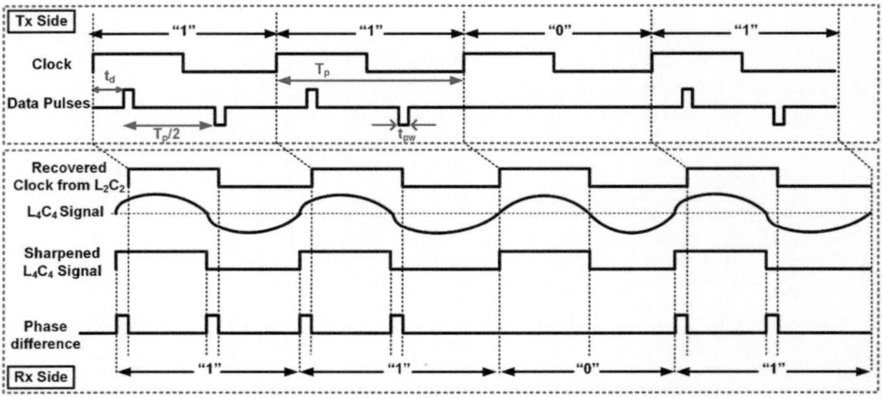

Fig. 13 PDM conceptual waveforms including their key parameters. Presence of data pulses within a bit period creates phase shift between the recovered clock extracted directly from the power carrier across L_2C_2-tank and the received and sharpened signal across L_4C_4-tank in Fig. 12. As a result, transmitting a "1" leads to a phase shift between the two square waveforms on the Rx side, the absence of which represents no phase shift and translates to a recovered "0" [63]

transmitted across the link. The first pulse is applied at the beginning of the bit period after a specific delay, t_d, to initiate a ringing across the low-Q L_4C_4-tank. t_d is selected in a way that the peak of the ringing coincides the original zero crossing onset of the interfering sinusoidal power carrier and shifts it in a certain direction. After $T_p/2$, a second pulse with equal but opposite amplitude is transmitted to add a similar time shift in the same direction as the first one to the next zero-crossing time of the received power carrier interference within that cycle.

Considering that L_1C_1 and L_2C_2 are both high-Q and tuned at f_p, the induced power carrier on the L_2C_2-tank is much stronger than that of the low-Q L_4C_4-tank. Therefore, the transmitted data pulses do not have any noticeable effects on the power transmission link. Since no pulses are transmitted for a bit "0", any delay between signals across L_4C_4 and L_2C_2 tanks represent a bit "1", which can be easily detected using a simple phase detector circuit after sharpening the received waveforms, as shown in Fig. 12.

The block diagram of the proposed fully-integrated PDM-based transceiver prototype in [62] has been shown in Fig. 12. The key aspects of this PDM transceiver include: (1) increasing data rate, (2) reducing both Tx and Rx power consumption, and (3) improving robustness against power carrier interference by reducing the required SIR to achieve the same BER. Inside the PDM Tx, shown in the left dashed box in Fig. 12, a clock generator block creates two non-overlapping clocks from an external master clock signal, Tx-Clk, at the desired carrier frequency, f_p, for a class-D PA, which generates the power carrier signal at a desired output power level. The power carrier is delivered to L_1 after passing through a matching circuit to induce current in L_2. Using the same Tx-Clk, a pulse pattern generator (PPG) generates two narrow pulses, which are in sync with the power carrier and spaced by half a

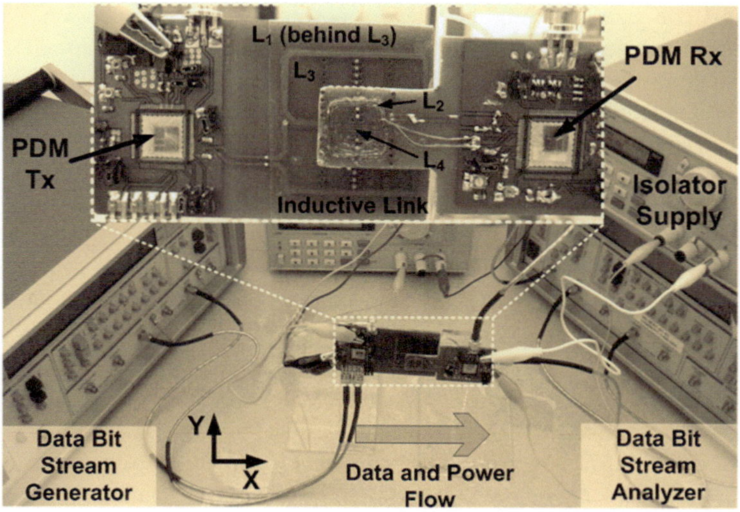

Fig. 14 PDM transceiver measurement setup. Inset: inductive links made of printed-spiral and wire-wound coils for power transmission (L_1 and L_2 in Fig. 12) and a pair of planar figure-8 coils on FR4 PCB for data transmission (L_3 and L_4 in Fig. 12) [26]. L_1 has been carefully aligned and glued behind L_3 to minimize k_{13}. L_2 has also glued over L_4 with careful alignment to minimize k_{24} [62]

power carrier cycle, $T_p/2$, for each Tx-Data bit "1". The LC driver circuit transmits each pulse across L_3-L_4 data link. Inside the Rx block, which is the right dashed box in Fig. 12, a passive full-wave rectifier is followed by a 1.8 V low drop-out (LDO) regulator to provide the IMD power supply. A clock recovery circuit extracts the internal IMD clock, CLK_R, from the received power carrier across L_2C_2-tank. V_R is the Rx input signal across L_4C_4-tank and the superposition of the power carrier interference through k_{14} and k_{24} and PDM pulses through k_{34}. V_R is amplified to create a square waveform, V_A, and then a pulse delay detector integrates the time shifts between V_A and CLK_R to recover the received data bit stream.

A typical measurement setup for dual-band power and data transmission based on PDM is shown in Fig. 14. Two PDM chips have been used in this experimental setup and wirebonded to QFN packages mounted on two-layer FR4 printed circuit boards (PCBs). Each PCB includes a planar figure-8 coil for data transmission as shown in Fig. 14 inset. The geometries of a printed spiral coil in the Tx and a wire-wound coil in the Rx were optimized at 13.56 MHz for power transmission. L_1 and L_2 were glued onto L_3 and L_4, respectively, following careful alignment to minimize undesired cross couplings between power and data coils, k_{13} and k_{24}. In measurements, the PDM transceiver can achieve a data rate of 13.56 Mbps with a BER of 4.3×10^{-7} across a 10 mm inductive link. Data Tx and Rx power consumptions under these conditions are only 960 and 162 pJ/bit, respectively, while a separate power link delivers 42 mW of regulated power to the load.

Table 1 compares different methods for power and data transmission. Although PHM results in lower power consumption, utilizing a PHM transceiver in an inductively-powered system is challenging. However, PDM is a pulse-based technique for simultaneous power and data transmission at high data rates with small power consumption. The work in [49] and [61] used two different carrier-based inductive links for power and data transmission at 2 and 20 MHz, respectively, called binary/differential phase-shift keying (BPSK/DPSK). Data is modulated on the phase of the 20 MHz carrier signal and transferred to the Rx, inside which an external high-pass filter (HPF) attenuates the 2 MHz power carrier to improve the SIR for $\sim$6–12 dB. Moreover, additional filtering and processing are done inside the Rx to reduce the interference effect. Therefore, data rate is limited to 2 MHz, which is 6.7 less than the one form PDM, the Rx consumed die area is large, and Rx power consumption is 19 times larger than PDM. This power reduction is important in inductively powered IMDs, where received regulated power is quite precious. The LSK data link in [37] achieved 2.8 Mbps with very small Tx power consumption. However, LSK can only be used for the uplink while other techniques such as PDM and BPSK can be used for both up and downlink. Moreover, LSK consumes low power for data transfer at the cost of reducing the delivered power by up to 50 % at high data rates.

3.6 Safety Issues for IMDs

Coil designs for IMD applications should consider the electromagnetic power absorbed in the human body to ensure that it meets international safety standards and do not pose health hazards [65]. Several guidelines for suggested absorption limits of electromagnetic energy are either expressed in terms of currents or specific absorption rate (SAR) of power induced in the human body, with the latter defined for a sinusoidal excitation in Watts/kg as,

$$SAR\,(x, y, z) = \frac{\sigma\,(x, y, z)\,E^2\,(x, y, z)}{2\rho\,(x, y, x)} \tag{23}$$

where ρ is the tissue density (in kg/m^3), σ is the conductivity (S/m), and E is the electric field amplitude (V/m) at point (x, y, z). Full-wave electromagnetic computational tools and experimental methods with phantoms filled with tissue simulants can be used to determine the field induced by telemetry or induction devices in the human body and compare them with standard safety limits.

Recommendations to prevent harmful effects in human beings exposed to electromagnetic energy have been issued by numerous organizations, such as IEEE [66]. For humans, the maximum permissible exposure (MPE) in terms of root-mean-square (RMS) electric (E) and magnetic (H) field strengths, and power densities are given in Table 2 as functions of frequency, f_0. Table 2 clearly shows that lower frequencies are more appropriate for the IMDs.

Table 1 Benchmarking of recent telemetry links

Ref	Mod	Distance (mm)	Data (MHz)	Power (MHz)	Data rate (Mbps)	Tx/Rx power (pJ/bit)	Tech (μm)	SIR (dB)	Area (mm^2) (data Tx/Rx)	V_{dd} (V)	BER
[34]	pcFSK	5	5/10	5/10	2.5	–/152	1.5	–	–/0.29	5	10^{-5}
[64]	BPSK	15	10	10	1.12	–/625	0.18	–	–/0.2	1.8	10^{-5}
[37]	LSK[d]	20	25	25	2.8	35.7/1,250	0.5	–	2.2/2.2[b]	2.8	10^{-6}
[58]	FSK	20	–/5	5	1.25	–	0.8	–	–	2.7	–
[58]	BPSK	20	48	5	3	1,962/–	0.8	–	2.3[b]	2.7	2×10^{-4}
[50]	QPSK	5	13.56	1	4.16	–	–	–	–	–	2×10^{-6}
[49]	BPSK	10–15	20	2	2	–/3,100	0.35	−12[a]	–/4.4	4.5	10^{-7}
[61][c]	DPSK	–	20	2	2	–	0.18	–	–	1.8	10^{-7}
[54]	PHM	10	66.6	–	20	345/294	0.35	–	0.1/0.5	1.8	8.7×10^{-7}
[62]	PDM	10	50	13.56	13.56	960/162	0.35	−18.5	0.34/0.37	1.8	4.3×10^{-7}

[a] A first order off-chip filter was used to improve SIR to −6 dB
[b] Including pads
[c] Second-order filter was used to improve SIR
[d] LSK is only used for uplink

Table 2 Maximum permissible exposure (MPE) limits to human body [66]

Frequency range (MHz)	Electric field strength, E (V/m)	Magnetic field strength, H (A/m)	Power density E-field, H-field (mW/cm^2)
0.003–0.1	614	163	100, 1,000,000
0.1–3.0	614	$16.3/f_0$	100, $10{,}000/f_0^2$
3–30	$1{,}842/f_0$	$16.3/f_0$	$900/f_0^2$, $10{,}000/f_0^2$
30–100	61.4	$16.3/f_0$	1, $10{,}000/f_0^2$

4 Conclusion

In this chapter, fundamental principles, modulation schemes, and practical considerations of wireless data by means of inductive coupling have been described. Fundamental near-field equations governing self- and mutual-inductance between magnetically coupled coils have been reviewed. Also important geometrical parameters that can affect the coupling coefficient of an inductive link have been specified. The time domain transfer function of an inductive link for data transmission was also derived.

Three major carrier modulation techniques for data transmission, namely ASK, FSK, and PSK, have been discussed and compared. It can be concluded that ASK provides the simplest solution for forward data transmission and would be an appropriate choice when low data rate and high PTE link with a single carrier is needed. FSK provides high data rates and a robust link at the expense of more complexity and reduced PTE. PSK can offer the highest bandwidth, however, synchronization issues might result in a high BER or high sensitivity to interference and artifacts.

Conflicting requirements for high PTE and high bandwidth through the same inductive link have led designers towards multi-carrier inductive links, in which two or three separate carrier signals are used for power transmission, uplink, and downlink. Pulse-based near-field data transmission has recently become popular due to its low power consumption, robustness, and high bandwidth, which are important in the IMD and FRID applications. Pulse-based PHM and PDM techniques and transceivers were also described. The PHM-based transceivers achieved high data rate, low-power consumption, and small die size. However, due to the on-off keying (OOK) nature of the PHM, the signal amplitude should be considerably larger than the power carrier interference at the Rx input for proper data recovery (e.g. SIR > 10 dB). Therefore, PDM is advantageous due to its robustness against large power carrier interference and offers reliable data communication in the presence of power interference that is one order of magnitude larger than the received data pulses. This is key in high performance IMDs, such as retinal prostheses that demand high power delivery. PHM, on the other hand, is suitable for low-power IMDs that are equipped with rechargeable batteries and do not need to be continuously powered.

Safety is paramount in design of every medical device, especially those that are meant to be implantable. In design of transcutaneous power and transmission links, the intensity of the magnetic field and its frequency of operation are the key factors that need to be chosen based on the electromagnetic safety standard guidelines. These guidelines are often expressed in terms of the SAR of the power induced in the human body.

References

1. D. Zhou, E. Greenbaum, *Implantable Neural Prostheses 1* (Springer, New York, 2009)
2. R. Allan, Medtronic sets the pace with implantable electronics. Electron. Des. **51**(24), 52–56 (2003)
3. M. Morrel, Responsive cortical stimulation for the treatment of medically intractable partial epilepsy. Neurology **77**, 1295–1304 (2011)
4. R. Fisher, Direct brain stimulation is an effective therapy for epilepsy. Neurology **77**, 1220–1221 (2011)
5. F. Zeng et al., Cochlear implants: system design, integration, and evaluation. IEEE Rev. Biomed. Eng. **1**, 115–142 (2008)
6. J. Weiland, M. Humayun, Visual prosthesis. Proc. IEEE **96**, 1076–1084 (2008)
7. K. Chen et al., An integrated 256-channel epiretinal prosthesis. IEEE J. Solid-State Circuits **45**(9), 1946–1956 (2010)
8. D.B. Shire et al., Development and implantation of a minimally invasive wireless subretinal neurostimulator. IEEE Trans. Biomed. Eng. **56**(10), 2502–2511 (2009)
9. A. Schwartz et al., Brain-controlled interfaces: movement restoration with neural prosthetics. Neuron **52**, 205–220 (2006)
10. T. Kuiken et al., Targeted reinnervation for enhanced prosthetic arm function in a woman with a proximal amputation: a case study. Lancet **369**, 371–380 (2007)
11. R. Fernandes et al., Artificial vision through neuronal stimulation. Neurosci. Lett. **519**, 122–128 (2012)
12. L. Theogarajan, Strategies for restoring vision to the blind: current and emerging technologies. Neurosci. Lett. **519**, 129–133 (2012)
13. M. Ghovanloo, K. Najafi, A modular 32-site wireless neural stimulation microsystem. IEEE J. Solid-State Circuits **39**, 2457–2466 (2004)
14. K. Finkenzeller, *RFID-Handbook*, 2nd edn. (Wiley, Hoboken, 2003)
15. N. Leavitt, Payment applications make E-commerce mobile. IEEE Comput. Soc. **43**, 19–22 (2010)
16. M. Sadiku, *Elements of Electromagnetics*, 4th edn. (Oxford University Press, Oxford, 2007)
17. C. Zierhofer, E. Hochmair, Geometric approach for coupling enhancement of magnetically coupled coils. IEEE Trans. Biomed. Eng. **43**, 708–714 (1996)
18. F. Grover, *Inductance Calculations Working Formulas and Tables* (D. Van Nostrand Company, New York, 1946)
19. F. Terman, *Radio Engineers Handbook* (McGraw-Hill, New York, 1943)
20. M. Soma et al., Radio-frequency coils in implantable devices: misalignment analysis and design procedure. IEEE Trans. Biomed. Eng. **34**, 276–282 (1987)
21. S. Venkatraman et al., A system for neural recording and closed-loop intracortical microstimulation in awake rodents. IEEE Trans. Biomed. Eng. **56**, 15–22 (2009)
22. J. Lee et al., A 64 channel programmable closed-loop neurostimulator with 8 channel neural amplifier and logarithmic ADC. IEEE J. Solid-State Circuits **45**, 1935–1945 (2010)
23. K. Arabi, M.A. Sawan, Electronic design of a multichannel programmable implant for neuromuscular electrical stimulation. IEEE Trans. Rehabil. Eng. **7**(2), 204–214 (1999)

24. S. Boyer et al., Implantable selective stimulator to improve bladder voiding: design and chronic experiment in dogs. IEEE Trans. Rehabil. Eng. **8**(4), 789–797 (2000)
25. B. Ziaie et al., A single-channel implantable microstimulator for functional neuromuscular stimulation. IEEE Trans. Biomed. Eng. **44**(10), 909–920 (1997)
26. B. Smith et al., An externally powered, multichannel, implantable stimulator-telemeter for control of paralyzed muscle. IEEE Trans. Biomed. Eng. **45**(4), 463–475 (1998)
27. W. Liu et al., A neuro-stimulus chip with telemetry unit for retinal prosthetic device. IEEE J. Solid-State Circuits **35**, 1487–1497 (2000)
28. G. Suaning, N. Lovell, CMOS neuro-stimulation ASIC with 100 channels, scalable output, and bidirectional radio-freq. telemetry. IEEE Trans. Biomed. Eng. **48**, 248–260 (2001)
29. P. Raker et al., Secure contactless smartcard ASIC with DPA protection. IEEE J. Solid-State Circuits **36**, 559–565 (2001)
30. U. Kaiser, W. Steinhaugen, A low-power transponder IC for high-performance identification systems. IEEE J. Solid-State Circuits **30**, 306–310 (1995)
31. A. Abrial et al., A new contactless smart card IC using an on-chip antenna and an asynchronous microcontroller. IEEE J. Solid-State Circuits **36**, 1101–1107 (2001)
32. D. Galbraith et al., A wide-band efficient inductive transdermal power and data link with coupling insensitive gain. IEEE Trans. Biomed. Eng. **34**, 265–275 (1987)
33. P. Troyk, G. DeMichele, Inductively-coupled power and data link for neural prostheses using a class-E oscillator and FSK modulation, in *Proc. IEEE 25th EMBS Conf.*, pp. 3376–3379, September 2003
34. M. Ghovanloo, K. Najafi, High data rate frequency shift keying demodulation for wireless biomedical implants. IEEE Trans. Circuits Syst. I **51**(12), 2374–2383 (2004)
35. M. Sawan et al., Wireless smart implants dedicated to multichannel monitoring and microstimulation. IEEE Circuits Syst. Mag. **5**, 21–39 (2005)
36. C. Marschner et al., A novel circuit concept for PSK-demodulation in passive telemetric systems. Microelectron. J. **33**, 69–75 (2002)
37. S. Mandal, R. Sarpeshkar, Power-efficient impedance-modulation wireless data links for biomedical implants. IEEE Trans. Biomed. Circuits Syst. **2**(4), 301–315 (2008)
38. Z. Tang et al., Data transmission from an implantable biotelemeter by load-shift keying using circuit configuration modulator. IEEE Trans. Biomed. Eng. **42**, 524–528 (1995)
39. L. Zhou, N. Donaldson, A fast passive data transmission method for eng telemetry. Neuromodulation **6**(2), 116–121 (2003)
40. M. Catrysse et al., An inductive power system with integrated bi-directional data-transmission. Sens. Actuators A **115**, 221–229 (2004)
41. G. Bawa, M. Ghovanloo, An active high power conversion efficiency rectifier with built-in dual-mode back telemetry in standard CMOS technology. IEEE Trans. Biomed. Circuits Syst. **2**(3), 184–192 (2008)
42. I. Obeid et al., Two multichannel integrated circuits for neural recording and signal processing. IEEE Trans. Biomed. Eng. **50**, 255–258 (2003)
43. E. Hawley et al., Telemetry system for reliable recording of action potentials from freely moving rats. Hippocampus **12**, 505–513 (2002)
44. P. Mohseni, K. Najafi, A fully integrated neural recording amplifier with dc input stabilization. IEEE Trans. Biomed. Eng. **51**, 832–837 (2004)
45. N. Neihart, R. Harrison, Micropower circuits for bidirectional wireless telemetry in neural recording applications. IEEE Trans. Biomed. Eng. **52**, 1950–1959 (2005)
46. K. Gosalia, G. Lazzi, M. Humayun, Investigation of a microwave data telemetry link for a retinal prosthesis. IEEE Trans. Microw. Theory Tech. **52**(8), 1925–1933 (2004)
47. M. Ghovanloo, S. Atluri, A wideband power-efficient inductive wireless link for implantable microelectronic devices using multiple carriers. IEEE Trans. Circuits Syst. I **54**(10), 2211–2221 (2007)
48. U. Jow, M. Ghovanloo, Optimization of data coils in a multiband wireless link for neuroprosthetic implantable devices. IEEE Trans. Biomed. Circuits Syst. **4**(5), 301–310 (2010)

49. M. Zhou et al., A non-coherent DPSK data receiver with interference cancellation for dual-band transcutaneous telemetries. IEEE J. Solid-State Circuits **43**, 2003–2012 (2008)
50. G. Simard et al., High-speed OQPSK and efficient power transfer through inductive link for biomedical implants. IEEE Trans. Biomed. Circuits Syst. **4**(3), 192–200 (2010)
51. U. Jow, M. Ghovanloo, Modeling and optimization of printed spiral coils in air, saline, and muscle tissue environments. IEEE Trans. Biomed. Circuits Syst. **3**, 339–347 (2009)
52. F. Inanlou, M. Ghovanloo, Wideband near-field data transmission using pulse harmonic modulation. IEEE Trans. Circuits Syst. I **58**(1), 186–195 (2011)
53. F. Inanlou et al., A 10.2 Mbps pulse harmonic modulation based transceiver for implantable medical devices. IEEE J. Solid-State Circuits **46**, 1296–1306 (2011)
54. M. Kiani, M. Ghovanloo, A 20 Mbps pulse harmonic modulation transceiver for wideband near-filed data transmission. IEEE Trans. Circuits Syst. II **60**, 382–386 (2013)
55. N. Miura et al., A 195-Gb/s 1.2-inductive inter-chip wireless superconnect with transmitter power control scheme for 3-D-stacked system in a package. IEEE J. Solid-State Circuits **41**(1), 23–33 (2006)
56. J. Yoo et al., A 1.12 pJ/b inductive transceiver with a fault tolerant network switch for multi-layer wearable body area network applications. IEEE J. Solid-State Circuits **44**(11), 2999–3010 (2009)
57. S. Lee et al., A low-energy inductive coupling transceiver with cm-range 50-Mbps data communication in mobile device applications. IEEE J. Solid-State Circuits **45**(11), 2366–2374 (2010)
58. A. Rush, P. Troyk, A power and data link for a wireless-implanted neural recording system. IEEE Trans. Biomed. Circuits Syst. **59**, 3255–3262 (2012)
59. G. Wang et al., Analysis of dual band power and data telemetry for biomedical implants. IEEE Trans. Biomed. Circuits Syst. **6**, 208–215 (2012)
60. G. Simard et al., Novel coils topology intended for biomedical implants with multiple carrier inductive link, in *Proc. IEEE Int. Symp. Cir. Syst.*, pp. 537–540, May 2009
61. K. Chen et al., A 37.6 mm^2 1024-channel high-compliance-voltage SoC for epiretinal prostheses, in *Digest of technical papers, IEEE Intl. Solid-State Cir. Conf.*, pp. 294–295, February 2013
62. M. Kiani, M. Ghovanloo, A 13.56-Mbps pulse delay modulation based transceiver for simultaneous near-field data and power transmission. IEEE Trans. Biomed. Circuits Syst. **9**(1), 1–11 (2014)
63. M. Kiani, M. Ghovanloo, Pulse delay modulation (PDM) a new wideband data transmission method to implantable medical devices in presence of a power link, in *IEEE Biomed. Cir. Syst. Conf.*, pp. 256–259, 2012
64. Y. Hu, M. Sawan, A fully integrated low-power BPSK demodulator for implantable medical devices. IEEE Trans. Circuits Syst. I **52**(12), 2552–2562 (2005)
65. J. Lin, Computer methods for field intensity predictions, in *CRC Handbook of Biological Effects of Electromagnetic Fields*, ch. 2, ed. by C. Polk, E. Postow (CRC Press, Boca Raton, 1986), pp. 273–313
66. IEEE standard for safety levels with respect to human exposure to radio frequency electromagnetic fields, 3 kHz to 300 GHz, 1999

Near-Field Wireless Power Transfer

Patrick P. Mercier and Anantha P. Chandrakasan

Abstract Wireless power transfer links are becoming increasingly important for consumer, industrial, and medical electronic devices. There are two principal applications for wireless power transfer that require different optimization criteria: continuous power deliver (e.g., cochlear implant) and periodic charging (e.g., cellular phone). In the former case, optimizing power transfer efficiency is a metric of great importance, while in the latter case, minimizing charging time by maximizing power transfer is important. This chapter presents analytical equations that predict optimal conditions for both applications through first-principals step-by-step reflected load analysis. These equations are then used as the basis for the design of a rapid wireless ultra-capacitor charging circuit that speeds up time-to-charge by 3.7×.

Keywords Wireless power transfer • Resonant power transfer • Inductive coupling • Near-field communication

1 Introduction

The short-range inductively coupled systems outlined in chapter "Centimeter-Range Inductive Radios" also offer the ability to wirelessly deliver power from the transmitter (often called the reader in RFID systems) to the receiver, eliminating the requirement for a battery or energy harvester on the receiver itself. This wireless power transfer (WPT) paradigm can be used in many applications, ranging from electric toothbrushes and cochlear implants, to smartwatches and electric vehicle chargers. The benefits of such an approach are many and include: convenience, ease-of-use, and full encapsulation for reduced exposure to unwanted environmental conditions. Adoption of wireless power transfer into consumer electronics has

P.P. Mercier (✉)
University of California San Diego, La Jolla, CA, USA
e-mail: pmercier@ucsd.edu

A.P. Chandrakasan
Massachusetts Institute of Technology, Cambridge, MA, USA
e-mail: anantha@mtl.mit.edu

P.P. Mercier, A.P. Chandrakasan (eds.), *Ultra-Low-Power Short-Range Radios*, Integrated Circuits and Systems, DOI 10.1007/978-3-319-14714-7_11

seen rapid recent growth (circa 2015), and is currently standardized by three competing organizations: the Alliance for Wireless Power (A4WP), the Power Matter Alliance (PMA), and the Wireless Power Consortium (WPC). Due to the varying requirements of each standard, this chapter will not delve into implementing a system for a specific standard; rather, this chapter will introduce the fundamental concepts necessary to understand and optimize near-field wireless power systems in general. In addition, a design example of a rapid wireless ultra-capacitor charger will be discussed [1].

2 Inductive Coupling

2.1 Overview

Near-field inductive coupling is the most popular approach to deliver wireless power between devices spaced up to a few centimeters apart. A schematic of a typical wirelessly-powered device is shown in Fig. 1.

The circuit operates as follows: a power amplifier, with an RF voltage-source input, sends power through a primary-side coil with N_1 turns. Some form of matching is used to tune-out the inherent loop inductance in order to decrease loading effects on the power amplifier. So long as the operational wavelength is much less than the physical dimension of the coils and their separation, energy will be contained primarily in near-field magnetics. In other words, the coils look like antennas that are *extremely* electrically small. Thus, the secondary-side coil, composed on N_2 turns and spaced a distance d from the primary-side coil, receives energy from the transmitter-produced time-varying magnetic field. The RF output of the secondary-side matching network then passes through a rectifier, converting the AC energy into DC energy used to power the load, in this case modeled by resistor $R_{L,DC}$. There are many excellent references that discuss the electromagnetic properties of such a system, as well as analytical formulae for predicting inductance and other parameters in more detail [2, 3].

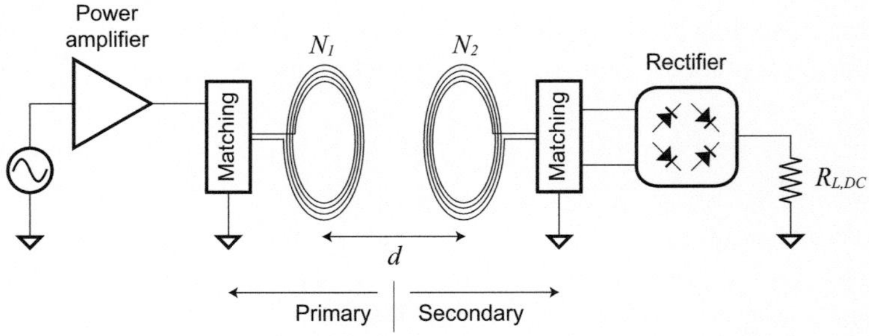

Fig. 1 An introductory schematic overview of a typical inductively-coupled system

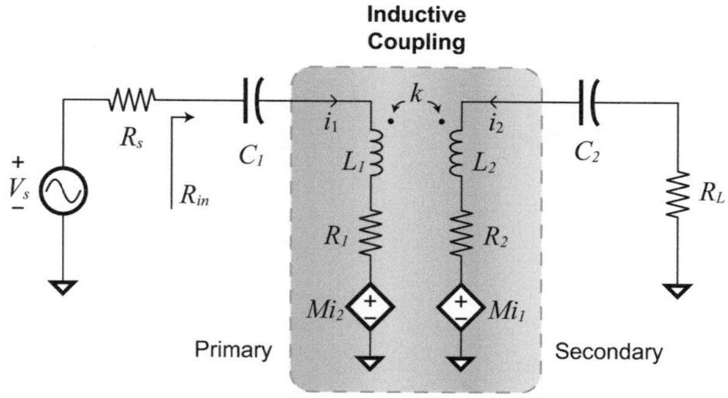

Fig. 2 Schematic model of an inductively coupled system. In this case, series capacitors C_1 and C_2 are used to resonate with inductors L_1 and L_2

A more detailed circuit diagram of a typical inductive coupled system is shown in Fig. 2. Inductors L_1 and L_2 model the primary and secondary loop reactances, respectively. The inductors are not perfect, however, as they both have finite quality factors. Specifically, the quality factor of each coil is given by $Q = \omega L/R$, where ω is the operating frequency and R represents the parasitic series loss resistance. Voltage sources Mi_2 and Mi_1 model the mutual coupling effects between the coils, where M is the mutual coupling factor, while capacitors C_1 and C_2 provide resonant matching with the inductors. The coil coupling coefficient, k, is defined by the following equation:

$$k \equiv \frac{M}{\sqrt{L_1 L_2}}. \tag{1}$$

It can be shown that k is dependent only on coil separation distance d and individual coil geometries [2]. Importantly, it should be noted that k varies between 0 (no coupling) and 1 (perfect coupling), and decreases with d. Most wireless power transfer systems have separation distances ranging from a few millimeters to 20 mm, resulting in $0.03 < k < 0.3$ for coils with centimeter-sized diameters.

The primary side is driven by a voltage source, V_s, that has a series resistance, R_s. The secondary side is loaded by a rectifier followed by the circuit to be wirelessly powered. Load circuits can often be modeled as an effective resistance, $R_{L,DC}$, in series with the rectifier. Since rectifiers are by definition non-linear, including them in this simplified linear model makes analysis overly difficult. Fortunately, it can be easily shown that a resistively-loaded rectifier can be approximated with a single resistance of half the size of the actual load resistor. In other words, with respect to the schematic in Fig. 1, $R_L \approx R_{L,DC}/2$ [2].

Finally, an important definition that will be useful for analysis is the turns ratio of an ideal transformer, assuming perfect coupling:

$$n \equiv \sqrt{\frac{L_2}{L_1}}. \tag{2}$$

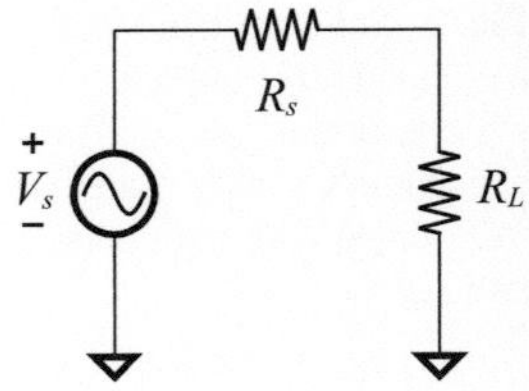

Fig. 3 Resistive divider circuit

Most wireless power transfer designs endeavor to maximize power transfer efficiency in order to minimize the size of the external power source. It can be shown that resonating the inductors with capacitors helps to achieve maximal power transfer efficiency. This has been shown using traditional circuit analysis techniques, preferred by electrical engineers, as well as coupled-mode theory, preferred by physicists [4]. As it turns out, both approaches provide the same results for high quality factor coils at large separation distances [5]. Circuit analysis, however, is more accurate over a wider range of cases and is arguably more convenient when dealing with more complicated circuits models. For this reason, we will use traditional circuit analysis techniques in this chapter.

In some cases, maximizing power transfer efficiency is not necessarily the overarching goal. For example, the most important metric in many charging systems is *charging time*, not charging efficiency. From a design perspective, minimizing charging time is equivalent to maximizing the amount of power delivered to the load circuit (e.g., a capacitor or battery) over a short period of time, given source and system constraints. It makes sense to refer to this scheme as wireless *energy* transfer, operating instantaneously at a maximum power transfer condition.

In many cases, maximizing efficiency is not the same as maximizing power transfer to the load. While most readers know this, it turns out that many good engineers do not fully appreciate this point. To better understand this, consider the classic resistive divider circuit in Fig. 3. The power transfer efficiency of this circuit is given by:

$$\eta = \frac{R_L}{R_L + R_s}, \tag{3}$$

which is maximized when the load resistance, R_L, is large relative to the source resistance, R_s. However, the power delivered to the load is given by:

$$P_L = \frac{V_s^2 R_L}{(R_L + R_s)^2}. \tag{4}$$

A large R_L implies low current from the source, which means low amounts of power are being delivered to the load. In the limit that R_L tends to infinity, η tends to 100 %, yet P_L tends to zero. The classic maximum power transfer theorem states that the load resistance should be matched to the source resistance in order to deliver the maximum amount of power possible to the load. In other words,

$R_L = R_s$ for maximum power transfer. Naturally, if the designer has the ability to make the source impedance arbitrarily small (as in many power amplifier design cases), both high efficiency and large output power can be achievable from a given voltage source. Maximum power transfer of course still applies, but if the source impedance can be made sufficiently low such that "enough" power is extracted from the source, high efficiency can still be achieved at the desired output power.

In the case of inductively-coupled links, however, it is often not possible to change the impedance seen before the load due to finite coil quality factors. Therefore, given a fixed voltage source, it is difficult to achieve both high efficiency and high power delivery. Consequently, it is worthwhile analyzing in greater detail what the fundamental limitations are in order to gain insight into any circuit solutions that can approach these limits.

2.2 *Reflected Load Analysis*

Inductive coupling theory has been well-studied in the past, and many excellent references describe analytical expressions that can accurately predict the maximum achievable efficiency of a given link [2, 3]. However, the analytical steps in previous work can, in some cases, be difficult to interpret correctly and apply to non-standard situations. Thus, this section carefully steps through the analysis necessary to determine maximum efficiency conditions, while also deriving the equations to predict maximum power transfer conditions under source constraints.

2.2.1 Transformer Model

The simplest way to analyze inductively-coupled circuits using circuit theory is by reflected load analysis. It is well-known that circuit elements can be reflected across the terminals of an ideal transformer, modulated by a factor of $1/a^2$, where a is the turns ratio of the transformer. A pair of inductively coupled coils as shown in Fig. 2, can be thought of as a loosely-coupled transformer (i.e., with $k < 1$), and can be modeled with an ideal transformer as one of its circuit elements. One such model is shown in Fig. 4, employing an ideal transformer with a turns ratio of $a = n/k$. Thus, any load on the secondary side can be reflected to the primary side, multiplied by a factor of $(k/n)^2$. It is important to know that reflected impedances must be seen in parallel to L_2 for reflection to be valid in this case.

2.2.2 Capacitor Tuning Options

Before reflecting impedances to compute maximum efficiency and power transfer conditions, it is beneficial to first discuss resonant tuning options. There are four primary topological options for resonating out inductors L_1 and L_2: the combi-

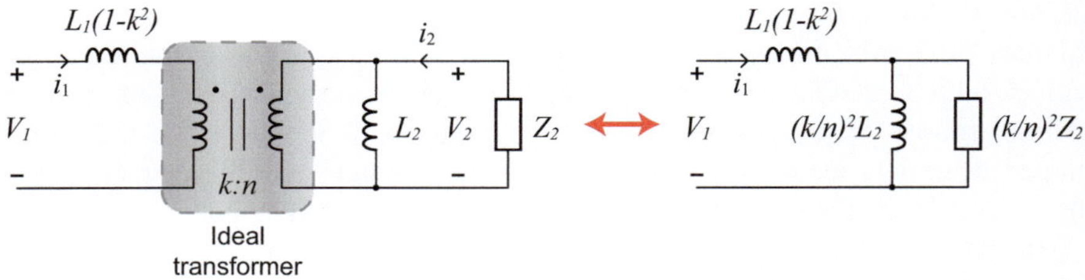

Fig. 4 Circuit model of a loosely-coupled transformer, and its load-reflected equivalent

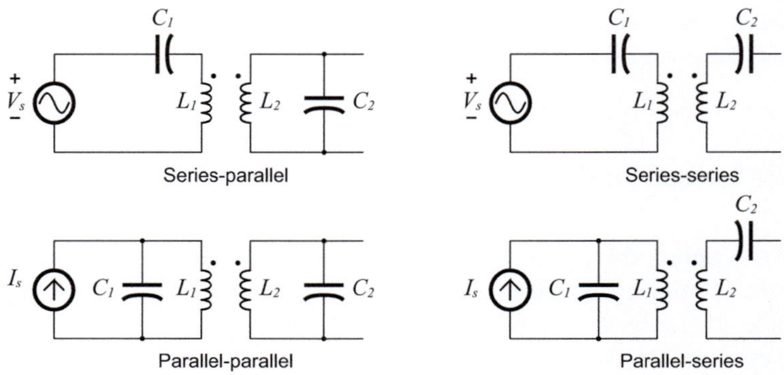

Fig. 5 The four basic options for resonant tuning of inductively coupled coils

nations of series or parallel capacitors on both the primary and secondary coils, as illustrated in Fig. 5. Series tuning on the primary is typically used to decrease loading effects of voltage-based sources, while parallel tuning is equivalently used for current-based sources. Since most drivers or power amplifiers effectively operate as voltage sources, series tuning in the primary is almost universally employed in the design of inductive power transfer links.

The choice of secondary-side tuning depends greatly on the application. It is well known that using a parallel secondary tuning capacitance induces a voltage multiplication factor, making rectifier diodes easier to turn-on.[1] This makes a series-parallel link configuration ideal for low-power (i.e., high R_L) applications, where it is difficult to generate sufficient voltage to activate non-linear rectifiers. At high powers (i.e., low R_L), the voltage multiplication factor may present voltages that go far beyond CMOS compatibility, limiting utility. Instead, a series-tuned secondary is preferred for high power applications, where increased power is generated through Q-multiplied current rather than voltage.

[1] This is somewhat confusing, since parallel *LC* resonators exhibit a *current* multiplicative factor. However, from the perspective of the dependent voltage source Mi_1 in Fig. 2, a parallel tuning capacitor ends up looking like it is in series with the inductor.

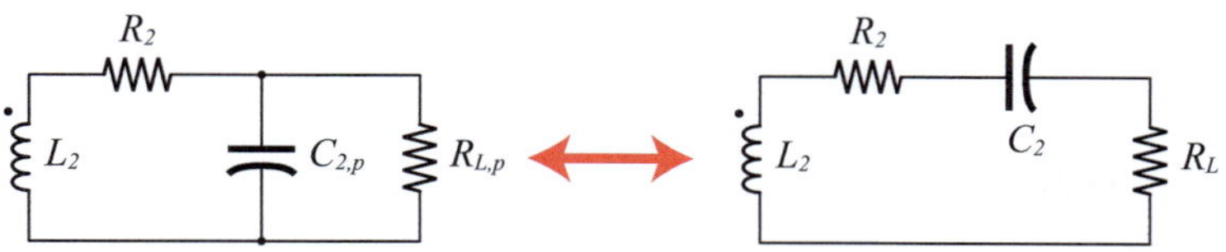

Fig. 6 Converting a parallel-tuned secondary to a series-tuned secondary

It turns out that a series-tuned secondary is the easier circuit to analyze, due to the fact that, as will described shortly, a parallel-tuned secondary does not necessarily always present a real impedance for arbitrary values of the constituent passive devices. Thus, the following analyzes a series-tuned secondary only. However, the following analysis can be easily applied to parallel-tuned secondaries by first transforming the parallel-tuned circuit to look like a series-tuned circuit as shown in Fig. 6 and through the following relationships:

$$Q_L = \omega C_{2,p} R_L \tag{5}$$

$$R_L = R_{L,p}(1 + Q_L^2) \tag{6}$$

$$C_2 = C_{2,p}\left(\frac{1 + Q_L^2}{Q_L^2}\right). \tag{7}$$

Note that the resonant frequency of a series-tuned load is given simply by:

$$\omega_o = \frac{1}{\sqrt{L_2 C_2}} = \frac{1}{\sqrt{L_1 C_1}}, \tag{8}$$

assuming the primary network is tuned to the same frequency. The case is not as simple for the parallel-tuned secondary, as the resonant frequency depends on the load:

$$\omega_o = \sqrt{\frac{1}{L_2 C_2} - \frac{1}{R_{L,p}^2 C_2^2}}. \tag{9}$$

This is in fact an additional reason to choose a series secondary for high power cases. If $R_{L,p}$ is small, then it's possible that:

$$\frac{1}{R_{L,p}^2 C_2^2} > \frac{1}{L_2 C_2}, \tag{10}$$

thereby indicating that there is no parallel resonate matching condition that will reflect as a real impedance to the primary. As a result, others have described primary circuits that sense this condition, and shift their driving frequency to compensate [6]. A series-tuned secondary avoids the need to track any resonance shifts in this case, thereby simplifying circuits, while also being inherently well suited for high power applications in the first place.

The following two subsections derive expressions for the maximum power transfer to the load under source constraints, and the maximum power transfer efficiency of an inductively coupled system. While maximum power transfer is derived first, most of the steps taken are directly related to computing the maximum efficiency, and thus the order is arbitrary.

2.2.3 Optimal Load Analysis for Maximum Power Transfer

In order to determine what conditions permit transfer of maximum power to the load (or maximum power transfer efficiency), it is necessary to obtain workable expressions for simplified, though equivalent circuits. For example, the circuit

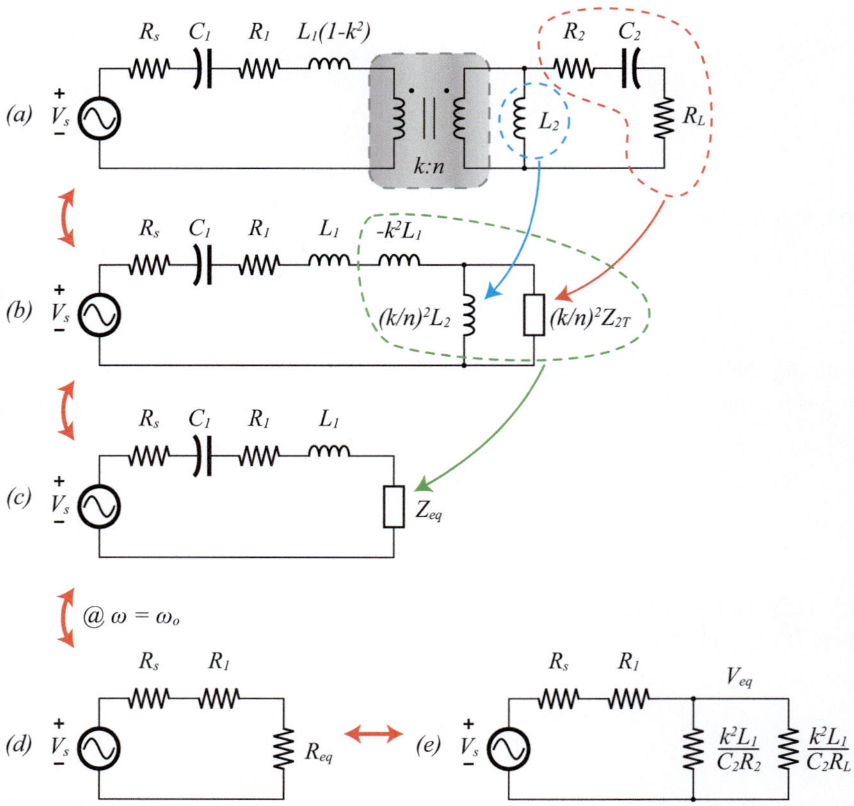

Fig. 7 Equivalent circuits for a series-turned secondary

in Fig. 7a is an accurate representation of an inductively coupled system with a series-resonant secondary. To gain design insight through simplified yet equivalent circuits, the total non-inductive secondary impedances are grouped together such that they exist in parallel to L_2 for reflection to the primary side. The total secondary impedance is given by Eq. 11 and is shown reflected to the primary side in Fig. 7b.

$$Z_{2T} = R_2 + R_L + \frac{1}{j\omega C_2} \tag{11}$$

Due to the parallel combination of an inductance and complex impedance, Fig. 7b is still too cumbersome for practical analysis. Mapping the values in Fig. 7b, c is the next step in enabling useful analysis. This can be done using equivalency mapping as follows:

$$\begin{aligned} Z_{eq} &= -j\omega k^2 L_1 + \frac{1}{\dfrac{1}{j\omega (k/n)^2 L_2} + \dfrac{1}{(k/n)^2 Z_{2T}}} \\ &= -j\omega k^2 L_1 + \frac{1}{\dfrac{1}{j\omega k^2 L_1} + \dfrac{1}{(k/n)^2 Z_{2T}}} \\ &= \frac{\omega^3 k^2 L_1 L_2 C_2 \left[\omega C_2 (R_L + R_2) + j(1 - \omega^2 L_2 C_2)\right]}{\omega^2 C_2^2 (R_L + R_2)^2 + (1 - \omega^2 L_2 C_2)^2} \end{aligned} \tag{12}$$

These results were shown in [3], and subsequently independently verified. However, Eq. 12 is complicated and not very useful for analysis. Fortunately, the expression simplifies considerably at resonance (i.e., $\omega = \omega_o = 1/\sqrt{L_2 C_2}$):

$$Z_{eq}|_{\omega_o} = \frac{k^2 L_1}{C_2 (R_L + R_2)} = R_{eq}. \tag{13}$$

What is critical to note here is that this is a purely real impedance. If the primary and secondary are set to resonate at the same frequency, the entire inductive coupling circuit looks purely resistive! The equivalent circuit under this resonant condition is shown in Fig. 7d. Note that in this series-tuned configuration, capacitors C_1 and C_2 must only be tuned with respect to their adjacent inductors, without regard for any other element in the circuit. The same cannot be said for parallel-tuned secondaries, where the tuning of C_2 must also take into account the load impedance per Eq. 9. If the load changes, as may be the case in many charging applications, either C_2 must be dynamically compensated, or the driving frequency must be modulated as described earlier.

All steps taken to this point are equally applicable to determine either maximum power or maximum efficiency conditions. The remainder of this section will now derive conditions for maximum power transfer, while Sect. 2.2.4 will discuss maximum efficiency.

The expression for R_{eq} can be split into two different components as follows:

$$\begin{aligned} R_{eq} &= \frac{k^2 L_1}{C_2(R_L + R_2)} \\ &= \frac{k^2 L_1}{C_2} \left(\frac{R_L R_2}{R_L + R_2} \right) \frac{1}{R_L R_2} \\ &= \left(\frac{k^2 L_1}{C_2 R_2} \right) // \left(\frac{k^2 L_1}{C_2 R_L} \right) \end{aligned} \tag{14}$$

A circuit with these two separated components is shown in Fig. 7e. Given that there is an isolated component proportional to the load resistor, the power consumption seen at the load can be found using the following set of equations:

$$\begin{aligned} P_{out} &= V_{eq}^2 \left(\frac{C_2 R_L}{k^2 L_1} \right) \\ V_{eq} &= V_s \left(\frac{R_{eq}}{R_s + R_1 + R_{eq}} \right) \\ &= V_s \left(\frac{k^2 L_1}{C_2(R_s + R_1)(R_L + R_2) + k^2 L_1} \right) \\ P_{out} &= V_s^2 \left(\frac{k^2 L_1 C_2 R_L}{(k^2 L_1 + C_2(R_s + R_1)(R_L + R_2))^2} \right). \end{aligned} \tag{15}$$

Solving Eq. 16 can lead to the load impedance that maximizes power transfer, given all other inductive coupling conditions.

$$\frac{\partial P_{out}}{\partial R_L} = 0. \tag{16}$$

The solution to this equation is shown here:

$$R_{L,opt} = R_2 + \frac{k^2 L_1}{C_2(R_s + R_1)} \tag{17}$$

Setting $R_s = 0$, which is a somewhat reasonable assumption given good practice power amplifier design[2], and substituting R's for Q's further simplifies the equation:

$$R_{L,opt} = \frac{\omega L_2}{Q_2} + \frac{k^2 Q_1}{\omega C_2}$$

$$= \frac{1}{Q_2}\sqrt{\frac{L_2}{C_2}} + k^2 Q_1 \sqrt{\frac{L2}{C_2}}$$

$$= \boxed{\sqrt{\frac{L_2}{C_2}}\left(\frac{1 + k^2 Q_1 Q_2}{Q_2}\right)} \tag{18}$$

Equation 18 shows that the ideal load impedance (i.e., the value of R_L that maximizes power transfer to itself) changes with the distance between coils (i.e., k). It is also dependent on the ratio of secondary reactance values and the coil quality factors. Thus, the optimal impedance depends on the coil separation distance, as well as their geometries. Figure 8a plots the optimal load impedances across all practical values of k for the particular inductive parameters indicated in the figure caption. In this example, R_L varies from a minimum of 3.6 Ω at $k \approx 0$, to a maximum of 17 kΩ at $k = 1$. Interestingly, the optimal R_L at low coupling coefficients is exactly equal to the parasitic resistance of the secondary coil, R_2, as can be inferred by Eq. 17. This makes good intuitive sense, since the state of the secondary circuit has almost no bearing on the primary circuit under low coupling conditions (i.e., its reflected impedance to the primary is negligible). Thus, a maximum power matching circuit on the secondary is the configuration that extracts the most power. At higher levels of coupling, reflected impedances are no longer negligible. Thus, matching R_L to R_2 would significantly load the primary circuit, thereby falling out of a maximum power transfer condition.

Figure 8b shows the extractable output power from the same inductive coupling setup as described in Fig. 8a, but this time plotted for varying output loads and coupling coefficients. Also shown is the theoretically maximum attainable output power and the associated load resistances for k varying from 0.001 to 0.5. At low coupling coefficients, the optimal R_L is small, as is the maximum extractable power. As k increases, the optimal R_L and associated extractable power increases. Although $R_{L,opt}$ continues to increase with k, $P_{out,max}$ sees diminishing returns, eventually plateauing beyond coupling coefficients greater than 0.2. This also matches intuition very well—at high values of k, the reflection coefficient becomes large, requiring a large R_L (due to the inverse relationship) to create an impedance that matches $R_1 + R_s$. At this point, the reflected $k^2 L_1/(C_2 R_2)$ becomes very large

[2] Alternatively, R_s can be lumped into the definition of the primary quality factor: $Q_{1,mod} = \omega L_1/(R_s + R_1)$.

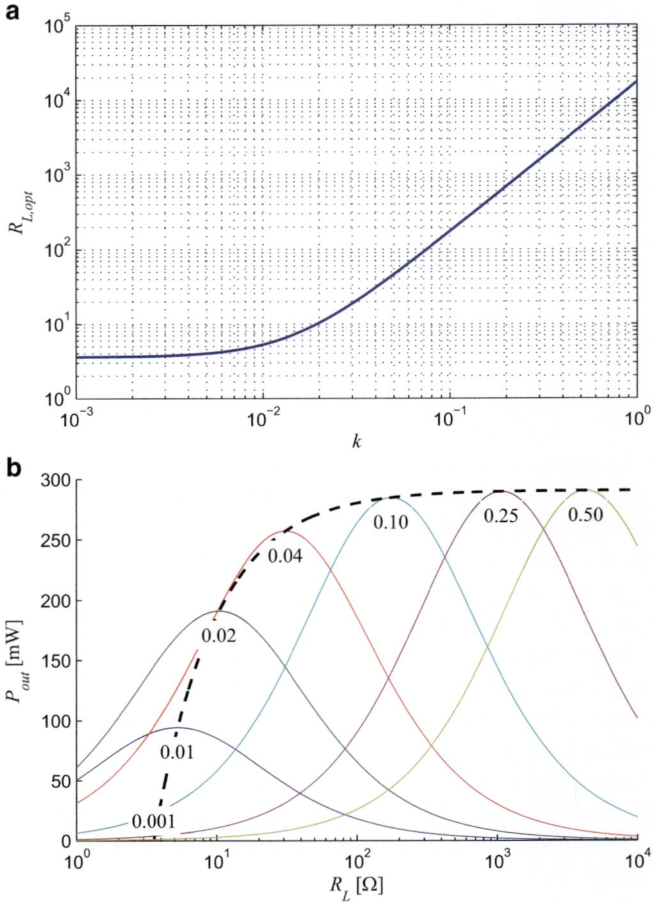

Fig. 8 Load impedances that maximize power transfer. In this example, $V_s = 3.3\ \mathrm{V}_{p-p}$, $f = 6.78\,\mathrm{MHz}$, $L_1 = 3.3\ \mu H$, $L_2 = 6.7\ \mu H$, $Q_1 = 60$, and $Q_2 = 80$. (**a**) Load impedances that maximize power transfer plotted across all practically useful coupling coefficients. (**b**) Amount of extracted output power plotted across a range of load resistances for various values of k (whose values are annotated in plain text on the figure). The *solid dashed line* represents the maximum extractable power for a given k, while also lining up with the associated optimal R_L for the given k

relative to $k^2 L_1/(C_2 R_L)$, making its effects negligible. Thus, the maximum power transfer theory can be simply applied here: $k^2 L_1/(C_2 R_L) = R_1 + R_s$, resulting in 50 % power transfer efficiency. In this example, 300 mW is thus the source-limited maximum power available to be delivered to the load.

It is important to note here that these curves are very setup-specific. For example, imagine designing an inductive link for a coupling coefficient of 0.1 and a load impedance of 170 Ω. At these conditions, 285 mW would be delivered to the load. If the coupling coefficient was increased to 0.25, perhaps because a different patient has a thinner skin thickness in a transcutaneous power transfer example, then for the same load impedance the extractable power drops by 2× to 137 mW. As a result of this sensitivity, inductive systems are typically either designed for a worst-case scenario, having on-board regulators dissipating any excess received power, or they are designed with some sort of feedback communication system in order to modulate the amount of delivered power. An example system that demonstrate modulation of power delivery by reconfiguring the secondary is discussed in Sect. 3.

2.2.4 Optimal Load Analysis for Maximum Power Transfer Efficiency

The maximum power transfer analysis developed in the preceding section can be easily leveraged to also derive optimal efficiency criteria. For example, it is certainly possible to take the P_{out} result in Eq. 15 and divide it by the input power to derive an expression for efficiency; however, this does not provide much analytical insight. Instead, total link efficiency can be broken out into two separate components: primary and secondary efficiencies. Specifically, the circuits in Fig. 7d and Fig. 7a can be used to calculate the primary and secondary efficiencies at resonance, respectively.

$$\begin{aligned}\eta_{tot} &= (\eta_1) \times (\eta_2) \\ &= \left(\frac{R_{eq}}{R_{eq} + R_s + R_1}\right) \times \left(\frac{R_L}{R_L + R_2}\right) \\ &= \ldots \text{(ignoring } R_s\text{, or lumping together with } Q_1\text{)} \ldots \\ &= \frac{k^2 Q_1 \omega C_2 R_L}{\left(\omega C_2 R_L + k^2 Q_1 + \frac{1}{Q_2}\right)\left(\omega C_2 R_L + \frac{1}{Q_2}\right)}\end{aligned} \tag{19}$$

Taking the derivative of efficiency with respect to R_L and equating to zero will yield the load impedance that maximizes power transfer efficiency:

$$\boxed{R_{L,\eta_{opt}} = \sqrt{\frac{L_2}{C_2}}\left(\frac{\sqrt{1 + k^2 Q_1 Q_2}}{Q_2}\right)} \tag{20}$$

Similar results were also derived in [2, 3], though the result in [2] is only applicable for reasonably large k. For completeness, the optimal efficiency given this result is shown here:

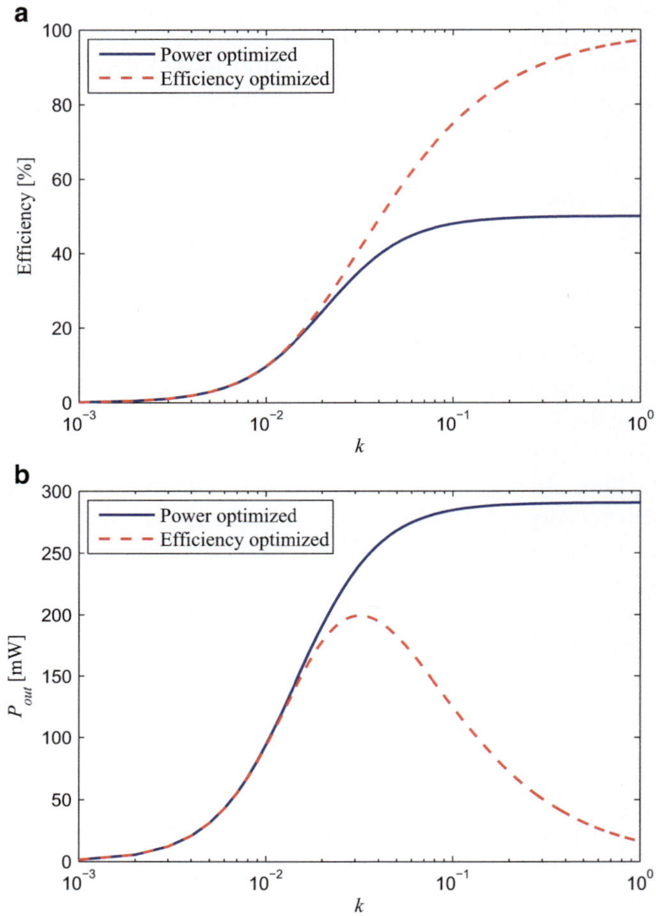

Fig. 9 Efficiency and load power delivery for a power-transfer-optimized link and an efficiency-optimized link. In this example, $V_s = 3.3\ \text{V}_{p-p}$, $f = 6.78$ MHz, $L_1 = 3.3\ \mu H$, $L_2 = 6.7\ \mu H$, $Q_1 = 60$, and $Q_2 = 80$. (**a**) Maximum achievable efficiency for all practical values of k, found by optimizing the load impedance for each value of k. (**b**) Maximum achievable load power for all practical values of k, found by optimizing the load impedance for each value of k

$$\eta_{opt} = \frac{k^2 Q_1 Q_2}{(1 + \sqrt{1 + k^2 Q_1 Q_2})^2}. \tag{21}$$

What is very interesting to note here is that the $R_{L,\eta_{opt}}$ expression is identical to the $R_{L,opt}$ expression for maximum power transfer, with the only difference being the square root of $(1 + k^2 Q_1 Q_2)$ in the numerator. It should also be mentioned that this expression is exact for both series- and parallel-tuned secondaries.

As shown in Fig. 9a, efficiency can theoretically approach 100 %, given large quality factors and high coupling. This is in contrast to links optimized for maximum power transfer, which in good agreement with theory, approach a maximum limit of 50 %. However, as shown in Fig. 9b, a highly-efficient link at high k delivers very little power compared to a link optimized for maximum power transfer. At low coupling coefficients, links designed for maximum efficiency or power transfer have the same behavior. This follows intuition nicely: at low coupling coefficients, reflected impedances are minimal, and thus, no matter how much power is consumed in the primary, it is always beneficial to match R_L to R_2 for low values of k.

2.2.5 Section Summary

This section has shown how to analyze inductively coupled links for the purposes of deriving theoretical maximal limits on efficiency and power transfer. It was shown that extracting maximum power or operating with optimal efficiency depends on the coil quality factors, the load impedance, the coupling coefficient, and the ratio of secondary reactances. Maximizing the coil quality factors improves both power transfer and efficiency, and should thus be attempted in all designs. The other parameters depend a great deal on the underlying application, and care must be taken to choose, or design around, these values. In particular, the designer should decide up-front if maximum power transfer or maximum power transfer efficiency is desired to help guide design decisions.

3 Design Example: Wireless Capacitor Charging

Since most previously published work discusses optimization of wireless power transfer links for maximum efficiency, this section will take the opportunity to instead optimize a link for maximum power transfer. This is an important condition for many charging applications—maximizing the power delivered to the load given source and coil parasitic impedances will ensure a minimal charging time. This section will specifically discuss charging an ultra-capacitor as a load, though other energy storage elements could easily be employed instead (with small caveats for battery chargers that require specific charging regimens).

3.1 Charging a Capacitor: Instantaneous Resistance

AC to DC converters employ the use of non-linear elements such as diodes to perform rectification. This makes rectifier modeling challenging, as large-signal analysis is typically required. Although solutions may exist under steady-state conditions, capacitor charging is by definition a transient event. Thus, it is very difficult to derive precise analytical expressions that provide design insight.

Fig. 10 Charging a capacitor at DC with a voltage step

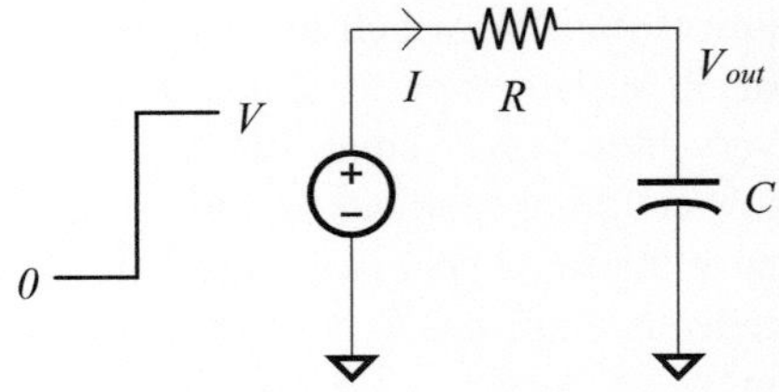

Instead, design insight can be attained by observing the behavior of analogous circuits. Consider, for example, a simple *RC* circuit in Fig. 10, where the capacitor is charged from a step voltage. If the input step occurs at time $t = 0$, the output voltage is given by:

$$V_{out}(t) = V\left(1 - e^{-t/RC}\right), \quad t \geq 0. \tag{22}$$

Similarly, the loop current is given by:

$$I(t) = \frac{Ve^{-t/RC}}{R}, \quad t \geq 0. \tag{23}$$

So at $t = 0$, $V_{out} = 0$ and $I = V/R$, while at $t = \infty$, $V_{out} = V$ and $I = 0$. The *instantaneous resistance* can then be found by dividing $V_{out}(t)$ by $I(t)$:

$$R_{C,inst}(t) = \frac{V_{out}(t)}{I(t)} = R\left(e^{t/RC} - 1\right), \quad t \geq 0. \tag{24}$$

At $t = 0$, the instantaneous resistance of the capacitor is zero; that is, the capacitor is able to sink as much current as limited by R. At $t = \infty$, the instantaneous resistance becomes infinite; that is, no current is able to enter the capacitor. Figure 11 shows the transient results of a representative example.

This situation is analogous to charging a capacitor from a rectified AC source: at $t = 0$ the capacitor is fully discharged, and can accept as much current as possible, limited only by rectifier losses. In other words, its instantaneous resistance is zero. Provided there is no resistive load in parallel, the capacitor cannot be further charged once steady-state is reached at $t = \infty$, and therefore its effective impedance is infinite.

Another way to think about this problem is by considering AC-charging a capacitor that is so large it can effectively be modeled by a DC voltage source over sufficiently short periods of time, as in Fig. 12. Then, by averaging the instantaneous voltage divided by the instantaneous current throughout a single cycle, the rectifier can be modeled with an effective resistance, R_D.[3]

For example, on average, the current flowing into the rectifier is given by:

[3]This resistance implicitly includes a conduction angle factor based on the fact that the diodes do not conduct all the time. A different model could include a resistance in series with a voltage

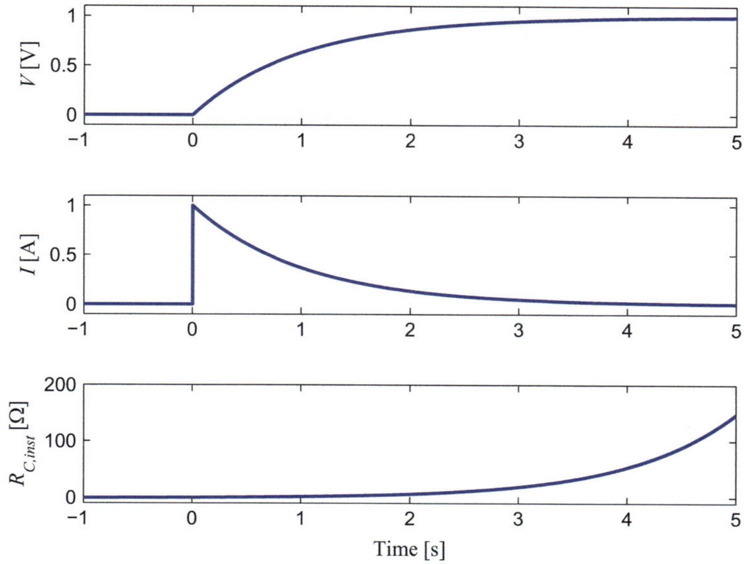

Fig. 11 Transient results of charging a 1 F capacitor from a 1 Ω source impedance to 1 V

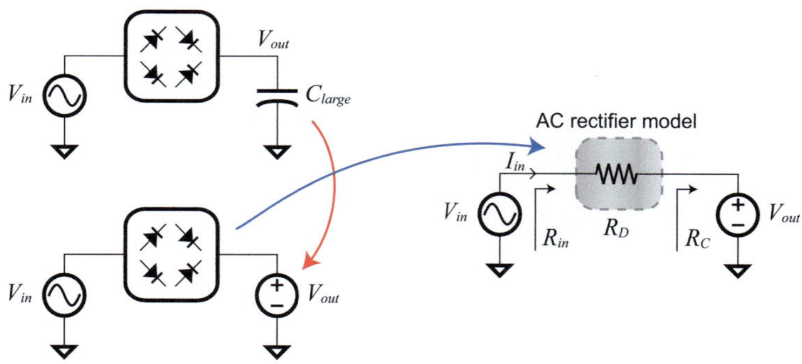

Fig. 12 Model used to calculate instantaneous charging capacitor resistance

$$I_{in} = \frac{V_{in} - V_{out}}{R_D}. \tag{25}$$

Thus, the average input resistance of the circuit is:

$$R_{in} = \frac{V_{in}}{I_{in}} = \frac{V_{in} R_D}{V_{in} - V_{out}}, \tag{26}$$

source, modeling the average diode drop. Regardless, this analysis is not meant to be quantitative or extremely precise, but is instead used to offer a more qualitative understanding.

and the instantaneous capacitor impedance can be given by:

$$R_C = R_{in} - R_D = \frac{V_{in} R_D}{V_{in} - V_{out}} - R_D = \frac{V_{out} R_D}{V_{in} - V_{out}}. \quad (27)$$

As with the step-response analysis, the instantaneous resistance of the capacitor is zero when its voltage is zero. Similarly, when its voltage is equal to V_{in}, its instantaneous resistance is infinite. In between, the capacitor's instantaneous resistance depends not only on the output voltage, but also the input voltage, and the effective diode resistance, which in itself will also likely change with voltage (or current). Further modeling of these effects are certainly possible, though for the following analysis, it is sufficient to only qualitatively understand this phenomenon.

3.1.1 The Main Issue for Inductive Coupling

As discussed in Sect. 2.2.3, given an inductive coupling setup fully defined in terms of L's, C's, R's, and k, there exists an optimal load resistance that maximizes the power transfer to said load. Unfortunately, the instantaneous resistance of a charging capacitor changes with its own voltage. As a result, there will only be a single point in time when conventional wireless energy delivery circuits are delivering maximum power to the implanted capacitor. To overcome this issue and provide maximum power for larger portions of the charging time, a proposed solution should either alter the effective load impedance in some way, or alter the setup of the inductive coupling circuit in some manner. Importantly, the ideal load impedance should be tunable not only to address changes in instantaneous capacitor resistance, but also to support robustness against coil separation distance (i.e., k).

3.2 *Modulating for Maximum Power Transfer*

There are many ways to attempt to modify the standard inductive coupling circuit in Fig. 1 to support varying instantaneous capacitor resistances and coupling coefficients. This section will suggest a few potential options that may suitable, though Sect. 3.3 will described the proposed solution that was actually implemented in this particular example.

3.2.1 DC/DC Converter as an Impedance Transformer

One of the most straightforward approaches is to attempt to change the effective ultra-capacitor resistance, at least as seen from the point of view of the inductive coupling circuit. The basic concept is shown in Fig. 13. Such an approach can be

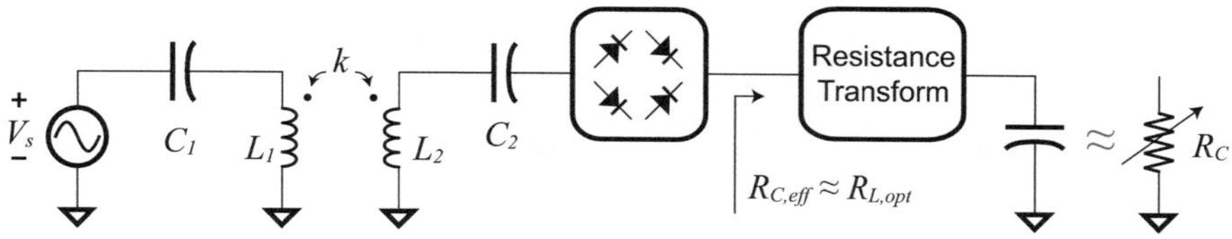

Fig. 13 Generalized circuit to modify the effective instantaneous capacitor resistance, $R_{C,eff}$ to look like the optimal load resistance, $R_{L,opt}$, for maximum power transfer

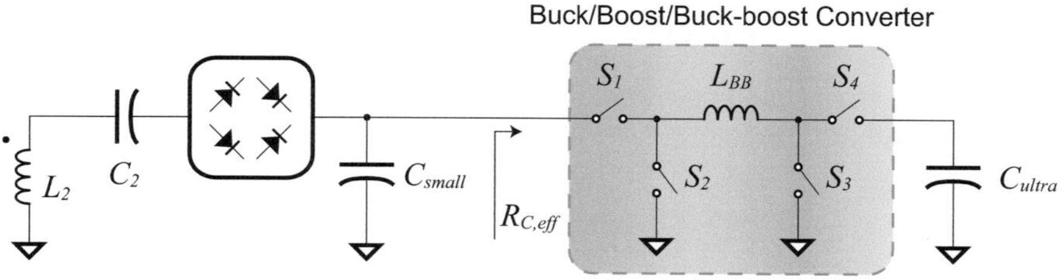

Fig. 14 Modulating $R_{C,eff}$ using a DC/DC converter

used either semi-statically, to provide robustness against varying k, or dynamically, to provide robustness against varying R_C.

One of the most straightforward approaches to modulate $R_{C,eff}$, irrespective of R_C, is to replace the resistance transform block in Fig. 13 with a DC/DC converter. The converter can be a switched capacitor converter, or a switched-inductor converter, the latter of which is illustrated in Fig. 14. In the illustrated case, it is possible to modulate the average input impedance in boost mode by simply switching the pulse width $\Phi 1$.

For efficiency and functionality reasons, it is necessary to ensure the voltage conversion ratio of the converter matches the voltages present at its input and output. For example, given a constant input voltage, the converter output voltage setting should ramp up as the capacitor charges. Depending on the way the inductive link is configured, this may require the use of buck, boost, and buck-boost methods of control.

Although this approach is directly able to set the input impedance, providing robustness against dynamically changing R_C and semi-static but potentially varying k, the circuit has high overhead. For instance, simplified linear expressions for input impedance only apply when the voltage conversion ratio is high in boost or buck modes. A more complex method of control would be necessary in buck-boost modes. Perhaps more importantly, it is exceedingly difficult to design a converter that operates efficiently over such a wide range of voltage conversion factors. Since the DC/DC converter is directly in-line with the output, any losses due to finite efficiency will directly affect charging time. On a positive note, the converter can be run in reverse to regulate load voltages after charging is complete. Others have described circuits with lower complexity to perform the same task [7].

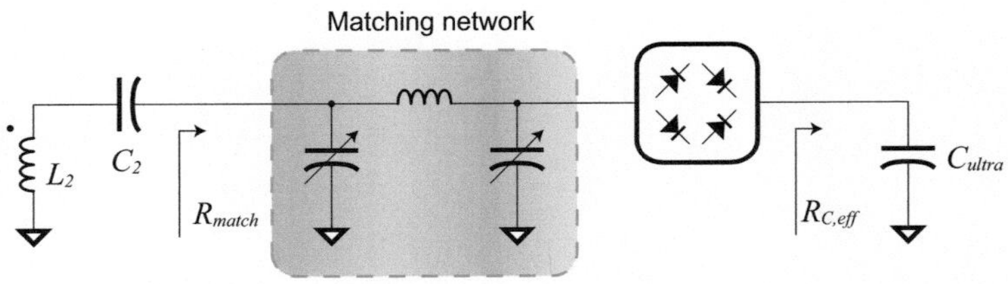

Fig. 15 Using a matching network to convert resistances

Although promising, an approach involving an in-line DC/DC converter suffers from several pitfalls and is therefore not further studied. It is certainly within the realm of possibility for future research to solve some of the outstanding issues, however.

3.2.2 Matching Network

It is well known in RF and microwave electronics that passive elements can provide impedance matching abilities at AC. Since the inductive link operates at AC prior to the rectifier, passive impedance matching circuits may be a logical choice to attempt to shape R_C. For example, the circuit in Fig. 15 uses a π network to convert the impedance seen at the input of the rectifier. The π-network capacitors can be varied in order to provide dynamic matching capabilities.

Unfortunately, passive matching networks have limited utility in this particular situation. The primary reason is that such matching network must use components that have finite quality factors, thereby limiting the achievable impedance transformation range. As shown in Fig. 8, the optimal impedance for maximum power transfer of the particular link example varies by several orders of magnitude for various k. Such a large impedance transformation range is simply not attainable with practical matching networks.

3.2.3 Resistance Compression Network

Resistance compression networks (RCNs) are a special class of matching networks that not only perform impedance matching, but also compress output impedance changes as seen at their input [8]. For example, a resistance compression network has been demonstrated to compress the input impedance of an energy recovery rectifier in an outphasing power amplifier design [9]. Figure 16 illustrates how a resistance compression network can be used for inductive coupling applications. Compression efficiencies can often approach 20×—meaning a 100:1 change in resistance at the output results in a 5:1 ratio at the input.

A resistance compression network can mitigate the effects of dynamically changing R_C in capacitor charging applications by compressing R_{RCN} to be closer to the optimal $R_{L,opt}$. However, the effective input impedance of the RCN is nominally

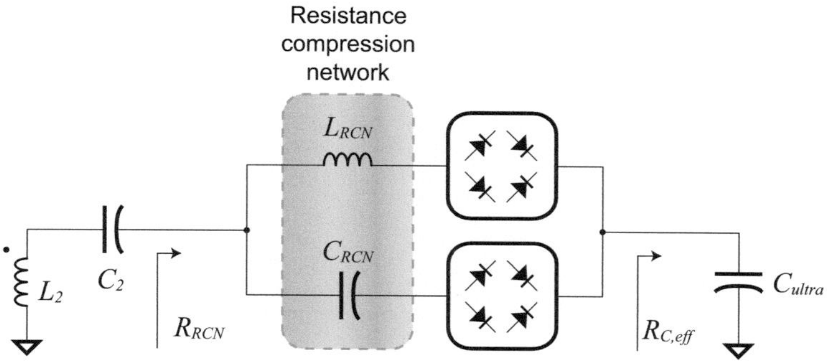

Fig. 16 Using a resistance compression network to convert and normalize resistances

centered around a single value: $R_{RCN,nom} = Z_0 = \sqrt{L_{RCN}/C_{RCN}}$. Thus, use of a resistance compression networks only benefits inductive coupling applications at a single value of k. The RCN can be made tunable to overcome this issue, but as was the case for matching networks, finite passive quality factors limit the range of achievable nominal resistances.

3.3 Proposed Impedance Modulation Solution: Multi-Tap Secondary

3.3.1 Theory

Equations 18 and 20 suggest that for given coil geometrical parameters and separation distances, there exists an optimal load impedance that either maximizes the total amount of power delivered to the load or maximizes power transfer efficiency. The solutions to modulate this impedance during charging regimens or changes in k as proposed in the preceding section have pitfalls in terms of efficiency, charging-time degradation, or practicality.

Fortunately, there is an alternate knob that can be used to adjust the system. Equations 18 and 20 show that the optimal impedance also depends on the ratio of secondary reactances: $\sqrt{L_2/C_2}$. Under resonant conditions, L_2 and C_2 are naturally related by the following relation: $\omega_o = 1/\sqrt{L_2C_2}$. Thus, the optimal impedance for maximum power transfer shown in Eq. 18 can be re-written in terms of either C_2 or L_2, the latter of which is shown here:

$$R_{L,opt} = \omega_o L_2 \left(\frac{1 + k^2 Q_1 Q_2}{Q_2} \right). \tag{28}$$

Changing the value of L_2 (and, correspondingly, C_2) can therefore change $R_{L,opt}$ for a given k. For example, Fig. 17 illustrates $R_{L,opt}$ plotted versus k for three

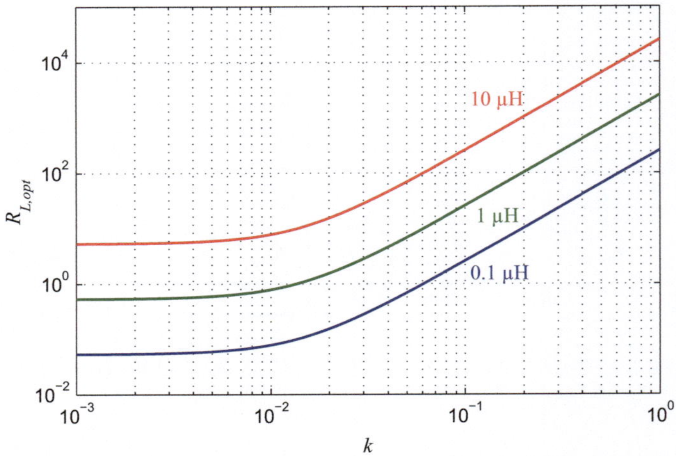

Fig. 17 Plot of $R_{L,opt}$ over all practical values of k for three different values of L_2. In this example, $V_s = 3.3\ V_{p-p}$, $f = 6.78$ MHz, $L_1 = 3.3\ \mu H$, $Q_1 = 60$, and $Q_2 = 80$. The maximum output power is identical to the power-optimized curve shown in Fig. 9b

separate values of L_2. With this design insight, a circuit that can change L_2 and C_2 can, at any point in time and for any k, change $R_{L,opt}$ to equal the charging capacitor's instantaneous resistance (R_C) at that precise instance in time. This directly solves the underlying issue—it would be possible to ensure the circuit is always delivering energy either at the maximum transfer point, or while operating with the highest possible efficiency.

Naturally, the complication here is that tunable inductors are difficult to implement. However, others have described tunable inductors for RFIC applications by using interesting bridge topologies [10] or multi-tap structures [11]. Even an inductor with three taps (i.e., three quantized inductance values) can provide substantial performance benefits. For example, Fig. 18 details analytical results of power delivered to R_L for various values of k typically found in transcutaneous charging applications. Specifically, each plot contains three curves for three separate values of L_2: 0.1 μH, 1 μH, and 10 μH in blue, green, and red, respectively. These results show that even with only three taps, a wide-range of $R_{L,opt}$ can be covered over the applicable range of k. Additionally, L_2 can be dynamically changed for a given k: as the capacitor charges, R_C increases, and the appropriate value of L_2 can be used such that instantaneously R_C is as close as possible to the selected value of $R_{L,opt}$.

3.3.2 Architecture

The architecture of the proposed multi-tap-secondary inductive coupling system for rapid wireless capacitor charging is shown in Fig. 19. The secondary coil is designed

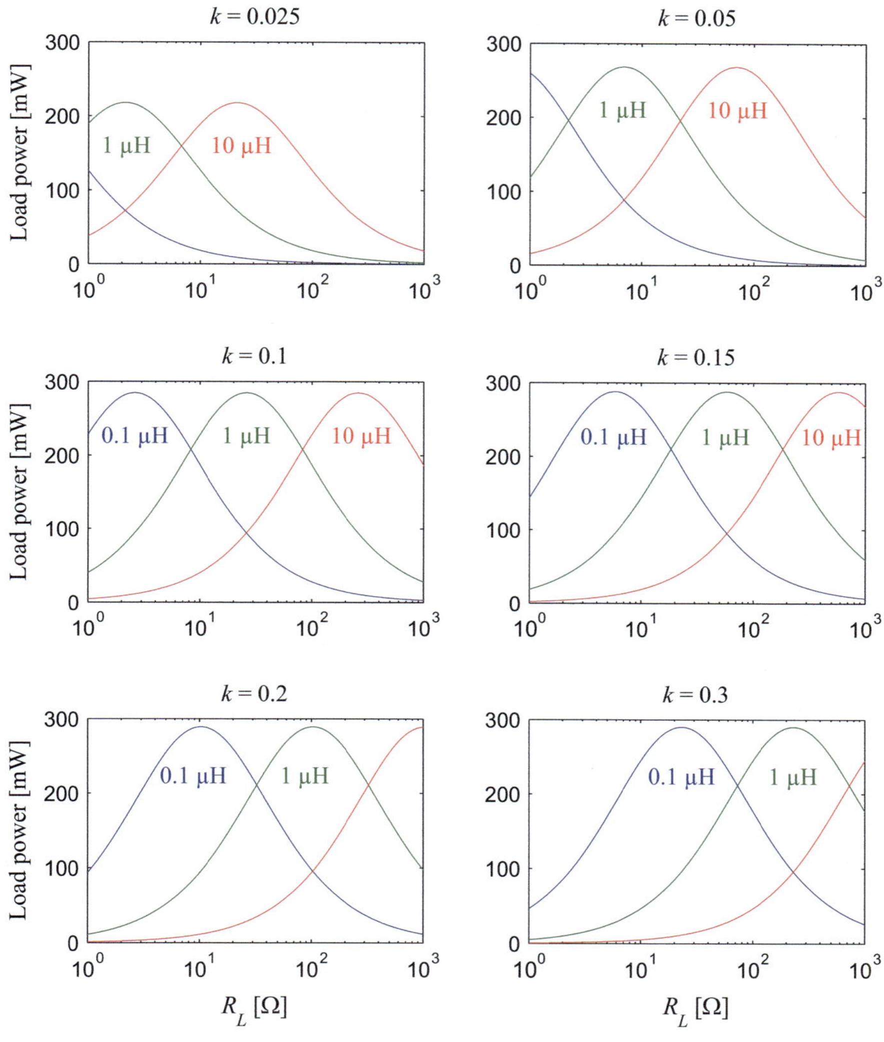

Fig. 18 Power delivered to R_L across various values of R_L and a few select values of k. In these plots, the *blue*, *green*, and *red curves* correspond to secondary inductances of 0.1 μH, 1 μH, and 10 μH, respectively. Additionally, $V_s = 3.3\ \mathrm{V}_{p-p}$, $f = 6.78$ MHz, $L_1 = 3.3\ \mu H$, $Q_1 = 60$, and $Q_2 = 80$

as a single large coil with inductance L_{23}. Two smaller inductances, L_{22} and L_{21}, are created by tapping into fewer turns of the coil. Each tap is allocated a single series capacitor used to resonate with the effective inductance seen at the output of the tap. The output of the series capacitors are connected to series switches, used to select a single tap configuration at a time. For an initial proof-of-concept demonstration, this architecture was implemented as a discrete prototype.

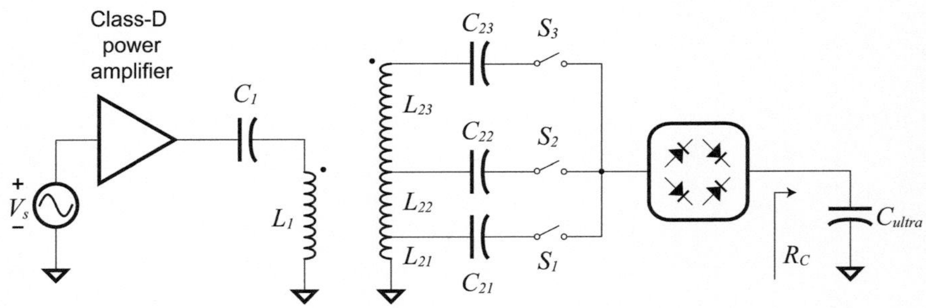

Fig. 19 Architecture of the proposed multi-tap-secondary rapid wireless capacitor charger

Design of the series switches is challenging. To have minimal effect on the inductor quality factors, the switch should be designed with as low on-resistance as possible. However, low-impedance switches typically have large associated parasitic capacitances. This presents a classic engineering trade-off: should resources be allocated to minimizing switch on-resistance or parasitic capacitance? Additionally complicating the matter is the need for high voltage-blocking capabilities. To appreciate this issue, consider the following scenario: switch S_1 is on, and switches S_2 and S_3 are off. Inductor L_{21} is resonating with capacitor C_{21}, and current is flowing through the rectifier, thereby charging C_{ultra}. Due to series resonance, the voltage at the node connecting L_{21} and C_{21} is Q multiplied. Since switches S_2 and S_3 are off, current in those branches are zero. As a result, the voltages at the inputs to S_2 and S_3 are at the same Q-multiplied level at the node connecting L_{21} and C_{21}. Since intrinsic Qs can be high—upwards of 100 for example—this appears to be a major issue. However, actual realized Qs are always much lower due to the loading effects of the rectifier and R_C. That being said, it is still important to maximize switch voltage-blocking capabilities for reliable operation.

As discussed in more detail in the following subsection, the discrete prototype design employs the use of a secondary inductor that is 25 mm in diameter. At this size, a total of 20 turns achieves an approximate inductance of less than 10 μH. At an operational frequency of 6.78 MHz, this results in a minimum series tuning capacitance of approximately 50 pF. To avoid disrupting resonant conditions, a switch should be selected such that its parasitic capacitance resides below this value. Thus, a Panasonic AQY221R2V solid-state optoelectronic relay was chosen as the switching element for this discrete prototype. Its specifications meet the requirement of this application: its on-resistance is 0.75 Ω, its parasitic capacitance is 12.5 pF, and it can safely block 40 V. When turned on, the relay requires approximately 5 mW to operate.

Many inductively-coupled applications employ the use of class-E power amplifiers on the primary circuit for high efficiency [12]. However, a class-E amplifier requires very precise knowledge of the load impedance in order to operate properly (and therefore at high efficiency) [13]. As previously discussed, inductive coupling systems operating with varying k and R_C conditions can present wildly varying impedances as seen by the primary, making the design of uncompensated class-E amplifiers impractical for these cases. To combat this issue, others have described interesting control loops to provide robustness against load variations that offer promising potential [14]. On the other hand, class-D amplifiers can operate reasonably efficiently without significant regard for load impedances. Consequently, such an amplifier was chosen for this initial proof-of-concept. Specifically, the amplifier was implemented as an inverter structure with NDS351AN and FDN360P NMOS and PMOS transistors, respectively.

The rectifier is composed of four Panasonic DB2S205 Schottky diodes in a bridge configuration.

3.3.3 Coil Design

The primary and secondary coils in the discrete prototype are both designed as printed inductors on an FR-4 substrate circuit board. The primary coil is an $N = 8$ turn design, while the secondary coil is an $N = 18$ turn design with an additional two taps after turns 2 and 5. Both coils are printed using 2-oz copper with a trace width/spacing of 0.2 mm. In terms of size, the primary coil is designed with a diameter of 30 mm, while the secondary coil is designed to be slightly smaller at 25 mm for robustness to mis-alignment. These types of sizes are often found in existing cochlear stimulation devices.

Electromagnetic simulations were performed using Mentor Graphics IE3D to extract inductance and mutual coupling information. A summary of parameters are shown in Table 1.

The quality factor is simulated to be higher for inductors with larger turns ratios. This matches theory well, as inductance increases with the square of the number of turns, while resistance increases only linearly with the number of turns [2]. Figure 20 shows the coupling coefficient of the structure, simulated across several different coil separation distances. Interestingly, the coupling coefficient is greater for the larger secondary inductances. This is purely due to the non-overlapping trace geometry occupying a larger area for a larger number of turns. Since the skin effect for such trace thickness is negligible, simulation results found that 2-oz copper increased the quality factor of the coils by 2× over 1-oz copper, as expected. A photograph of the manufactured secondary coil is shown in Fig. 21.

Table 1 Simulated coil parameters

	L_1	L_{21}	L_{22}	L_{23}
N	8	2	5	18
L [μH]	3.31	0.28	1.25	6.73
R_{series} [Ω]	2.02	0.47	1.10	3.07
Q	70	26	48	93
C_{tune} [pF]	167	1954	443	82

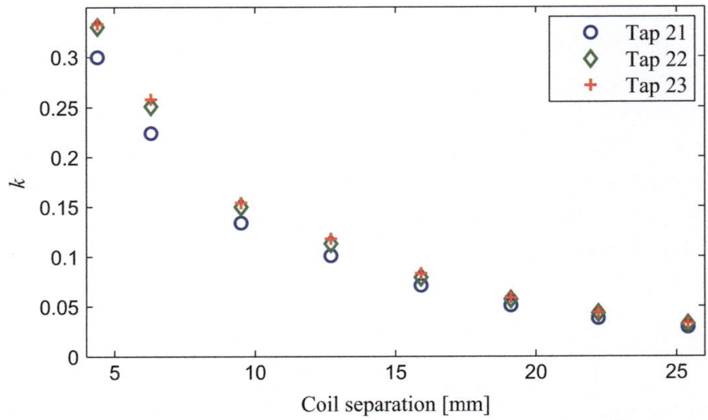

Fig. 20 Simulated coupling coefficients for various coil separation distances

Fig. 21 Photograph of the secondary three-tap coil

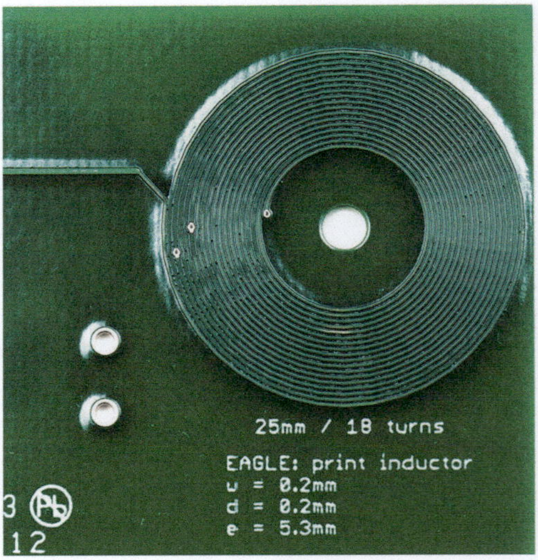

To set more realistic expectations, the results originally shown in Fig. 18 are re-calculated using the actual realized inductance values and quality factors. In addition, a 0.9 Ω source resistance was included in the calculation of Q_1, while a 0.75 Ω switch resistance and 3 Ω equivalent rectifier resistance was included in the calculation of Q_2. The results of these analytical calculations are shown in Fig. 22. At low values of R_L, I^2R losses in the secondary parasitics consume a large fraction of the total available power to the load, thereby limiting the effectiveness of this technique. However, this effect is most pronounced at low k with low secondary inductances; the small amount of passive gain from the inductive turns ratio limit the usefulness of this configuration anyway. Thus, at low-k, it is almost always beneficial to use a high value of L_2, resulting in a higher $R_{L,opt}$ and therefore a lower proportion of I^2R losses relative to delivered load power. On the other hand,

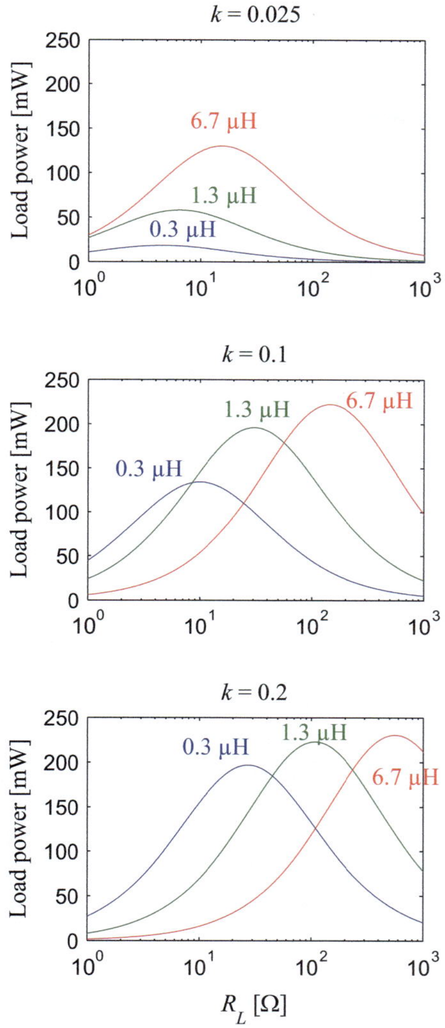

Fig. 22 Power delivered to R_L across various values of R_L and a few select values of k. In these plots, the *blue*, *green*, and *red curves* correspond to secondary inductances of 0.28 μH, 1.25 μH, and 6.73 μH, respectively. Additionally, $V_s = 3.3\ V_{p-p}$, $f = 6.78\,\text{MHz}$, $L_1 = 3.3\ \mu H$. $Q_{1,eff} = 48$, $Q_{21,eff} = 2.8$, $Q_{22,eff} = 11$, and $Q_{23,eff} = 42$

this technique permits much higher power delivery at high-k configurations over a wide range of load impedances.

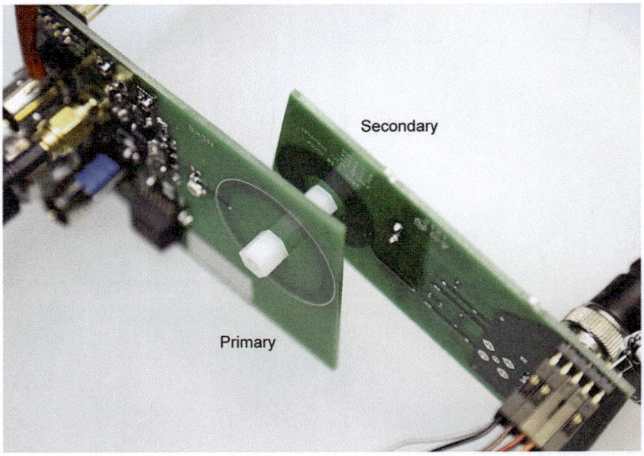

Fig. 23 Photograph of the testing setup for the rapid wireless capacitor charger

3.3.4 Discrete Prototype Measurement Results

The discrete prototype was tested in a regular electronics lab environment using nylon board-spacers of various lengths to separate the primary and secondary coils. A photograph of the testing setup is shown in Fig. 23 [1]. Voltage and power monitoring are simultaneously performed using a Keithley 2400 sourcemeter at the output of the rectifier.

Figure 24 demonstrates the benefits of using a multi-tapped secondary coil. Across various distances and output capacitor voltages, it is clear that not a single tap configuration offers the best performance. At close coil separation distances (i.e., high-k), inductor L_{21} delivers the most power to the load, while inductor L_{23} delivers the most power to the load at long distances (low-k). At intermediate distances, dynamically switching between secondary coil configurations as the output capacitor voltage increases can achieve superior power delivery results.

Also shown in Fig. 24 are measurement results from a board that did not include the series switches; instead, the output of the second tap (i.e., L_{22} and C_{22}) was directly connected to the rectifier. Since the overall extractable power is not significantly higher in this configuration, these measurement results demonstrate that a multi-tap secondary is still beneficial, even when taking into account series switch losses and activation power consumption.

Since it is difficult to analytically model a rectifier feeding a DC voltage source, the x-axis shown in Fig. 24 is converted to resistance values by dividing the output capacitor voltage by its incoming current. These results are then compared in Fig. 25 to analytical predictions using the equations derived in Section 2.2.3, under the same conditions as stated in Fig. 22. Measured results match the theoretically predicted behavior very closely for all tap configurations. Note that the measured output resistance is limited under certain tap and distance configurations due to output capacitor voltage limits.

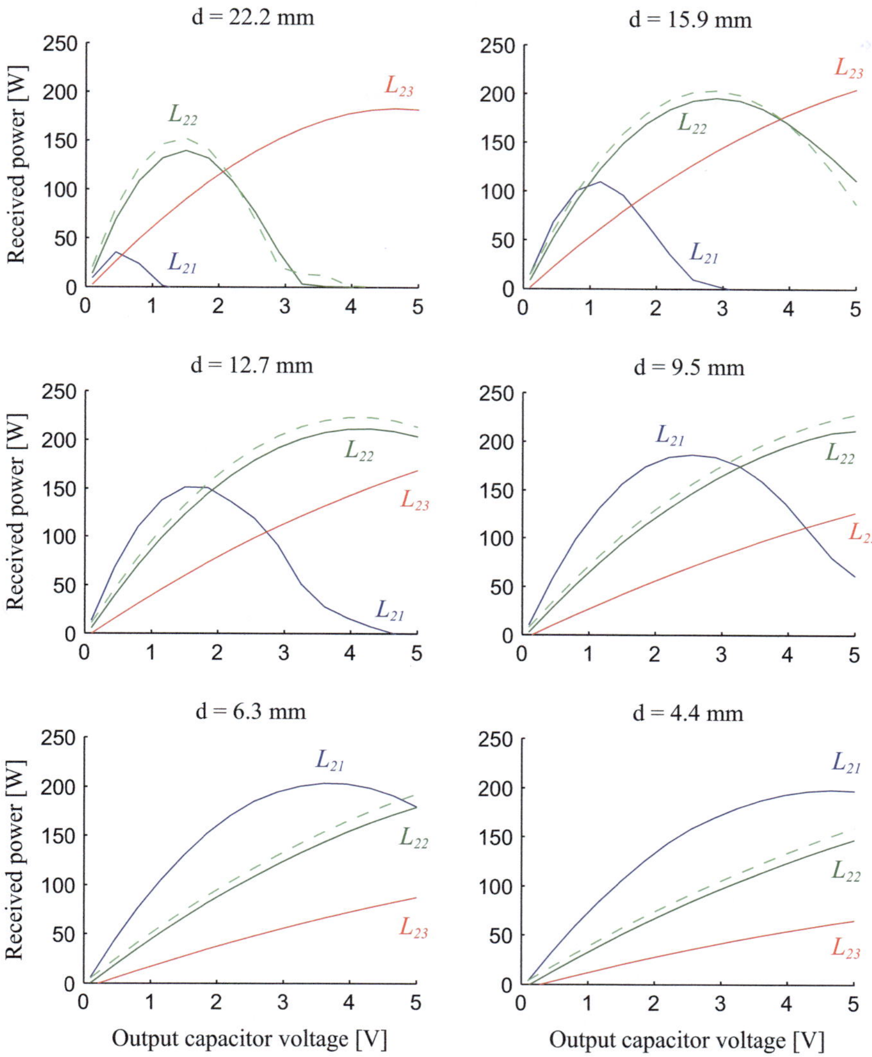

Fig. 24 Power delivered to a Keithley 2400 sourcemeter set at fixed voltage values for a few select distances. In these plots, the overlaid *solid blue*, *green*, and *red curves* correspond to inductors L_{21}, L_{22}, and L_{23}, respectively. The *light dashed green line* corresponds to a fixed secondary tap configuration (i.e., without any series switches). Here, $V_s = 3.3\ \mathrm{V}_{p-p}$ and $f = 6.78\,\mathrm{MHz}$

The discrete prototype was also used to demonstrate improved capacitor charging time and range. Figure 26 shows measured transient results of charging a 2.5 F ultra-capacitor to 5 V for two separate coil separation distances. The colored thin lines correspond to static individual tap configurations, while the thick black line

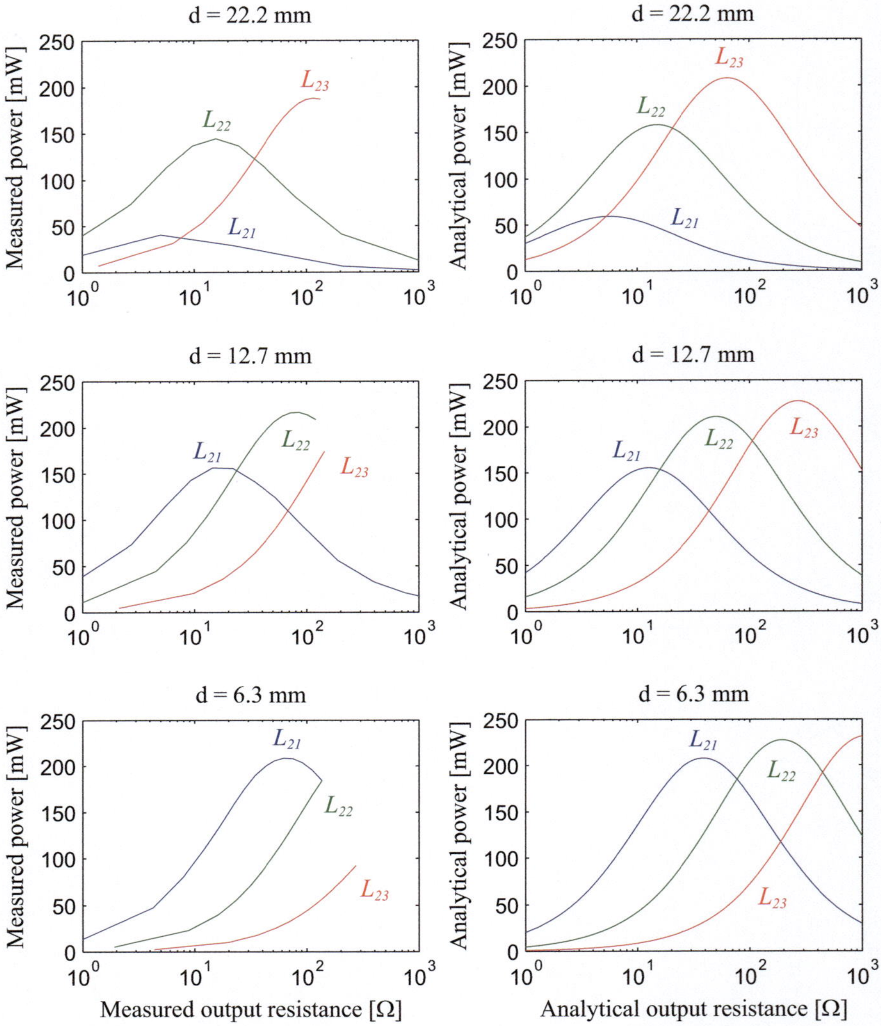

Fig. 25 Measured power delivery (left) compared to analytically predicted power delivery (right) under the same conditions as in Figs. 22 and 24

corresponds to dynamic switching between taps.[4] At a distance of 4.4 mm, inductor L_{21} offers the same level of performance as the dynamic configuration, both achieving a charging time that is 3.7× faster than using inductor L_{23}. At a distance

[4]A Keithley 2400 sourcemeter was used to measure the power of each tap at output intervals of 0.5 V to dynamically determine the optimal configuration.

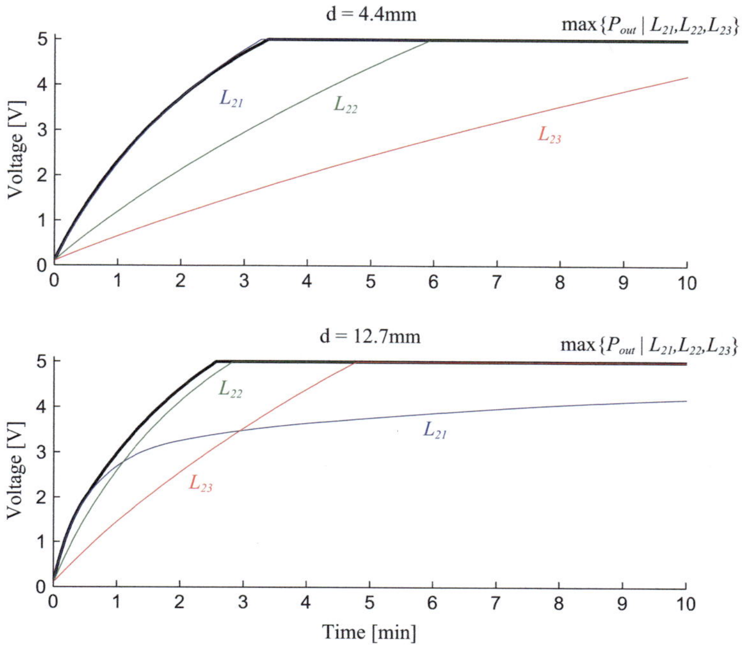

Fig. 26 Transient measurement of charging a 2.5 F ultra-capacitor to 5 V. The *thin blue*, *green*, and *red lines* corresponds to exclusive use of inductors L_{21}, L_{22}, and L_{23}, respectively, while the *thick black line* corresponds dynamic switching between them

of 12.7 mm, however, inductor L_{21} requires an estimated 26 min to reach 5 V, which is 10.3× slower than the dynamic configuration.

Figure 27 summarizes charging time results plotted versus coil separation distance. As was observed from Fig. 24, inductor L_{21} provides the best performance at small coil separations distances, inductor L_{22} at medium distances, and inductor L_{23} are long distances. Dynamically switching between taps offers the fastest charging time at essentially all distances, while also expanding the operational range by up to 2.5×.

4 Conclusions

Wireless power transfer links are becoming increasingly important for consumer, industrial, and medical electronic devices. There are two principal applications for wireless power transfer that require different optimization criteria: continuous power deliver (e.g., cochlear implant), and periodic charging (e.g., cellular phone). In the former case, optimizing power transfer efficiency is a metric of great

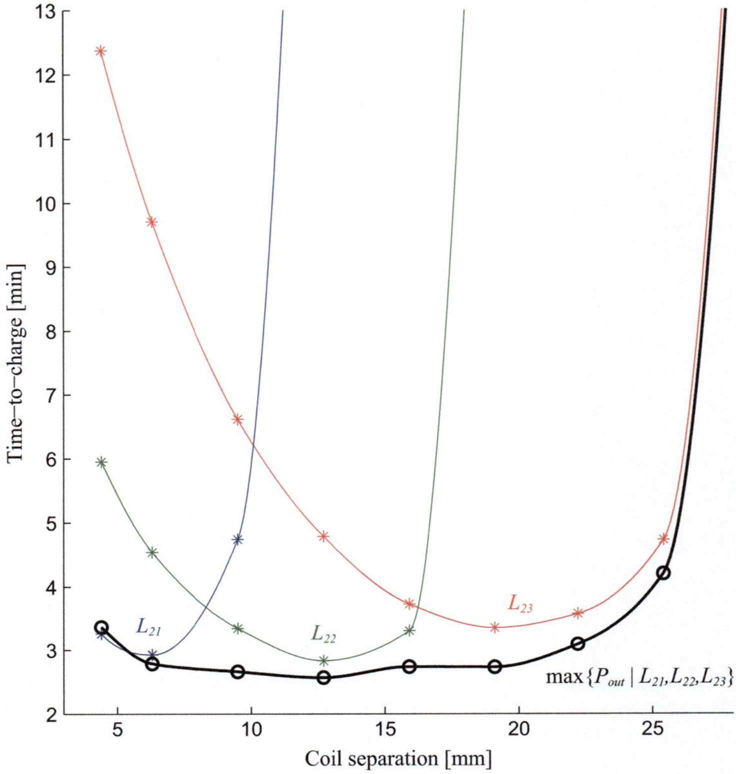

Fig. 27 Charging time for a 2.5 F, 5 V ultra-capacitor. The *thin blue*, *green*, and *red dashed lines* corresponds to exclusive use of inductors L_{21}, L_{22}, and L_{23}, respectively, while the *thick black line* corresponds dynamic switching between them

importance, while in the latter case, minimizing charging time by maximizing power transfer is important.

This chapter has developed analytical circuit theory used to predict the value of load impedance that maximizes either power transfer or power transfer efficiency in a near-field inductive link. Engineers can then use this analytical framework to help optimize their designs given application requirements.

Since most literature in wireless power transfer focuses on optimizing efficiency, this chapter took the opportunity to discuss a wireless charging application that required maximization of delivered power under source and finite coil quality factor constraints. The main challenge was ensuring a near-optimal link impedance for maximum power transfer under wildly variable load impedances (inherent in a charging ultra-capacitor) and coupling coefficients. With insight derived from the equations developed in this chapter, a multi-tap secondary circuit was presented in order to alter the optimal load condition. A rapid wireless capacitor charging system was designed using the proposed techniques, demonstrating charging times that are up to 3.7× faster and operate over a 2.5× larger coil separation distance than conventional approaches.

References

1. P.P. Mercier, A.P. Chandrakasan, Rapid wireless capacitor charging using a multi-tapped inductively-coupled secondary coil. IEEE Trans. Circuits Syst. I **60**(9), 2263–2272 (2013)
2. R. Sarpeshkar, *Ultra Low Power Bioelectronics: Fundamental, Biomedical Applications, and Bio-inspired Systems* (Cambridge University Press, Cambridge, 2010)
3. K. Van Schuylenbergh, R. Puers, *Inductive Powering: Basic Theory and Application to Biomedical Systems* (Springer, Dordrecht, 2009)
4. A. Kurs, A. Karalis, R. Moffatt, J. Joannopoulos, P. Fisher, M. Soljačić, Wireless power transfer via strongly coupled magnetic resonances. Science **317**(5834), 83–86 (2007)
5. M. Kiani, M. Ghovanloo, The circuit theory behind coupled-mode magnetic resonance-based wireless power transmission. IEEE Trans. Circuits Syst. I **59**(9), 2065–2074 (2012)
6. K. Van Schuylenbergh, R. Puers, Self-tuning inductive powering for implantable telemetric monitoring systems. Sens. Actuators A Phys. **52**(1), 1–7 (1996)
7. W. Sanchez, C. Sodini, J. Dawson, An energy management IC for bio-implants using ultracapacitors for energy storage, in Proc. IEEE Symp. VLSI Circuits (2010), pp. 63–64
8. Y. Han, O. Leitermann, D. Jackson, J. Rivas, D. Perreault, Resistance compression networks for radio-frequency power conversion. IEEE Trans. Power Electron. **22**(1), 41–53 (2007)
9. P. Godoy, D. Perreault, J. Dawson, Outphasing energy recovery amplifier with resistance compression for improved efficiency. IEEE Trans. Microw. Theory Tech. **57**(12), 2895–2906 (2009)
10. A. Tanabe, K. Hijioka, H. Nagase, Y. Hayashi, A novel variable inductor using a bridge circuit and its application to a 5–20 GHz Tunable LC-VCO. IEEE J. Solid-State Circuits **46**(4), 883–893 (2011)
11. C. Fu, C. Ko, C. Kuo, Y. Juang, A 2.4–5.4-GHz wide tuning-range CMOS reconfigurable low-noise amplifier. IEEE Trans. Microw. Theory Tech. **56**(12), 2754–2763 (2008)
12. M. Paernel, High-efficiency transmission for medical implants. IEEE Solid-State Circuits Mag. **3**(1), 47–59 (2011)
13. T.H. Lee, *The Design of CMOS Radio-Frequency Integrated Circuits*, 2nd edn (Cambridge University Press, Cambridge, 2004)
14. M. Baker, R. Sarpeshkar, Feedback analysis and design of RF power links for low-power bionic systems. IEEE Trans. Biomed. Ciruits Syst. **1**(1), 28–38 (2007)

Energy Harvesting Opportunities for Low-Power Radios

Saurav Bandyopadhyay and Yogesh K. Ramadass

Abstract Advancements in integrated circuit design have made it possible to have ultra-low-power wireless sensor nodes for health monitoring, smart buildings, industrial automation and for the automotive industry. These low power circuits generally have an Analog Front End (AFE) to sense weak signals, ADCs to digitize the sensed signals, microcontrollers for processing and low power radios for transmitting the low data rate information to a base station. These wireless sensors may be deployed in remote locations or may be in large numbers making battery replacement challenging. By harvesting the ambient energy, it is possible to power these systems and achieve near perpetual operation making battery replacement unnecessary. However, in order for these systems to extract energy from harvesters, these circuits need to not only be ultra-low-power themselves but they also need to ensure maximum available power is always extracted from the energy harvester. In this chapter, the basics of energy harvesting systems will be discussed with a focus on low power design techniques, maximum power extraction and battery management in these systems.

Keywords Energy harvesting • Wireless sensors • IoT • Photovoltaic • Thermoelectric • Piezoelectric • BQ25570 • Battery charger

1 Introduction

Sophisticated battery operated electronic systems and self-powered devices have found diverse applications in autonomous wireless sensors for industrial automation, wearable or implantable biomedical sensors or sensors in smart buildings. In all of these systems, long battery or operational lifetimes are highly important. Energy efficiency of integrated circuits within these systems plays a major factor

S. Bandyopadhyay (✉)
Texas Instruments, Dallas, TX, USA
e-mail: s-bandyopadhyay@ti.com

Y.K. Ramadass
Texas Instruments, Santa Clara, CA, USA
e-mail: yogesh-ramadass@ti.com

P.P. Mercier, A.P. Chandrakasan (eds.), *Ultra-Low-Power Short-Range Radios*, Integrated Circuits and Systems, DOI 10.1007/978-3-319-14714-7_12

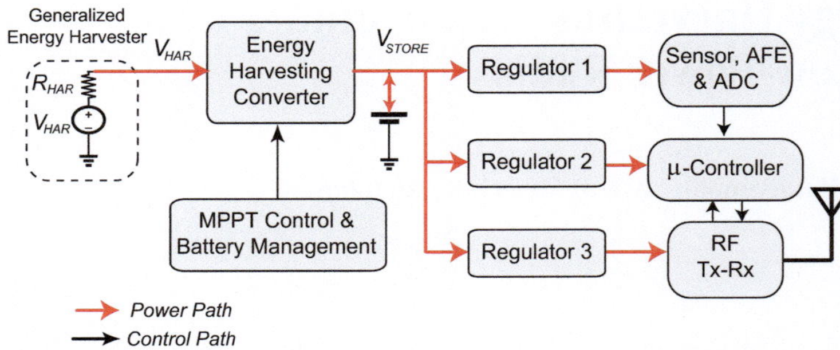

Fig. 1 Block diagram of a typical wireless sensor with sensor-AFE, microcontroller, RF Tx–Rx along with energy harvesting system and storage

Table 1 Typical numbers for power consumption of various circuit blocks

Circuit block	Power consumption	Comments
AFE[1]	2.9 μW	0.6 V operation
ADC[2]	0.6 nW	10b SAR with LSB-first successive approximation, 0.5–1 V operation
μController[3]	2.72 μW	0.5 V, 16b μController, 128 kb SRAM, 100 kHz
RF Tx–Rx[4, 5]	Tx-440 pJ/bit and Rx-180 pJ/bit	0.7 V operation, 2.4 GHz, 1 Mbps with FBAR

in determining the size, weight, cost and lifetime of portable electronics. Highly aggressive low-power circuit design and efficient power delivery is essential to meet the power constraints in such systems.

Typical wireless sensor node found in most low power systems consist of a sensor, AFE, microcontroller to locally process the data and a radio to power the information out to receivers as shown in Fig. 1. The entire system is powered by an energy subsystem. Table 1 shows the typical power numbers of these subsystems. They have been steadily decreasing through low voltage operation. The overall system is duty cycled. With sensor node electronics power decreasing so much, a new class of self powered systems are made possible. These self-powered systems all share the same qualities of low data rate, low duty cycle and ultimately low power consumption.

Through the course of this chapter we will look at the energy harvesting sources and their characteristics, along with energy storage options used in these systems. We will then discuss the architecture of an energy harvesting system with details of the energy management circuits.

Table 2 Harvestable power from different energy harvesters [9]

Energy harvesters	Power generated per unit area
Photovoltaic	$10\,\mu W/cm^2$ indoor, $10\,mW/cm^2$ outdoor
Thermoelectric	$30\,\mu W/cm^2$ body worn, $1–10\,mW/cm^2$ industrial
Piezoelectric	$4\,\mu W/cm^2$ body worn, $100\,\mu W/cm^2$ industrial

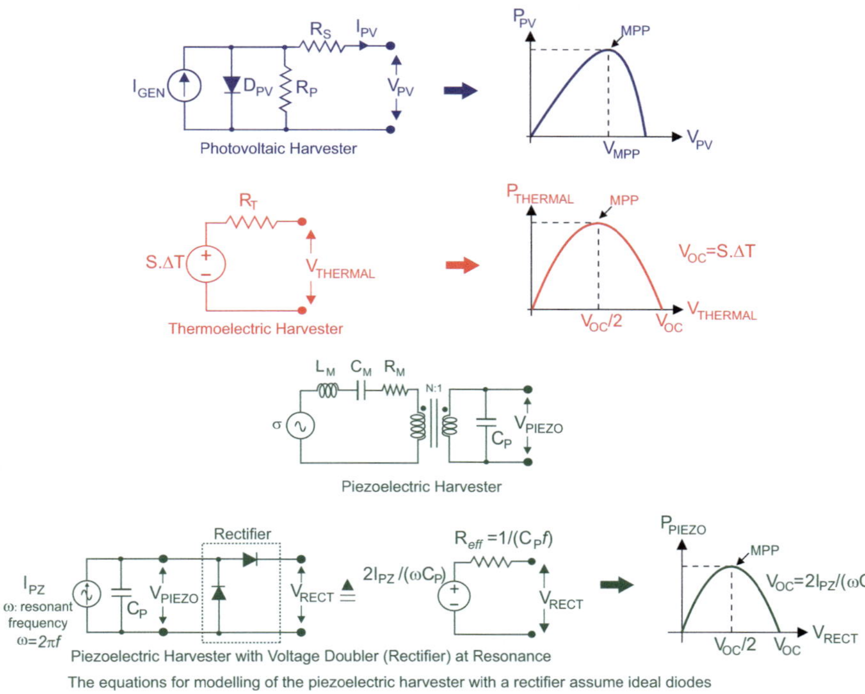

Fig. 2 Harvester electrical models

2 Energy Harvesting Sources

Energy harvesters like photovoltaic cells[6], thermoelectric generators[7] and piezoelectric harvesters[8] have become popular due to their suitable power density for the applications alluded to earlier (Table 2).

Most energy harvesters can be modeled as either a voltage or current source and a circuit element limiting the maximum extractable power from the harvester. Figure 2 shows representative electrical models along with the plot of power versus closed circuit voltage for individual harvesters. For photovoltaic cells, the current source I_{GEN} represents a light intensity dependent energy source. For a given light intensity, the amount of extractable power depends on the cell voltage (V_{PV}). For low cell voltages, the current supplied by the photovoltaic cell (I_{PV}) is close to I_{GEN}, but the output power is low because of low V_{PV}. For higher cell voltages, the diode parallel to I_{GEN} starts to shunt part of I_{GEN} to ground, thereby reducing I_{PV}. Therefore, there is an optimum voltage for power extraction from the photovoltaic cells. This is the maximum power point voltage (V_{MPP}). Depending on the type of cell, the series and shunt resistors (Rs and Rp in Fig. 2) may also affect the V_{MPP}. It must be noted that this point varies with the light intensity.

In the case of thermoelectric harvesters [7], as shown in Fig. 2, the voltage being generated is given by S.ΔT, where S is the Seebeck coefficient of the thermoelectric material and ΔT is the temperature differential applied across the thermoelectric material. In this case, the internal resistance (R_T) in series with the voltage source limits the maximum extractable power.

Piezoelectric harvesters can be modeled as an ac circuit with a transformer that converts the mechanical vibrations into electrical energy. Before connecting the harvester to a DC–DC converter, the voltage output must be rectified. There are many different techniques to rectify the voltage (full bridge rectifiers, voltage doublers, bias-flip rectifier [10], etc.). It is possible to show by an analysis similar to the one in [11], that the ac model of the piezoelectric harvester at resonance, along with the rectifier, is equivalent to a dc voltage source in series with a resistor as shown in Fig. 2. Depending on the type of rectifier, this resistor may be variable or fixed. However, the important point to note is that the harvester would have a maximum power plot similar to the ones shown in Fig. 2.

3 Energy Storage

Until this point, we have discussed some of the basics of the energy sources. In this section, we will look at the energy storage elements that are generally used in energy harvesting systems. Let us first see why we need to store the energy instead of directly powering the load circuits in such systems. Most of the electronic systems do not have a constant power draw. In a sensor node, the sensor wakes up periodically, takes a measurement, transmits the data to a base station and then goes to sleep. Figure 3 has been used to conceptually describe the sensor's energy storage requirement. The power drawn by the load circuits looks like that shown in the red curve. On the other hand, the power obtained from a harvesting element looks like that shown in green. Under this case, we need the storage element to act as a buffer accumulating the power obtained from the input and providing the peak currents required by the load. The storage element acts as an energy buffer. Besides, there may be large periods of time where no power is available from the harvesting element. An example is using solar cell systems in dark surroundings. During these periods of time, the storage element acts as the only source of energy in the systems.

The key requirements for the storage element are that it needs to hold onto the charge accumulated effectively, have a large enough capacity in a small form factor and provide peak output currents with minimal droop. Table 3 shows a comparison of the commonly used energy storage elements. Conventional rechargeable batteries like Li-ion, NiMH and NiCd have a large capacity going up to 2.5 A hours though they have a large form factor. The bigger AA, AAA batteries can support high load

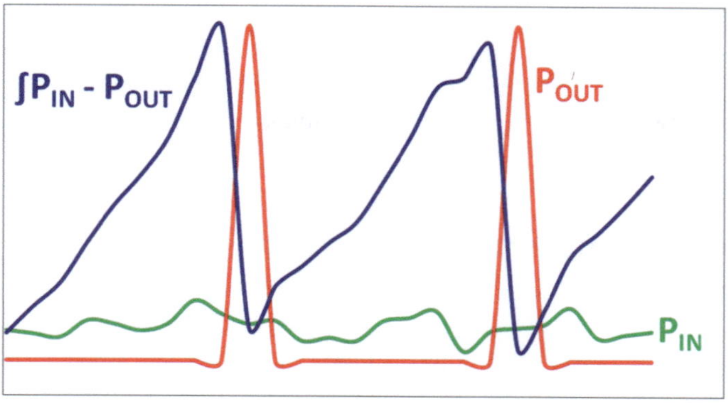

Fig. 3 Different P_{IN} and P_{OUT} profiles making energy storage necessary in energy harvesting system

Table 3 Comparison of energy storage capabilities of conventional batteries, thin film batteries and supercapacitors

	Conventional batteries	Thin film batteries	Supercapacitors
Recharge cycles	100 s	5–10 K	Millions
Self discharge	Moderate	Negligible	High
Charge time	Hours	Minutes	Sec-minutes
Impedance	Low-high	Low	Low
Physical size	Large	Small	Medium
Capacity	0.3–3,500 mAh	12–2,200 μAh	10–100 μAh

currents of 100's of mA owing to their internal impedance of a few 10's of milliohms whereas the smaller coin cell batteries have very poor constant current drain of only 100's of μA, limited by their high internal impedance. This is not useful when powering duty-cycled radio loads. To overcome these challenges, a special newer battery technology that uses solid-state electrolytes is become increasingly popular now. These batteries have a solid-state LiPON electrolyte and operate like conventional chemical batteries. However owing to the solid electrolyte they have extremely low self-discharge of only a few 1 % a year. While their impedance is higher than AA batteries, at 10's of ohms they are much better than coin cell batteries. This enables them to support up to 50 mA of load current which is more than sufficient for most radio loads in sensor nodes. The other impressive thing about these batteries is that they are extremely small in form factor and can withstand high temperatures owing to their solid electrolyte.

Supercapacitors, unlike batteries, do not store energy chemically making them much safer to operate. This also enables them to be charged and recharged many times without losing capacity. Supercapacitors can support very high load currents on par with AA batteries but suffer from poor self discharge and have only limited capacity. As can be seen, different storage options each have their advantages and disadvantages. Energy harvesting systems may also incorporate more than one storage mechanism to overcome some of the issues. For eg., a small supercapacitor in parallel with a solid-state battery combines the benefits of both in terms of high load current, low self discharge and long life times with high temperature operation.

4 General System Architecture of Energy Harvesting Systems

Energy harvesting systems generally consists of two power converter stages [12, 13], as shown in Fig. 4. Since the energy harvested may not match the load profile at any given time, the first converter harvests energy and stores it onto a storage element (battery or supercapacitor) while the second converter regulates the output voltage of the load circuits. In addition to charging the intermediate storage element, the first converter also performs MPPT (Maximum Power Point Tracking). Moreover, for system with batteries, we also need to incorporate battery management with Over Voltage (OV) and Under Voltage (UV) and Over Temperature (OT) control [14]. In addition to these circuits, the system may also have a Cold Start Circuit to ensure the system can start up even when the battery is fully discharged. In this section, we will discuss the details of circuit techniques used to design these blocks.

4.1 Battery Charger Power Stage and Maximum Power Tracking Circuits

Depending on the nature of the harvester, a DC–DC converter (photovoltaic and thermoelectric harvesters) or an AC–DC converter (piezoelectric harvesters) is used to charge the battery. In order to process power, different switching regulator topologies (buck, boost, buck-boost, etc.) may be used. For harvesters like photovoltaic (one or two cells) or thermoelectric harvesters, a boost converter is usually used.

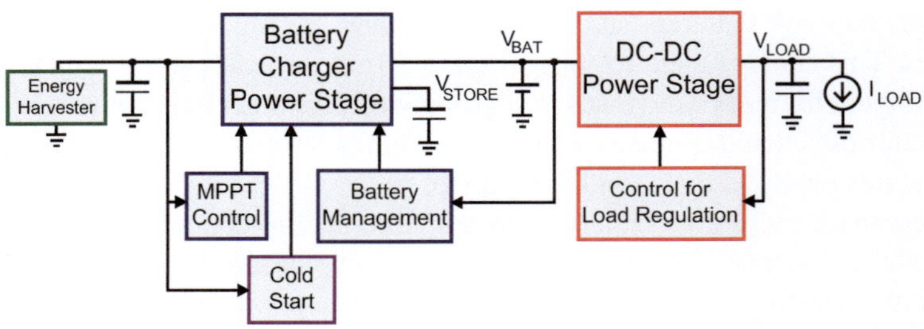

Fig. 4 Energy harvesting system architecture

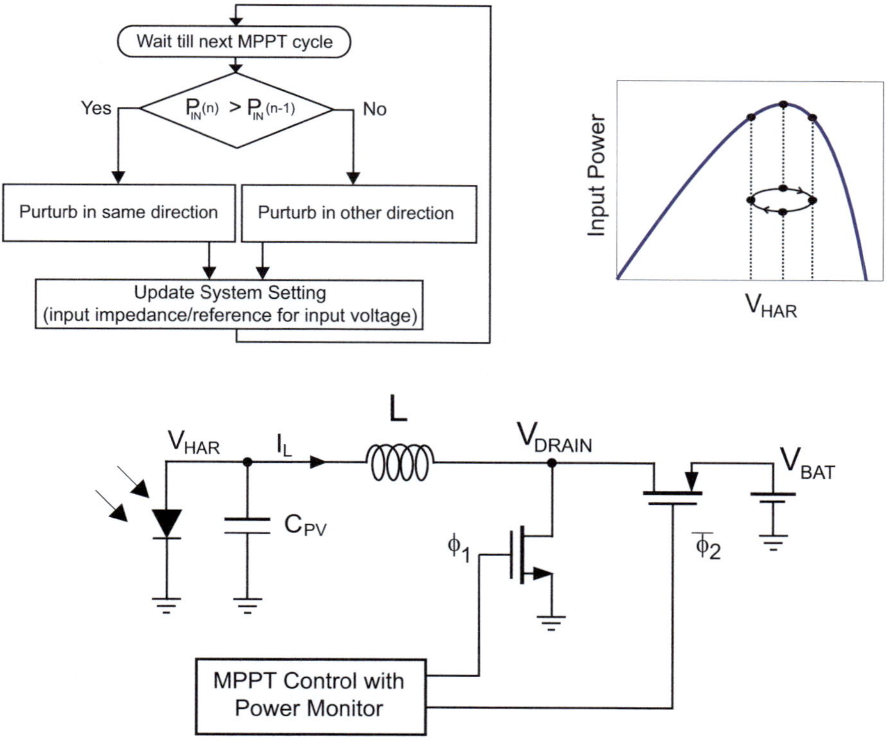

Fig. 5 Perturb and observe MPPT in the context of boost converter

However, for piezoelectric harvesters, a rectifier along with a buck or buck-boost converter may be used.

This power converter stage is also required to ensure maximum power is always extracted from the harvesters. This essentially means this converter must be controlled to either present the optimal input impedance [13, 15] necessary for maximum power transfer or to regulate the input voltage to the converter such that the converter extracts maximum power from the harvester [14]. Both these techniques are equivalent and one may be favored over the other depending on the accuracy of input voltage detection (for boost converter implementations, the input voltage may be of the order of 10's of mV). For switched-mode power converters, the input impedance can be viewed as the ratio of the input voltage to the average input current calculated over a number of switching periods. In most energy harvesters, the maximum power point may depend on the environmental conditions (light intensity, temperature, etc.) therefore, the system needs to be dynamically biased such that MPPT is achieved. This requires a power monitor circuit to detect when the system is extracting maximum power from the harvester. Figure 5 shows the generic MPPT using a *perturb and observe* technique. The MPPT operation consists of first *perturbing* the input impedance or the input voltage

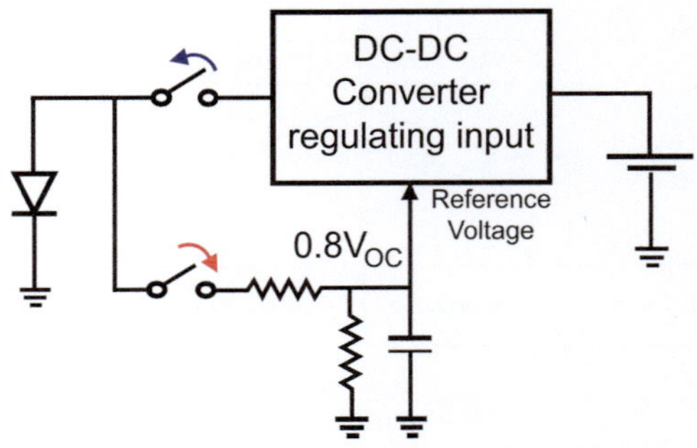

Fig. 6 Fractional open circuit voltage MPPT for photovoltaics

to the converter and then *observing* whether the output from the power monitor increases or decreases. If a power increase is detected, then the system is perturbed in the same direction in the next MPPT cycle. On the other hand, if the power decreases, then the perturbation direction is reversed. This way in steady state the system oscillates between three states where the two corner cases are for lower input power and the middle state is for maximum power. In some systems however, MPPT may not be required as the ratio of the maximum power point voltage to open circuit voltage of the harvester is approximately known [14]. In such cases, the input voltage is regulated to a fraction of the open circuit voltage where the fraction depends on the energy harvester (0.75–0.8 for most photovoltaic cells and 0.5 for thermoelectric harvesters). The control for maximum power extraction reduces to periodically sampling the open circuit voltage of the harvester and then regulating the converter input to the predefined fraction of the sampled open circuit voltage (as shown in Fig. 6).

4.2 Battery Management

The battery management in energy harvesting systems essentially consists of Over-Voltage (OV), Under-Voltage (UV) and Over-Temperature (OT) protection. Unlike battery chargers in portable devices where the current charging the battery needs to be regulated depending on the state of charge of the battery [16], in energy harvesting systems since the power from the harvesters are generally in the μW level, the battery capacity is generally much higher compared to the charging current. Therefore, we can afford to trickle charge the battery without degrading it. The OV, UV and OT control is done by comparators that detect when the battery voltage is higher than its OV threshold, lower than its UV threshold and when the temperature has exceeded the maximum acceptable temperature for charging a battery. Battery management implementations in some energy harvesting products is going to be highlighted in Sect. 5.

4.3 Load Regulation

In order to provide a regulated output to the load circuits (sensors, AFE, μController and RF Tx–Rx in Fig. 1), a second power converter is used as shown in Fig. 4. These low power load circuits generally operate at voltages around 0.6–1 V [1–5]. Therefore a buck converter is used to power these circuits off the battery. It must be noted that due to the low-power nature of the system, the quiescent power of the regulator control needs to be kept as low as possible to ensure the system sustains itself. However, due to the two power converters in series, the energy from the harvester is processed twice before reaching the load circuits, thereby limiting the overall efficiency. Section 5 will discuss a technique that addresses this issue and increases the end to end efficiency by reducing the number of power conversion stages depending on the power coming from the harvester and the load conditions.

4.4 Startup Circuits

In addition to the power converters, in systems where the input voltage (or harvester voltage) is in the 10 to a few 100's of mV, a cold start up circuit may be required for cases where the battery is fully discharged. This entails extra circuits that can operate at low voltages and can provide some initial charge to the battery. Following this, the system can operate normally and work of the battery. Section 5 will highlight a few low voltage circuits that aid in the cold startup.

5 Overview of Some Previously Published Energy Harvesting Systems

5.1 BQ25570, Energy Harvesting Battery Charger and Regulator

BQ25570 [17] is a boost battery charger with a buck regulator with integrated MPPT, Battery Management and a PFM control for the Buck Regulator as shown in Fig. 7. The system starts with a DC–DC boost converter/charger. Once started, the boost charger extracts power from low voltage output harvesters such as thermoelectric generators or single or dual cell solar panels. The boost charger can be started with input voltages as low as 330 mV, and once started, can continue to harvest energy down to input voltages of 100 mV.

The BQ25570 implements a fractional open circuit maximum power point tracking scheme to extract maximum available power from the harvesters. The fraction of open circuit voltage that is sampled and held can be controlled by pulling

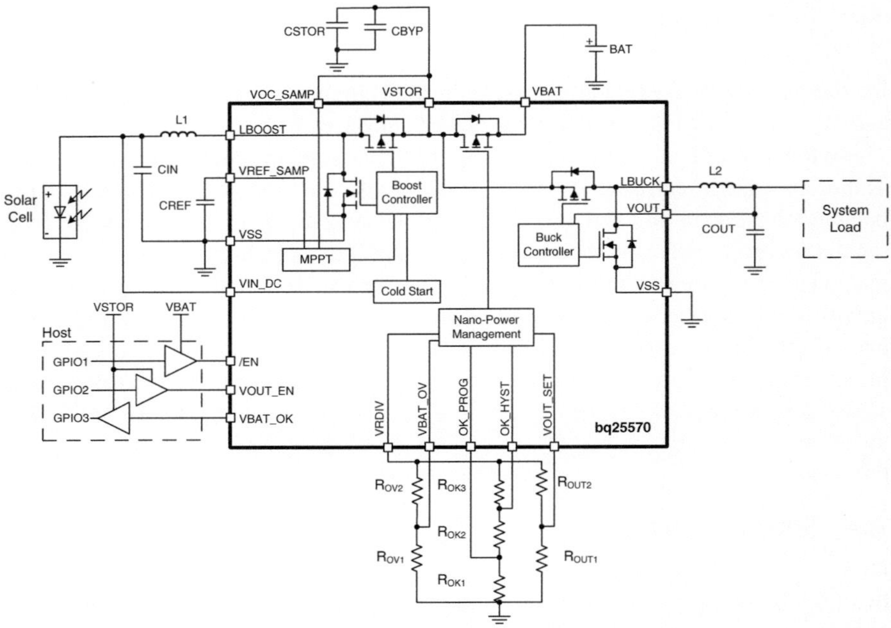

Fig. 7 Block diagram of BQ25570 [17]

the VOC_SAMP pin high or low (80 %or 50 % respectively) or by using external resistors. This sampled voltage is held with an external capacitor (CREF) on the VREF_SAMPpin. Connecting VOC_SAMP to VSTOR sets the MPPT threshold to 80 % and results in the IC regulating the voltage on the attached harvester to 80 % of its open circuit voltage. Alternatively, an external reference voltage can be provided by a MCU running a more complex MPPT algorithm. In addition to the boost charging front end, the BQ25570provides the system with an externally programmable regulated supply via the buck converter with 93 % peak efficiency. The regulated output has been optimized to provide high efficiency across low output currents (less than 10 μA) to high currents (around 110 mA). To prevent damage to the system's storage element, the voltage on the storage element is monitored against internally set under-voltage (UV) and programmable over-voltage (OV) levels.

The boost charger is intended to be powered from a high impedance DC source, such as a solar panel, TEG or piezoelectric module; therefore, it regulates its input voltage (VIN_DC) in order to prevent the input source from collapsing. The boost charger monitors its output voltage (VSTOR) and stops switching when VSTOR reaches a resistor programmable threshold level. The buck converter is powered from VSTOR. Both converters are based on a switching regulator architecture which maximizes efficiency while minimizing start-up and operation power. Both use pulse frequency modulation (PFM) to maintain efficiency, even

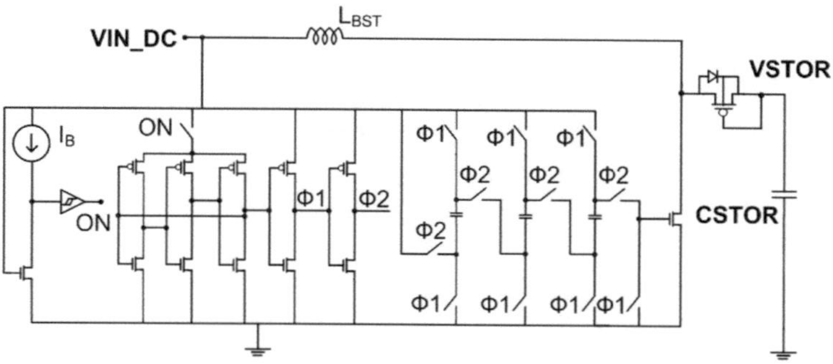

Fig. 8 Low voltage oscillator with charge pump for cold start

under light load conditions. In addition, the boost charger implements battery protection features so that either rechargeable batteries or capacitors can be used as energy storage elements at the storage element output (VBAT). The high efficiency of the BQ25570 boost charger and buck converter is achieved with nano-power management circuitry. These circuits essentially samples and holds all references in order to reduce the average quiescent current. That is, the internal circuitry is only active for a short period of time and then off for the remaining period of time at the lowest feasible duty cycle.

In order to cold start from 330 mV, the system uses a low voltage oscillator and a charge pump to generate pulses of higher than VIN_DC voltage swing (Fig. 8). Using a low V_T low side device in the boost converter further helps with the startup process.

5.2 *LTC3108, A Low Input Voltage with Transformer Assisted Startup Circuit*

LTC3108 uses a slightly different topology for boosting up low voltages to higher voltages as compared to some of the techniques that have been discussed until now. Figure 9 shows the block diagram [18]. By using a transformer with a large turns ratio (1:100 in this example) and a low V_T MOSFET on the SW pin, an oscillator is with a high voltage swing. An internal rectifier converts this ac voltage to an unregulated dc voltage on the V_{STORE} pin. An LDO is then used to regulate this down to V_{OUT}. To support multiple voltage domains, another output V_{OUT2} is also generated. Due to the transformer, the design can startup from voltages as low as 20 mV without any other assisted circuits.

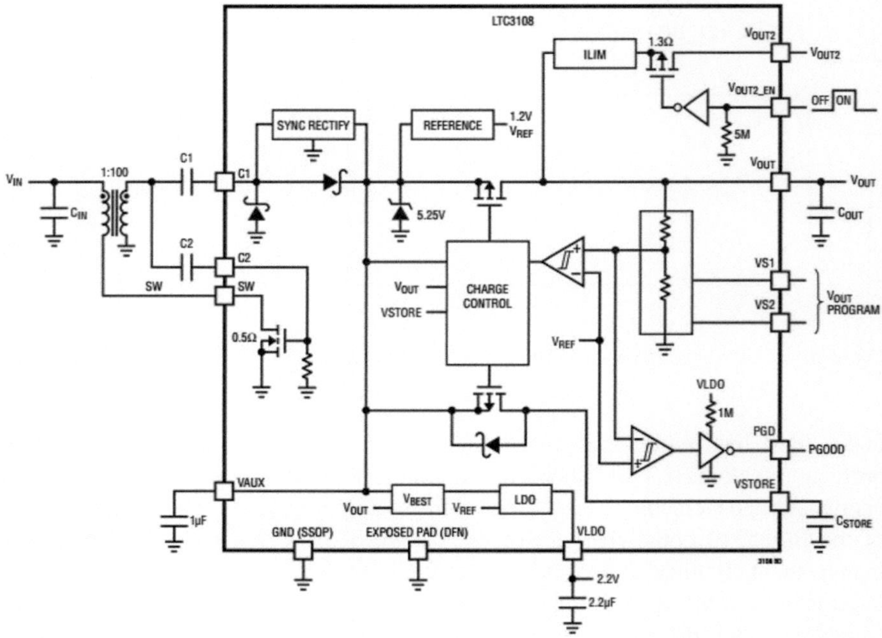

Fig. 9 LTC 3108 block diagram [18]

5.3 *MAX17710, A Boost Charger with Integrated LDO for Energy Harvesting Sources*

The MAX17710 has two main features- a battery charger that charges up a low-capacity cell with overcharge protection and an LDO regulator output with over discharge protection. The block diagram has been shown in Fig. 10 [19]. The system starts up utilizing some initial energy from the battery. In this state, the device pulls only 1 nA (typ) from the cell and LDO functions are disabled. The important aspect to note is that the design can support either a high voltage energy harvesting source that connects directly to the CHG pin of the device and charges up the battery in a linear fashion, or the system may use a boost converter to charge the battery from low voltage sources. The OV protection for the battery is provided by an integrated zener diode. An integrated LDO is then powers the load circuits off the battery.

5.4 *Efficient Rectifier for Piezoelectric Harvesters*

For piezoelectric harvesters, traditionally diode rectifiers are used. These rectifiers suffer from poor energy extraction capabilities from piezoelectric harvesters. To understand this, consider the full-bridge rectifier shown in Fig. 11 with ideal

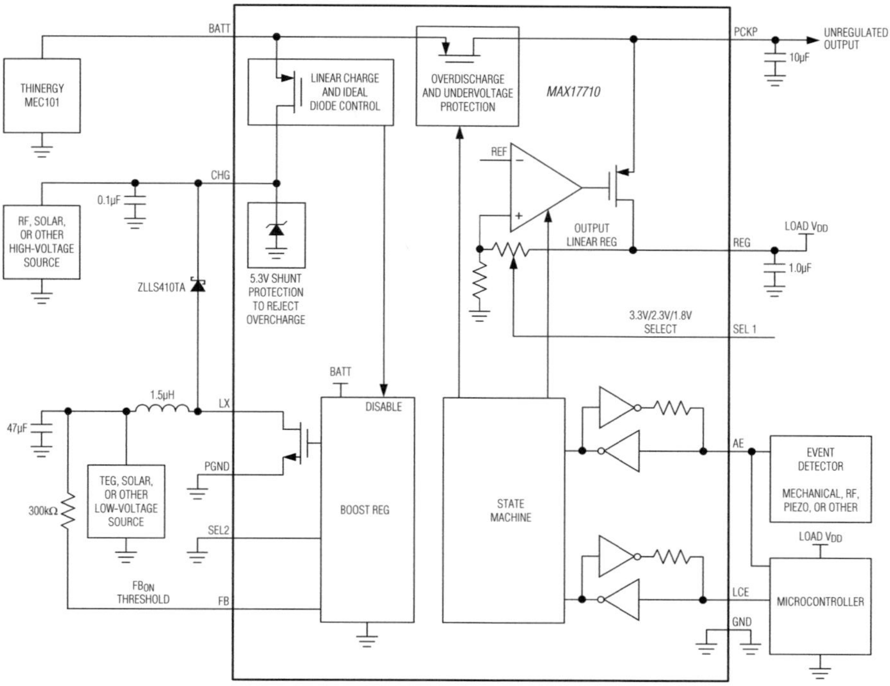

Fig. 10 MAX17710 block diagram [19]

diodes. During interval 1 when current IP is positive, diodes D1 and D4 are ON, I_P flows into the output capacitor and the voltage across the PE harvester is clamped at $+V_{RECT}$. During interval 2, I_P changes direction and diodes D1 and D4 turn OFF. However, before the diodes D2 and D3 can turn ON for the current to flow to the output, the voltage across C_P has to change from $+V_{RECT}$ to $-V_{RECT}$. Hence, during this phase, I_P instead of flowing to the output, flows into the capacitor C_P to discharge it. Only after C_P has discharged to V_{RECT} can the diodes D2 and D3 turn ON for I_P to start flowing into the output. The shaded portion of the current waveform shows the amount of charge lost in discharging C_P during this process. The process repeats itself every half-cycle and the loss in charge limits the amount of actual electrical power that can be extracted using the full-bridge rectifier. The total charge available from the harvester can be given by $4C_P V_P$ where V_P is the open-circuit voltage amplitude of the harvester. The charge lost is dependent on the rectifier output voltage and can be given by $4C_P V_{RECT}$. At low values of V_{RECT}, most of the charge available flows into the output but the voltage is low. At high values of V_{RECT}, very little charge flows into the output. These opposing trends causes the full-bridge rectifiers output power to reach a maximum of $C_P V_P^2 f$. The maximum occurs at an optimal rectifier voltage of one-half V_P.

The first stage of improvement in power extracted can be obtained by connecting a simple switch across the piezoelectric harvester and turning it ON at the right

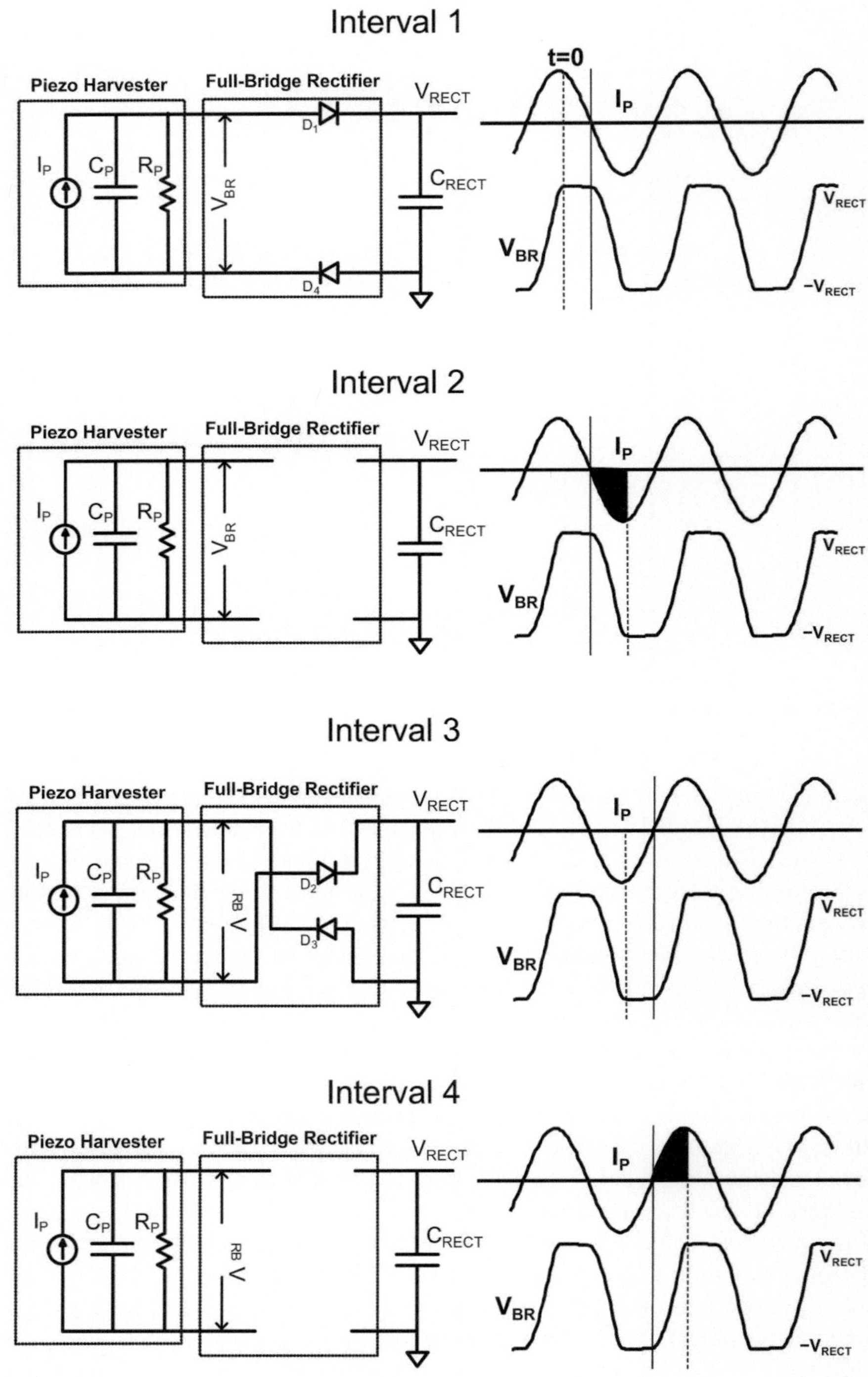

Fig. 11 Limited energy extraction capability of full bridge rectifiers [10]

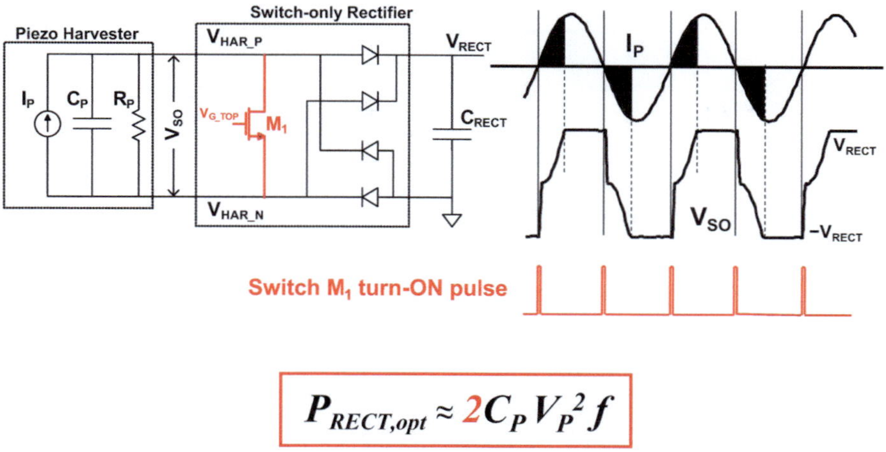

Fig. 12 Switch only implementation of efficient rectifier [10]

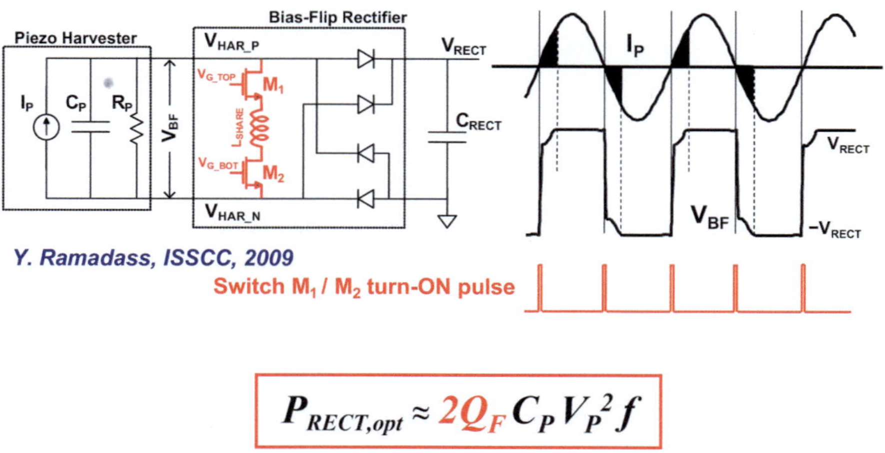

Fig. 13 Bias-flip implementation of efficient rectifier [10]

instant (Fig. 12). At every half-cycle, when I_P changes direction, the switch M1 is turned ON briefly to discharge the voltage across C_P. Now, the piezoelectric current only has to charge up the capacitor C_P from 0 to $+V_{RECT}$ or $-V_{RECT}$ before it can flow into the output. This reduces the charge lost by a factor of 2 as can be seen by the decrease in area of the shaded portion of the current waveform. This reduction in charge lost improves the extractable power by a factor of 2. It also increases the optimal output voltage by 2×.

We can go one-step further and build the bias-flip rectifier (Fig. 13) where the switch is replaced with an inductor L_{SHARE} which is connected in parallel with the piezoelectric harvester. The switches M1 and M2 of the bias-flip rectifier are turned ON for a brief time when I_P crosses zero in either direction. When the switches

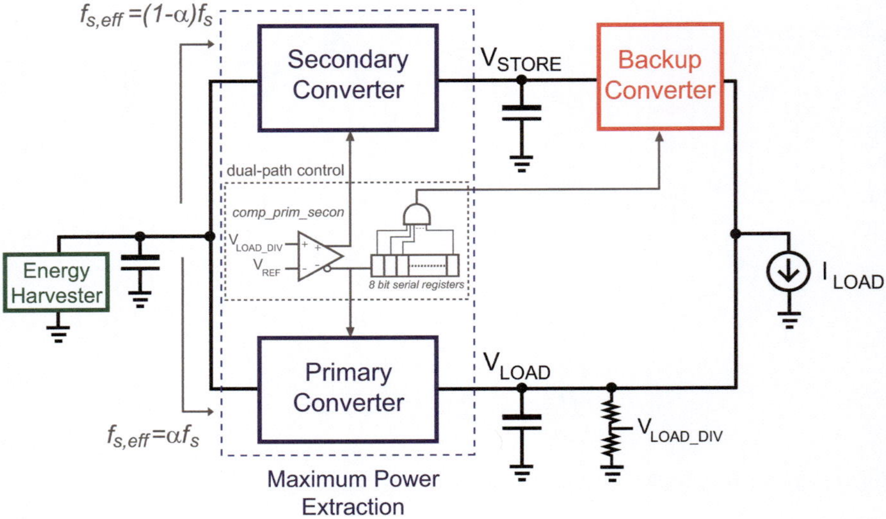

Fig. 14 *Dual-path* architecture [15]

are ON, the inductor helps in flipping the voltage across C_P. The series resistance along the L_{SHARE}, C_P resonant path limits the magnitude of this voltage inversion. After the switches open, the piezoelectric current I_P needs to supply only a small amount of charge to C_P to bring it up to $+V_{RECT}$ or $-V_{RECT}$. This significantly reduces the charge lost and improves the power extractable from the harvester. The output power that can be obtained with the bias-flip rectifier is greater than that obtained by the full-bridge rectifier by a factor of $2Q_F$. Q_F can be thought of as a parallel combination of the Q-factor of the piezoelectric element and that of the L_{SHARE}, C_P resonant path. For most piezoelectric harvesters this corresponds to a 4–8× improvement in the output power compared to the full-bridge rectifier [10].

5.5 Architectural Improvements to Improve System Efficiency

Until now, we have looked at different energy harvesting systems with essentially the standard *charger followed by regulator* architecture. A previously published work [15] has looked into an improvement over this standard architecture. Figure 14 shows the *dual-path* architecture that achieves higher efficiency than the traditional architecture. Here, the Maximum Power Extraction stage is split into two parallel converters- primary and secondary converters. When the ambient energy is available, the primary converter directly powers V_{LOAD}. Therefore, energy is transferred from the harvester to the load using a single power converter stage. However, as the load requires regulation, the primary converter is enabled only when the voltage $V_{LOAD,DIV}$ (from a tunable resistive division of V_{LOAD}), is below an internal reference

V_{REF}. When $V_{LOAD,DIV}$ is higher than V_{REF}, the secondary converter is enabled, storing the excess energy from the harvester on to V_{STORE} (battery or supercapacitor). The control circuit for enabling these converters consists of a clocked comparator. This comparator enables the gate drive signals either to the primary converter or to the secondary converter power stages, depending upon whether $V_{LOAD,DIV}$ is lower or higher than V_{REF}. Therefore, due to the complementary switching of the primary and secondary converters, V_{LOAD} is regulated and V_{STORE} is charged at the same time.

In the *dual-path* scheme, the primary converter adequately delivers power to the load when the harvester is able to meet the load requirement. When the ambient energy is not enough for the load, the primary converter alone is not able to regulate the output. Therefore, backup converter is activated, transferring the previously stored energy from V_{STORE} to V_{LOAD}. Conceptually, this can be done by detecting when $V_{LOAD,DIV}$ falls below $V_{REF} - \Delta V$, another reference slightly lower than V_{REF}. The ΔV offset is required to ensure that the backup converter is active only when the primary converter alone is not able to regulate the load.

The dual-path architecture provides an efficiency improvement over the traditional architecture. It bypasses the second stage of the traditional two stage architecture (Regulation Block in Fig. 4) when the input energy is able to meet the load requirement. Therefore, the system reduces to a single power converter supplying energy directly to the load. When the load is much higher than the input power, the dual-path architecture functions exactly like the traditional architecture transferring the previously stored energy from V_{STORE} to V_{LOAD}. Therefore, we can see that by arranging the DC-DC converters appropriately, it is possible to have one converter between the harvester and the load in the best case and two converters (similar to traditional architectures) in the worst case. For the implementation discussed in [15], the dual-path architecture provides a peak efficiency improvement of close to 11–13 %.

6 Conclusions

In this chapter, we looked at energy harvesting systems, harvester sources and storage options, general architectures of these systems and then finally some implementations that highlight the circuit/architectural techniques used to improve these systems. Energy harvesting is becoming increasingly popular for systems where the user can deploy a wireless sensor node and does not have to be concerned about replacing batteries periodically. Whether its for smart buildings, industrial automation or biomedical applications, wireless sensor nodes are becoming more and more ubiquitous and energy harvesting is helping alleviate the energy self-sufficiency issues in these systems.

References

1. M. Yip, J.L. Bohorquez, A.P. Chandrakasan, A 0.6 V 2.9μW mixed-signal front-end for ECG monitoring, in IEEE Symposium on VLSI Circuits (June 2012)
2. F.M. Yaul, A.P. Chandrakasan, A 10b 0.6 nW SAR ADC with data-dependent energy savings using LSB-first successive approximation, in IEEE International Solid State Circuits Conference (February 2014)
3. J.Y. Kwong, Y.K. Ramadass, N. Verma, M. Koesler, K. Huber, H. Moormann, A.P. Chandrakasan, A 65 nm Sub-Vt microcontroller with integrated SRAM and switched-capacitor DC-DC converter, in IEEE International Solid State Circuits Conference (February 2008)
4. A. Paidimarri, P. Nadeau, P. Mercier, A. Chandrakasan, A 440 pJ/bit 1 Mb/s 2.4 GHz multi-channel FBAR-based TX and an integrated pulse-shaping PA, in IEEE Symposium on VLSI Circuits (June 2012)
5. P. Nadeau, A. Paidimarri, P. Mercier, A. Chandrakasan, Multi-channel 180 pJ/bit 2.4 GHz FBAR-based receiver, IEEE Radio Frequency Integrated Circuits (RFIC) Symposium (June 2012)
6. M. Gratzel, Photovoltaic and photoelectrochemical conversion of solar energy. Phil. Trans. R. Soc. A **365**, 993–1005 (2007)
7. J. Lim, C.-K. Huang, M. Ryan, G.J. Snyder, J. Herman, J.-P. Fleurial, MEMS/ECD methods for making $Bi_{2-x}Sb_xTe_3$ thermoelectric devices. NASA Technical Reports (July 2008)
8. N.S. Shenck, J.A. Paradiso, Energy harvesting with shoe-mounted piezoelectrics. IEEE Micro **21**, 30–42 (2001)
9. R.J.M. Vuller, R. van Schaijk, I. Doms, C. Van Hoof, R. Mertens, Miropower energy harvesting. Solid State Electron. **53**, 684–693 (2009)
10. Y.K. Ramadass , A.P. Chandrakasan, An efficient piezoelectric energy harvesting interface circuit using a bias-flip rectifier and shared inductor, IEEE J. Solid State Circuits **45**(1), 189–204 (2010)
11. G.K. Ottman, H.F. Hofmann, A.C. Bhatt and G.A. Lesieutre, Adaptive piezoelectric energy harvesting circuit for wireless remote power supply. IEEE Trans. Power Electron. **17**(5), 669–676 (2002)
12. N.J. Guilar, R. Amirtharajah, P.J. Hurst, S.H. Lewis, An energy-aware multiple-input power supply with charge recovery for energy harvesting applications. IEEE ISSCC Digest of Technical Papers (February 2009), pp. 298–299
13. Y.K. Ramadass, A.P. Chandrakasan, A battery-less thermoelectric energy harvesting interface circuit with 35 mV startup voltage. IEEE J. Solid State Circuits **46**(1), 333–341 (2011)
14. K. Kadirvel, Y. Ramadass, U. Lyles, J. Carpenter, V. Ivanov, V. McNeil, A. Chandrakasan, B. Lum-Shue-Chan, A 330 nA energy-harvesting charger with battery management for solar and thermoelectric energy harvesting, in IEEE International Solid-State Circuits Conference (ISSCC) (February 2012)
15. S. Bandyopadhyay, A.P. Chandarkasan, Platform architecture for solar, thermal, and vibration energy combining with MPPT and single inductor. IEEE J. Solid-State Circuits **47**(9), 2199–2215 (2012)
16. M. Chen, G.A. Rincon-Mora, Accurate, compact, and power-efficient Li-ion battery charger circuit. IEEE Trans. Circuits Syst. Express Briefs **53**(11), 1180,1184 (2006)
17. Texas Instruments Datasheet BQ25570. Available Online-http://www.ti.com/product/bq25570
18. Linear Technology Datasheet LTC3108. Available Online-http://www.linear.com/product/LTC3108
19. Maxim Integrated Datasheet MAX17710. Available Online-http://datasheets.maximintegrated.com/en/ds/MAX17710.pdf

MIX
Papier aus verantwortungsvollen Quellen
Paper from responsible sources
FSC® C105338

If you have any concerns about our products,
you can contact us on
ProductSafety@springernature.com

In case Publisher is established outside the EU,
the EU authorized representative is:
Springer Nature Customer Service Center GmbH
Europaplatz 3, 69115 Heidelberg, Germany

Printed by Libri Plureos GmbH
in Hamburg, Germany